C000120858

Interdisciplinary Applied Mathematics

Volume 55

Series Editors

Anthony Bloch, University of Michigan, Ann Arbor, MI, USA

Charles L. Epstein, University of Pennsylvania, Philadelphia, PA, USA

Alain Goriely, University of Oxford, Oxford, UK

Leslie Greengard, New York University, New York, USA

Advisory Editors

Rick Durrett, Duke University, Durham, NC, USA

Andrew Fowler, University of Oxford, Oxford, UK

L. Glass, McGill University, Montreal, QC, Canada

R. Kohn, New York University, New York, NY, USA

P. S. Krishnaprasad, University of Maryland, College Park, MD, USA

C. Peskin, New York University, New York, NY, USA

S. S. Sastry, University of California, Berkeley, CA, USA

J. Sneyd, University of Auckland, Auckland, New Zealand

Problems in engineering, computational science, and the physical and biological sciences are using increasingly sophisticated mathematical techniques. Thus, the bridge between the mathematical sciences and other disciplines is heavily traveled. The correspondingly increased dialog between the disciplines has led to the establishment of the series: Interdisciplinary Applied Mathematics.

The purpose of this series is to meet the current and future needs for the interaction between various science and technology areas on the one hand and mathematics on the other. This is done, firstly, by encouraging the ways that mathematics may be applied in traditional areas, as well as point towards new and innovative areas of applications; and secondly, by encouraging other scientific disciplines to engage in a dialog with mathematicians outlining their problems to both access new methods as well as to suggest innovative developments within mathematics itself.

The series will consist of monographs and high-level texts from researchers working on the interplay between mathematics and other fields of science and technology.

L. Angela Mihai

Stochastic Elasticity

A Nondeterministic Approach to the
Nonlinear Field Theory

 Springer

L. Angela Mihai
Mathematics
Cardiff University
Cardiff, UK

ISSN 0939-6047   ISSN 2196-9973   (electronic)
Interdisciplinary Applied Mathematics
ISBN 978-3-031-06691-7   ISBN 978-3-031-06692-4   (eBook)
https://doi.org/10.1007/978-3-031-06692-4

Mathematics Subject Classification: 33B15, 34C15, 60G50, 60G60, 60H15, 60H30, 70K50, 74B20, 74C15, 74E05, 74E10, 76A15, 82D60, 94A15

© Springer Nature Switzerland AG 2022
This work is subject to copyright. All rights are reserved by the Publisher, whether the whole or part of the material is concerned, specifically the rights of translation, reprinting, reuse of illustrations, recitation, broadcasting, reproduction on microfilms or in any other physical way, and transmission or information storage and retrieval, electronic adaptation, computer software, or by similar or dissimilar methodology now known or hereafter developed.
The use of general descriptive names, registered names, trademarks, service marks, etc. in this publication does not imply, even in the absence of a specific statement, that such names are exempt from the relevant protective laws and regulations and therefore free for general use.
The publisher, the authors and the editors are safe to assume that the advice and information in this book are believed to be true and accurate at the date of publication. Neither the publisher nor the authors or the editors give a warranty, expressed or implied, with respect to the material contained herein or for any errors or omissions that may have been made. The publisher remains neutral with regard to jurisdictional claims in published maps and institutional affiliations.

This Springer imprint is published by the registered company Springer Nature Switzerland AG
The registered company address is: Gewerbestrasse 11, 6330 Cham, Switzerland

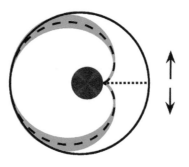

In memory of Iulian Beju, mathematical mechanician and professor at Bucharest University

Preface

Set at the interface between analytical continuum mechanics and advances in probability theory, this book combines in a coherent and unified manner nonlinear elasticity and applied probability to develop a nondeterministic approach for the quantification of uncertainties in the macroscopic elastic responses of materials at large strains.

At a continuum level, the elasticity of many solid materials can be described by phenomenological relations with the input parameters derived from carefully designed experiments. These are the so-called "hyperelastic models" characterised by strain-energy densities. They include the classical Hooke's law of linear elasticity at small strains and may take different forms at large strains. When performing a physical experiment multiple times on a test sample or on different samples of a material, the measurements, although qualitatively similar, are quantitatively different, and more so under different testing protocols. Traditionally, deterministic approaches have been employed to obtain model parameters from average values of the experimental measurements. However, as predictions rely on constitutive modelling, it may not be adequate for a model to depend on a single set of fixed parameters, regardless of how well they reproduce certain experiments. Recently, the use of information about uncertainties in the elastic behaviour of materials, stemming from uncertainties in the acquired data, begun to be explored.

The nondeterministic methods in this book place random variables at the foundation of stochastic hyperelastic models where the input parameters are defined by probability densities. Such models rely on the notion of entropy and on the maximum entropy principle, and are able to propagate uncertainties from input data to output physical quantities. Specific geometries include cuboids, spheres, and cylindrical tubes, which are capable of universal deformations that can be sustained for a wide range of materials. However, under particular loading conditions, a nonlinear hyperelastic material may cause the deformation to become unstable, and these instabilities must be characterised. Although synthetic and natural structures are typically irregular, regular geometries can be studied systematically to identify the independent influence of material properties on the elastic response. For each problem, the elastic solution is firstly derived analytically, then the effect of fluctuating model parameters is quantified in terms of probabilities. In this way, the propagation of uncertainties from input data to output quantities of interest is mathematically tractable, uncovering key aspects of how probabilistic approaches are incorporated into the nonlinear field theory. As a by-product, addressing well-known problems from a nondeterministic perspective creates opportunities for fresh insights and refinements of previously established results.

Stochastic elasticity is a fast-developing field that combines nonlinear elasticity and stochastic theories in order to significantly improve model predictions by accounting for uncertainties in the mechanical responses of materials. However, in contrast to the tremendous development of computational methods for large-scale problems, which have been proposed and implemented extensively in recent years, at the fundamental level, there is very little understanding of the uncertainties in the behaviour of elastic materials under large strains. In this case, small sample tests often present situations that might be overlooked by large-scale stochastic techniques.

Based on the idea that every large-scale project starts as a small-scale data problem, this book aims to combine fundamental aspects of large-strain finite elasticity and probability theories, which are prerequisites for the quantification of uncertainties in the elastic responses of soft materials. In practice, this type of material can be found under both load-carrying and non-load-carrying conditions, and form the subject of intensive research in industrial and biomedical applications where the advent of additive manufacturing has led to increased interest in the optimal design of controlled reproducible geometries. For these materials, the deformations are inherently nonlinear, and their elastic properties, which change with the deformation, can be independently optimised for improved serviceability. A central challenge in predicting mechanical performance of many engineering and natural materials is the lack of quantitative characterisation of the uncertainties inherent in the experimental data and in the mathematical models derived from them. By recasting specific problems, for which the deformation is known explicitly, within the nondeterministic framework, crucial insights can be gained into the effects of fluctuating parameters at a continuum level, which cannot be captured by a deterministic approach. Such effects have important implications for the optimal design of soft elastic materials and their model prediction in various applications.

There are many great books on nonlinear elasticity, including: Biot [61], Ciarlet [101–103], Green and Zerna [218], Ogden [415] and Truesdell and Noll [558] where the fundamental theory is presented and explained in detail; Treloar [550] focusing on phenomenological models for rubber-like materials; Antman [23], Atkin and Fox [31], Bigoni [60], Holzapfel [256], Landau & Lifshitz [315], Marsden and Hughes [343], and Steigmann [525] where classical and modern results are made accessible to a wider audience; Goriely [205] advancing the phenomenological modelling of biological structures; Dorfmann and Ogden [146] where the nonlinear elasticity framework is extended to electroelasticity and magnetoelasticity; Anand and Govindjee [16] and Volokh [575] providing a modern introduction to elasticity, plasticity, viscoelasticity and coupled field theories; Warner & Terentjev [594] where the theory of rubber elasticity is adapted to liquid crystal elastomers; and

Belytschko et al. [55], Bonet and Wood [66] and Oden [407] on the finite element representation and numerical approximation of soft solids.

Fewer books have been written on uncertainty quantification in solid mechanics, namely: Wiener [607] on statistical mechanics in elasticity; Soize [510] setting the foundation for stochastic modelling and uncertainty quantification of large-scale computational models for elastic solids; Elishakoff [150, 151] where probability theory is applied to problems in structural mechanics underpinned by classical linear elasticity; Sobcczyk and Kirkner [505] presenting a concise introduction to stochastic modelling of material microstructures; Ostoja-Starzewski [420] offering a comprehensive analysis of stochastic models and methods in the mechanics of random media; and Malyarenko and Ostoja-Starzewski [338] and Malyarenko et al. [339] introducing stochastic continuum models for elastic materials coupled with electric or magnetic fields.

Stochastic Elasticity is the first book to combine fundamental large strain nonlinear elasticity and probability theories. The aim is to make both theories accessible to those wishing to incorporate uncertainty quantification in phenomenological studies of soft materials. In this sense, the book should be of interest to workers and research students in applied mathematics, engineering, biomechanics and materials science who need to gain an insight into fundamental aspects of nonlinear elasticity from a continuum point of view, within a nondeterministic framework. This is the more obvious audience, as the book is written on an advanced and detailed level, with the more elementary background deferred to appendices. There is also an expectation that the book should reach further to those with no previous exposure to the subject, who would like to adopt a similar nondeterministic mind-set and incorporate uncertainties in their own scientific work. Whether with the intention to keep them under control, as is the case of manufactured and functional materials, or just to observe them, as in natural and living biological matter, taking uncertainties into account can both increase the immediate significance of many new or existing results, and render them more likely in the future. Of course, there is always an element of surprise when dealing with uncertainties: One never quite knows what more they can bring. That makes every study of uncertainty quantification unique.

Cardiff, UK   L. Angela Mihai
March 21, 2022

Acknowledgements

My understanding of stochastic elasticity, such as it is, would not have been possible without the complementary expertise and abundant support of my close collaborators, Alain Goriely (University of Oxford) and Thomas Woolley (Cardiff University). Alain and I have enjoyed solving elasticity problems for over a decade. We learned together and from each other, and the exchange was never dull. The work with Thomas on stochastic modelling, although more recent, was so effective as if we had known each other for years.

For illuminating discussions on uncertainty quantification of elastic materials, I am indebted to Johann Guilleminot (Duke University), whose original work inspired my own.

On the mechanics of liquid crystal elastomers, I learned enormously from my collaborators Haoran Wang (Utah State University), with whom I published my first cover image, depicting a molecular dynamics simulation, and Devesh Mistry and Helen Gleeson (University of Leeds), who created auxetic nematic elastomers.

I was fortunate also to have very committed students: Khulud Alayyash, Alex Safar, Danielle Fitt, Manal Alamoudi, Aaron English and James Scaife, and research associates: Hayley Wyatt, Maciej Buze and Tom Raistrick, who contributed to many of the publications referenced in this book.

My work on the mathematics of solid mechanics has received generous funding over the years from the Engineering and Physical Sciences Research Council of Great Britain, for which I am most grateful.

The final thanks go to my family and close friends for their cheerful support and for giving me plenty of time to write and think. Their names are written in my heart, and this book is in a way their reward.

Contents

1 **Introduction** 1

2 **Finite Elasticity as Prior Information** 7
  2.1   Finite Elastic Deformations 7
        2.1.1 Hyperelastic Materials 11
        2.1.2 Strains 12
        2.1.3 Stresses 12
        2.1.4 Adscititious Inequalities 16
  2.2   Nonlinear Elastic Moduli in Isotropic Elasticity 19
        2.2.1 Incremental Elastic Moduli 21
        2.2.2 Stretch Moduli 23
        2.2.3 Poisson Functions 28
        2.2.4 Bulk Moduli 29
        2.2.5 Shear Moduli 30
        2.2.6 Universal Relations 36
        2.2.7 Torsion Moduli 37
        2.2.8 Poynting Moduli 40
  2.3   Nonlinear Elastic Moduli in Anisotropic Elasticity 42
  2.4   Quasi-Equilibrated Motion 45

3 **Are Elastic Materials Like Gambling Machines?** 49
  3.1   Stochastic Hyperelastic Models 50
        3.1.1 Model Calibration 51
        3.1.2 Multiple-Term Models 55
        3.1.3 One-Term Models 56
  3.2   Stochastic Modelling of Rubber-Like Materials 57
        3.2.1 Hypothesis Testing 59
        3.2.2 Parameter Estimation 60
        3.2.3 Bayesian Model Selection 60

4 **Elastic Instabilities** 67
  4.1   Necking 68

4.2   Equilibria of an Elastic Cube Under Equitriaxial Surface Loads.....   72
      4.2.1   The Rivlin Cube   73
      4.2.2   The Stochastic neo-Hookean Cube   78
      4.2.3   The Stochastic Mooney–Rivlin Cube   81
4.3   Cavitation of Spheres Under Radially Symmetric Tensile Loads   86
4.4   Inflation and Perversion of Fibre-Reinforced Tubes   95
      4.4.1   Stochastic Anisotropic Hyperelastic Tubes...................   98
      4.4.2   Inflation of Stochastic Anisotropic Tubes 102
      4.4.3   Perversion of Stochastic Anisotropic Tubes 104

5   Oscillatory Motions 111
5.1   Cavitation and Radial Quasi-equilibrated Motion of
      Homogeneous Spheres 115
      5.1.1   Oscillatory Motion of Stochastic Neo-Hookean
              Spheres Under Dead-Load Traction 117
      5.1.2   Static Deformation of Stochastic Neo-Hookean
              Spheres Under Dead-Load Traction 123
      5.1.3   Non-oscillatory Motion of Stochastic Neo-Hookean
              Spheres Under Impulse Traction 124
      5.1.4   Static Deformation of Stochastic Neo-Hookean
              Spheres Under Impulse Traction 126
5.2   Cavitation and Radial Quasi-equilibrated Motion of
      Concentric Homogeneous Spheres 127
      5.2.1   Oscillatory Motion of Spheres with Two Stochastic
              Neo-Hookean Phases Under Dead-Load Traction 128
      5.2.2   Static Deformation of Spheres with Two Stochastic
              Neo-Hookean Phases Under Dead-Load Traction 132
      5.2.3   Non-oscillatory Motion of Spheres with Two
              Stochastic Neo-Hookean Phases Under Impulse Traction ... 135
      5.2.4   Static Deformation of Spheres with Two Stochastic
              Neo-Hookean Phases Under Impulse Traction 137
5.3   Cavitation and Radial Quasi-equilibrated Motion of
      Inhomogeneous Spheres... 140
      5.3.1   Oscillatory Motion of Stochastic Radially
              Inhomogeneous Spheres Under Dead-Load Traction 141
      5.3.2   Static Deformation of Stochastic Radially
              Inhomogeneous Spheres Under Dead-Load Traction 145
      5.3.3   Alternative Modelling of Radially Inhomogeneous
              Spheres 148
      5.3.4   Non-oscillatory Motion of Stochastic Radially
              Inhomogeneous Spheres Under Impulse Traction............ 151
      5.3.5   Static Deformation of Stochastic Radially
              Inhomogeneous Spheres Under Impulse Traction............ 152

5.4 Oscillatory Motion of Circular Cylindrical Tubes 153
    5.4.1 Tubes of Stochastic Mooney–Rivlin Material Under
        Impulse Traction 159
    5.4.2 Tubes with Two Stochastic Neo-Hookean Phases
        Under Impulse Traction 167
    5.4.3 Stochastic Radially Inhomogeneous Tubes Under
        Impulse Traction 173
5.5 Generalised Shear Motion of a Stochastic Cuboid 177
    5.5.1 Shear Oscillations of a Cuboid of Stochastic
        Neo-Hookean Material 178

6 Liquid Crystal Elastomers 183
6.1 Uniaxial Elastic Models 186
6.2 Stresses in Liquid Crystal Elastomers 191
    6.2.1 The Case of Free Nematic Director 191
    6.2.2 The Case of Frozen Nematic Director 193
6.3 A Continuum Elastic–Nematic Model 193
6.4 Shear Striping Instability 195
6.5 Inflation of a Nematic Spherical Shell 205
6.6 Necking of a Nematic Elastomer 206
6.7 Auxeticity and Biaxiality 208

7 Conclusion 217

A Notation 219

B Fundamental Concepts 223
B.1 Tensors and Tensor Fields 223
B.2 Polar Systems of Coordinates 229
    B.2.1 Cylindrical Coordinate System 229
    B.2.2 Spherical Coordinate System 230
B.3 Random Variables and Random Fields 232
B.4 Normal Distribution as Limiting Case of the Gamma Distribution .. 236
B.5 Linear Combination of Independent Gamma Distributions 239
B.6 Maximum Entropy Probability 239

Bibliography 241

Index 271

Chapter 1
Introduction

We have become complete determinists, and even those who wish to reserve the right of human free will at least allow determinism to reign undisputed in the inorganic world.

—H. Poincaré [437]

Finite elasticity is the study of elastic responses of solid materials under reversible deformations. When geometrical changes are small, such that deformations are not detectable by the human eye, a material configuration can be considered as fixed. Then any geometrical changes can be neglected. This so-called small strain assumption is at the basis of classical theory of linear elasticity. A brief but illuminating historical overview of the theory of elasticity is provided in [201].

While the linear elasticity theory is successfully used in structural mechanics and many engineering applications, many modern applications are concerned with soft solids (inflatable structures, polymers, synthetic rubbers) or biological structures (plants, vital organs), which involve visibly large deformations. For the study of such deformations, finite elasticity covers the simplest case where internal forces only depend on the present state of the body and not on its history, that is, it excludes plasticity, viscosity, and damage. Here is a concise description of the main difference between classical elasticity theory, which is limited to infinitesimally small strains, and finite strain elasticity:

> With reference to problems in elasticity, the terms 'small displacement', 'finite displacement', 'small strain' and 'finite strain' have the following meanings which will be made more precise later. If the displacement of every particle in a strained elastic body is much smaller than the greatest linear dimension of the body, the displacement is small; otherwise it is finite. If the elongation per unit length of every element in the body is much smaller than unity the strain is small; otherwise it is finite . . . In most problems the strain is small only if the displacement is also small. However, in the case of thin rods and plates the displacement may be finite and yet the resulting strain may be small.—G.E. Hay [240]

In particular, a homogeneous isotropic linearly elastic material, for which the material properties are the same in all directions, is characterised by two physical constants, for example, the Young's modulus and Poisson's ratio. Any other material constant, such as the shear modulus or the bulk modulus, is then derived from

© Springer Nature Switzerland AG 2022
L. A. Mihai, *Stochastic Elasticity*, Interdisciplinary Applied Mathematics 55,
https://doi.org/10.1007/978-3-031-06692-4_1

established universal relations. In contrast, nonlinear elastic materials cannot be represented by constants but are generally described by parameters that change with the deformation. To be effective in estimating elastic material behaviours, these parameters must satisfy certain conditions, namely [362]:

1. To be generally applicable, they must be obtainable for all materials in a class (e.g., all incompressible or compressible homogeneous isotropic hyperelastic materials).
2. Ideally, they should be measurable under multiaxial deformations, which, in principle, are closer to real physical situations.
3. For mechanical consistency with the classical theory, they must be equal to the corresponding linear elastic parameters at small strains.

An important parameter that satisfies the above conditions is the *nonlinear shear modulus*. This was defined for isotropic materials in [362] and for anisotropic materials in [373]. In its most general form, the shear modulus incorporates information from both axial and shear deformations. Then universal relations with other elastic parameters, such as the nonlinear stretch modulus and the Poisson function, can be established.

On the importance of shear modulus, Treloar et al. [551] remarked that, for a theory to be helpful in explaining the elastic responses of a material, it should take into account its properties *"not only in simple extension and compression, but also in other types of strain"*. This led to the phenomenological theory for vulcanised rubber [550, p. 65] (read also [188]):

> ...the theory is based on the concept of a vulcanised rubber as an assembly of long-chain molecules linked together at a relatively small number of points so as to form an irregular three-dimensional network.—L.R.G. Treloar [550, p. 60]

Statistical treatment of rubber networks led to an elegant formula for the shear modulus, depending on the molecular structure. Importantly, this modulus can be inferred from macroscopic deformations as well:

> At first sight it seems surprising that a single elastic constant should be sufficient to define the properties of a rubber, in which the strains may be large, whereas in the classical theory of elasticity, which is limited to small strains, two independent elastic constants ...are required. This apparent inconsistency is a result of the assumption of constancy of volume or incompressibility ...As a result the response to a stress is effectively determined solely by the shear modulus ...—L.R.G. Treloar [550, pp. 67–68]

The first experimental measurements for macroscopic load-deformation responses of rubber-like material at large strains were recorded by Rivlin and Saunders [470]. But taking these measurements had its challenges:

> Now, it is well known that rubber vulcanizates exhibit in greater or lesser degree the effects of hysteresis, internal friction and permanent set. These effects represent departures from the ideal elasticity envisaged by the theory, and it is therefore important to design the experiments in such a manner that the influence of these departures is minimized.—R.S. Rivlin and D.W. Saunders [470]

Furthermore:

> This task is made more difficult than it otherwise would be by the fact that some of the test-pieces used have to be moulded individually, and it is difficult to make two rubber specimens having identical properties even if nominally identical procedures are followed in preparing them.—R.S. Rivlin and D.W. Saunders [470]

On subsequent theoretical results, Treloar again remarked:

> The accuracy of these results is limited by the accuracy of the original data, which are themselves limited not so much by the accuracy of the measurements of stress and strain in themselves, but rather by the inherent lack of complete reproducibility (reversibility) of the rubber ... Finally, it should be pointed out that the general validity of Rivlin and Saunders' conclusions ... must be subject to some uncertainty, on account of the experimental limitations, referred to above, and also, of course, to possible variations between different rubber vulcanizates.—L.R.G. Treloar [550, pp. 225–227]

Indeed, the original data show some variability (albeit small) between the measured responses of different samples (see also [550, p. 224], or [558, p. 181]). Following the traditional deterministic approaches, many constitutive models calibrated to the average value of these data have been proposed (e.g., in [80, 117, 120, 138, 237, 289, 342, 362, 417, 526, 559, 567]).

In fact, the elastic properties of solid materials are rarely deterministic. Uncertainties in the measurement of these properties typically arise from:

- Sample to sample variability (due to the inherent material inhomogeneity)
- Observational data, which may be sparse (samples in short supply, or easily damaged), indirect (measured through some sophistic equipment), and polluted by noise
- Imperfect reversibility of elastic deformations (especially at large strains)

To capture elastic properties, material models are either completely phenomenological, i.e., based only on physical observations at the macroscopic level, or involve a transfer of information from the study of underlying physical properties at a smaller scale. In the latter case, statistical approaches can be employed to characterise a material at the microscopic or mesoscopic scale, as, for example, in the case of polymer networks. For the mechanical analysis of rubber-like networks, statistical approaches are reviewed in [300]. An overview of constitutive equations for biological soft tissues, including statistical models for collagen networks, is provided in [92]. Many multiscale models are devised also through homogenisation techniques, which may include stochastic[1] techniques for random media [420]. For heterogeneous linearly elastic solids, a statistical theory was introduced by McCoy (1973) [347]. Stochastic strategies for the investigation of mesoscopic mechanical effects in random materials were also proposed by Huet (1990) [269]. Further

---

[1] The word "stochastic" originates from the Greek word "στόχος" (stokhos) meaning "aim" or "guess" and refers to a random variable, or function, characterised by probabilities, from the Latin word *probābilis* (likely), or analysed statistically, but not known precisely (e.g., stochastic processes are random functions of time).

developments in the stochastic modelling of heterogeneous solids are reviewed by [420]. There are, however, many challenges introduced by the consideration and quantification of uncertainties in mathematical models and their use in making predictions [33, 34, 161, 220, 270, 334, 335, 408–411, 608].

The focus of this book is on stochastic hyperelastic models that are phenomenological models built on a minimal set of fundamental assumptions, with the parameters characterised by probability distributions that can be calibrated to macroscopic experimental measurements. These models reduce to the usual deterministic functions when the parameters are single-valued constants. For the stochastic versions, an important question is: *What a priori distribution should be used to characterise an elastic modulus?*

In [183], experimental strain values for liver soft tissue under compressive stress were assumed to vary according to a normal (Gaussian) distribution. A nondeterministic model was then calibrated numerically to the mean stress–strain curve. As the computed results were somewhat unsatisfactory, it was concluded that the normal distribution might not adequately capture the variability in the observed dataset.

Advanced stochastic strategies based on information theory, which aid with the calibration of hyperelastic models for isotropic elastic materials, were introduced by Staber and Guilleminot [519–521]. Extensions to anisotropic material models, presenting different stochastic properties in different directions, were treated in [95, 522, 523]. Before that, in [225, 226], a similar strategy was applied to the stochastic representation of tensor-valued random variables and random fields in linear elasticity. These strategies, which rely on the principle of maximum entropy introduced by E.T. Jaynes [280, 281] (see also [282]), are reviewed in [224, 227]. The measure of entropy (or uncertainty) of a probability distribution was first defined in the context of information theory by C.E. Shannon [492] (see also [514]).

In this book, the nonlinear shear modulus is central to the stochastic elastic framework where it is cast as a fluctuating parameter described by a maximum entropy probability. Within this framework, a range of interesting theoretical problems are treated, which are well-known due to the anomalous behaviour of their nonlinear elastic solutions (not observed at small strains). These problems have been previously formulated and treated deterministically by other authors. The challenge then is to "engraft" the probabilistic parameters onto the elastic solution and observe how their uncertainty propagates to the final results. At a first glance, this may seem rather straightforward: For every realisation of the probabilistic moduli, the deterministic solution remains valid, while the stochastic solution consists of the ensemble solutions thus obtained. Except that letting parameters follow probability distributions raises some fundamental questions about the stability of the probabilistic solution in each case. Even for well-behaved but nonlinear solutions, when the variance increases, the mean value becomes less significant from the physical point of view. Such problems exemplify how stochastic elastic models offer a rich description of physical effects and can thus be integrated into the nonlinear field theory.

The main content of the book is organised as follows: Chap. 2 presents an overview of nonlinear finite elasticity that is at the foundation of the stochastic elastic theory developed in this book. The discussion of stochastic elasticity begins in Chap. 3 where its general principles are displayed through a calibration procedure for the construction of stochastic hyperelastic models from small datasets. Experimental measurements of rubber-like material in uniaxial tension are used to illustrate the technique. In addition to the standard relative error criterion, an explicit lower bound on Bayes factor is also employed to select the best performing model. To illustrate the general principles, Chaps. 4 and 5 include carefully chosen instability problems from the analytical continuum mechanics literature, but with the material parameters characterised by probability distributions. The focus of Chap. 4 is on quasi-static problems and homogeneous materials, while in Chap. 5, quasi-equilibrated motions and spatially dependent material properties are treated as well. Chapter 6 is devoted to the mathematical modelling of nematic liquid crystal elastomers and their striking mechanical behaviour to which stochastic elasticity is applied. Nematic elastomers are at the forefront of materials research and a topic of growing interest to applied mathematicians, engineers, physicists, chemists, and materials scientists. Their theoretical investigation requires a solid background in nonlinear elasticity, and by including them here, it is hoped that some readers will become attracted to working in this area where much is yet to be discovered and understood. The key question in Chaps. 4, 5 and 6 is: *How do the probabilistic input parameters influence the output mechanical responses?* Throughout the book, physical quantities are treated symbolically, and units of measure only appear in Chaps. 3 and 6 where experimental data are required. The final chapter concludes with a further outlook. Notation and selected classical results that are particularly relevant are summarised in appendices.

Chapter 2
Finite Elasticity as Prior Information

> *Mechanics as a whole is non-linear; the special parts of*
> *mechanics which are linear may seem nearer to common sense,*
> *but this shows merely that good sense in mechanics is*
> *uncommon. We should not be resentful if materials show*
> *character instead of docile obedience.*

—C. Truesdell [557]

This chapter presents a summary of fundamental concepts in finite elasticity that will be useful for the stochastic methodology developed in the following chapters. Special attention is given to the *nonlinear elastic moduli*, which characterise the elastic properties of materials at large strains.

2.1 Finite Elastic Deformations

A material body $\mathcal{B}$ is a three-dimensional (3D) differential manifold, the elements of which are called material points or continuum particles.[1] A system of coordinates establishes a one-to-one correspondence between the body's particles and triples of real numbers $(x_1, x_2, x_3) \in \mathbb{R}^3$. For the position of a particle, either Cartesian or polar coordinates can be used. The number of coordinates is called the number of degrees of freedom of the physical system. Any two systems of coordinates are related through a continuously differentiable transformation. When a material point has first the position P and then the position Q, the point is said to have undergone a change of position, or a *displacement*, which is determined by the vector $\overrightarrow{PQ}$.

Denoting by $\mathbb{R}^3$ the 3D Euclidean space, we consider a continuous material body $\mathcal{B}$ that occupies a compact domain $\overline{\Omega} \in \mathbb{R}^3$. This body is made of particles whose positions at a time instant t define the configuration of the body. A configuration is a

---

[1] As noted in [607], it is important to distinguish between continuum particles and single atoms or molecules.

© Springer Nature Switzerland AG 2022
L. A. Mihai, *Stochastic Elasticity*, Interdisciplinary Applied Mathematics 55,
https://doi.org/10.1007/978-3-031-06692-4_2

smooth mapping of $\overline{\Omega}$ onto a region of $\mathbb{R}^3$. Among all configurations, we choose one that does not vary in time as a *reference configuration* and identify each particle of the body with its position $\mathbf{X}$ in the reference configuration. The interior of the body is identified as an open, bounded, connected subset $\Omega \subset \mathbb{R}^3$ with boundary $\Gamma = \partial\Omega = \overline{\Omega} \setminus \Omega$. We further assume that Γ is Lipschitz-continuous and, in particular, that a unit normal vector $\mathbf{n}$ exists almost everywhere along Γ.

A finite elastic deformation of the body from the reference configuration $\mathcal{B}_0$ to the current configuration $\mathcal{B}$ is defined by a one-to-one, orientation-preserving mapping

$$\chi : \Omega \rightarrow \mathbb{R}^3 \tag{2.1}$$

such that its derivative with respect to the reference state [343], defined by

$$\mathbf{A} = \nabla\chi = \text{Grad } \chi, \tag{2.2}$$

and commonly known as "the gradient tensor," has positive determinant, i.e.,

$$J = \det \mathbf{A} > 0 \qquad \text{on} \qquad \Omega. \tag{2.3}$$

The one-to-one condition on Ω guarantees that interpenetration of matter is avoided and that a body occupying a finite non-zero volume cannot be compressed to zero. However, self-contact on the body's surface is permitted; hence, this transformation need not be injective on $\overline{\Omega}$ (see Fig. 2.1).

- In the Lagrangian (material, reference) representation, one fixes a particle $\mathbf{X}$ (Lagrangian variable) in the reference configuration and observes its distortion $\chi(\mathbf{X})$.
- In the Eulerian (spatial, current) representation, one fixes a point $\mathbf{x} = \chi(\mathbf{X})$ (Eulerian variable) in the current configuration, which corresponds to the place occupied by the particle $\mathbf{X}$ after deformation, and studies what happens at that point as the time progresses.
- Under a finite strain deformation with gradient tensor $\mathbf{A}$, the following relations hold for line, volume, and surface changes: a volume element $d\mathcal{V}$, a line vector $d\mathbf{X}$, and a vector area element $\mathbf{d}\mathcal{A}$ in the reference configuration transform into the corresponding elements dv, dx, and $\mathbf{d}a$ in the deformed configuration, by the formulæ:

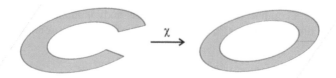

Fig. 2.1 Schematic of an elastic body in the reference state (left) and the deformed state with self-contact (right)

$$dv = J d\mathcal{V}, \qquad d\mathbf{x} = \mathbf{A} d\mathbf{X}, \qquad d\mathbf{a} = (\text{Cof} \mathbf{A}) \, d\mathcal{A}, \qquad (2.4)$$

where $J = \det \mathbf{A}$ and $\text{Cof} \mathbf{A} = J \mathbf{A}^{-T}$ is the cofactor of $\mathbf{A}$, with "T" denoting the transpose and "$-T$" the inverse transpose.

- The displacement field is defined by

$$\mathbf{u}(\mathbf{X}) = \mathbf{x} - \mathbf{X}. \qquad (2.5)$$

The gradient of the displacement is equal to

$$\nabla \mathbf{u} = \text{Grad } \mathbf{u} = \mathbf{A} - \mathbf{I}, \qquad (2.6)$$

where

$$\mathbf{I} = \text{diag}(1, 1, 1) \qquad (2.7)$$

is the second-order tensor identity, with $\text{diag}(a, b, c)$ denoting the diagonal second-order tensor having the entries a, b, c along its first principal diagonal. Equivalently, $\mathbf{I} = (\delta_{ij})_{i, j = 1, 2, 3}$, with the entries given by the Kronecker delta: $\delta_{ii} = 1$ for $i = 1, 2, 3$ and $\delta_{ij} = 0$ for $i \neq j, i, j = 1, 2, 3$.

Polar decomposition is central for the geometrical interpretation of the deformation in elasticity [558, pp. 52–53] and shows that every deformation gradient can be expressed as the product between a rotation and a stretch tensor[2] (see Fig. 2.2). The *polar decomposition theorem* states that $\mathbf{A}$ has two unique multiplicative decompositions, namely,

$$\mathbf{A} = \mathbf{R} \mathbf{U} \qquad \text{and} \qquad \mathbf{A} = \mathbf{V} \mathbf{R}, \qquad (2.8)$$

Fig. 2.2 Polar decomposition of the deformation gradient for an elastic solid

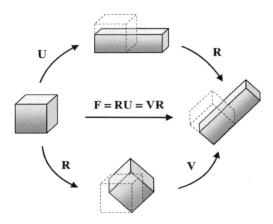

$$F = RU = VR$$

---

[2] An efficient algorithm for the computation of the polar decomposition of a 3×3 matrix is described in [251].

where

$$\mathbf{U} = \left(\mathbf{A}^T\mathbf{A}\right)^{1/2} \quad \text{and} \quad \mathbf{V} = \left(\mathbf{A}\mathbf{A}^T\right)^{1/2} \tag{2.9}$$

are symmetric and positive definite, representing the right and left stretch tensors, respectively, and $\mathbf{R}$ is proper orthogonal (i.e., $\mathbf{R}^{-1} = \mathbf{R}^T$ and $\det \mathbf{R} = 1$), representing the rotation tensor (see Appendix B.1 for a detailed proof).

The following tensors are symmetric and positive definite by construction: the right Cauchy–Green tensor

$$\mathbf{C} = \mathbf{U}^2 = \mathbf{A}^T\mathbf{A} \tag{2.10}$$

and the left Cauchy–Green tensor

$$\mathbf{B} = \mathbf{V}^2 = \mathbf{A}\mathbf{A}^T. \tag{2.11}$$

By (2.8), the identity $\mathbf{V} = \mathbf{R}\mathbf{U}\mathbf{R}^T$ holds, and therefore, both the right and left stretch tensors $\mathbf{U}$ and $\mathbf{V}$ have as eigenvalues the singular values of the tensor $\mathbf{A}$,

$$\mathrm{eig}(\mathbf{U}) = \mathrm{eig}(\mathbf{V}) = (\alpha_1, \alpha_2, \alpha_3), \tag{2.12}$$

such that $\alpha_1 \geq \alpha_2 \geq \alpha_3 > 0$. These values represent the *stretch ratios* in the principal directions and are usually referred to as "the principal stretches" [558, p. 53]. It follows that

$$\mathbf{B} = \mathbf{V}^2 = \mathbf{R}\mathbf{U}^2\mathbf{R}^T = \mathbf{R}\mathbf{C}\mathbf{R}^T, \tag{2.13}$$

and hence, the eigenvalues of both the elastic Cauchy–Green tensors are

$$\mathrm{eig}(\mathbf{C}) = \mathrm{eig}(\mathbf{B}) = (\alpha_1^2, \alpha_2^2, \alpha_3^2). \tag{2.14}$$

The principal invariants of these tensors take the form [517]

$$I_1 = \mathrm{tr}\,\mathbf{B} = \mathrm{tr}\,\mathbf{C}, \qquad I_2 = \mathrm{tr}(\mathrm{Cof}\mathbf{B}) = \mathrm{tr}(\mathrm{Cof}\mathbf{C}), \qquad I_3 = \det \mathbf{B} = \det \mathbf{C}. \tag{2.15}$$

Equivalently, in terms of the principal stretches,

$$I_1 = \alpha_1^2 + \alpha_2^2 + \alpha_3^2, \qquad I_2 = \alpha_1^2\alpha_2^2 + \alpha_2^2\alpha_3^2 + \alpha_3^2\alpha_1^2, \qquad I_3 = \alpha_1^2\alpha_2^2\alpha_3^2. \tag{2.16}$$

Note that

$$I_1 \geq 3I_3^{1/3} \quad \text{and} \quad I_2 \geq 3I_3^{2/3}, \tag{2.17}$$

where equality holds if and only if $\alpha_1 = \alpha_2 = \alpha_3$ [176]. Relations (2.17) are useful in the construction of constitutive models for compressible hyperelastic materials.

2.1.1 Hyperelastic Materials

A material is *homogeneous* if there exists a reference configuration such that all material particles respond in the same way to the deformations described with respect to this configuration, i.e., the material properties are independent of position. Additionally, it is *isotropic* if the material behaviour is the same in all directions.

We are interested in hyperelastic (Green[3] elastic) materials, that is, materials characterised by a strain-energy density function. For homogeneous isotropic hyperelastic materials, the strain-energy function depends only on the deformation gradient $\mathbf{A}$, i.e.,

$$W = W(\mathbf{A}). \tag{2.18}$$

Strain-energy functions can be assumed equal to zero in the reference configuration, i.e., $W(\mathbf{I}) = 0$, and they are based on the following fundamental assumptions:

Material Objectivity (Frame-Indifference) This condition states that constitutive equations must be invariant under changes of frame of reference [558, p. 44]. It requires that the scalar strain-energy function, $W = W(\mathbf{A})$, depending only on the deformation gradient $\mathbf{A}$, with respect to the reference configuration, is unaffected by a superimposed rigid-body transformation, which involves a change of position after deformation, i.e., $W(\mathbf{R}^T \mathbf{A}) = W(\mathbf{A})$, where $\mathbf{R} \in SO(3)$ is a proper orthogonal tensor (rotation). Material objectivity holds if and only if $W(\mathbf{A}) = W(\mathbf{U})$, where $\mathbf{U}^2 = \mathbf{A}^T \mathbf{A}$, and it is guaranteed by defining strain-energy functions in terms of the principal invariants, I_1, I_2, I_3, of the Cauchy–Green tensors.

Material Isotropy This property requires that the strain-energy function is unaffected by a rigid-body transformation prior to deformation, i.e., $W(\mathbf{A}\mathbf{R}) = W(\mathbf{A})$, where $\mathbf{R} \in SO(3)$ [213, 461], [558, p. 139]. For isotropic materials, the strain-energy function is a symmetric function of the principal stretches, $\alpha_1, \alpha_2, \alpha_3$ [566].

Although isotropic strain-energy functions are the main focus of this book, some anisotropic relations, having different constitutive properties in different directions, will also be considered.[4]

---

[3] The notion of "potential function" was introduced, and the properties of this function were considered and applied, by George Green (1828) in "An Essay on the Application of Mathematical Analysis to the Theories of Electricity and Magnetism" [214] (see also [86, 162, 518]).

[4] A review article on isotropic and anisotropic hyperelastic models is [92].

2.1.2   Strains

For an elastic deformation with gradient tensor $\mathbf{A}$, we define the following general
elastic strain tensors [362] (see also [415, pp. 89–91,118,156,159]):

$$
\mathbf{e}_n = \begin{cases} (\mathbf{B}^{n/2} - \mathbf{I})/n & \text{if } n < 0, \\ \ln \mathbf{C}^{1/2} & \text{if } n = 0, \\ (\mathbf{C}^{n/2} - \mathbf{I})/n & \text{if } n > 0. \end{cases} \tag{2.19}
$$

Some particular cases are often encountered in engineering applications, namely:
for $n = 0$, $\mathbf{e}_0 = \mathbf{e}^{(H)}$ is known as the Hencky[5] (logarithmic, true) strain tensor
[248]; for $n = 1$, $\mathbf{e}_1 = \mathbf{e}^{(B)}$ is the so-called Biot strain tensor [61]; for $n = 2$,
$\mathbf{e}_2 = \mathbf{e}^{(G)}$ is the (Lagrangian) Green–St Venant strain tensor [139, 215]; and for
$n = -2$, $\mathbf{e}_{-2} = \mathbf{e}^{(A)}$ is the (Eulerian) Almansi–Hamel strain tensor [12, 233] (see
also the review article [552]).

If the deformations are sufficiently small, then every strain tensor defined
by (2.19) takes the form

$$
\bar{\mathbf{e}} = \frac{1}{2}\left(\nabla \mathbf{u} + \nabla \mathbf{u}^T\right) \tag{2.20}
$$

and coincides with the linear strain tensor from the classical theory.

2.1.3   Stresses

To simplify the notation, we write a strain-energy function as W and infer its
argument from the context. We define the following stress tensors:

Cauchy Stress The Cauchy (or true) stress tensor represents the internal force
acting within the deformed solid on a unit of deformed area and is equal to

$$
\mathbf{T} = J^{-1}\frac{\partial W}{\partial \mathbf{A}}\mathbf{A}^T - p\mathbf{I} = 2J^{-1}\mathbf{A}\frac{\partial W}{\partial \mathbf{C}}\mathbf{A}^T - p\mathbf{I} = 2J^{-1}\frac{\partial W}{\partial \mathbf{B}}\mathbf{B} - p\mathbf{I}, \tag{2.21}
$$

where $p = 0$ for unconstrained materials, and $J = 1$, with $J = \det \mathbf{A}$ (i.e.,
deformations are isochoric [558, pp. 71–72]), for incompressible materials. This
stress tensor is symmetric, i.e., $\mathbf{T}^T = \mathbf{T}$.

For materials constrained by the internal condition $C(\mathbf{A}) = 0$, where $C(\mathbf{A}) =
\det \mathbf{A} - c = J - c$, with c a positive constant and $J = \det \mathbf{A}$, the scalar p represents
the Lagrange multiplier associated with the internal constraint and is commonly
referred to as the arbitrary hydrostatic pressure [205, pp. 286–287], [415, pp. 198–
201].

---

[5] See biographical article [534].

For unconstrained materials, the Cauchy stress tensor defined by (2.21) can be represented equivalently as follows [558, p. 140]:

$$\mathbf{T} = 2J^{-1} \left(\frac{\partial W}{\partial I_1} \frac{\partial I_1}{\partial \mathbf{B}} + \frac{\partial W}{\partial I_2} \frac{\partial I_2}{\partial \mathbf{B}} + \frac{\partial W}{\partial I_3} \frac{\partial I_3}{\partial \mathbf{B}} \right) \mathbf{B} = \beta_0 \, \mathbf{I} + \beta_1 \, \mathbf{B} + \beta_{-1} \, \mathbf{B}^{-1},$$
$$(2.22)$$

where

$$\beta_0 = \frac{2}{\sqrt{I_3}} \left(I_2 \frac{\partial W}{\partial I_2} + I_3 \frac{\partial W}{\partial I_3} \right), \qquad \beta_1 = \frac{2}{\sqrt{I_3}} \frac{\partial W}{\partial I_1}, \qquad \beta_{-1} = -2\sqrt{I_3} \frac{\partial W}{\partial I_2}$$
$$(2.23)$$

are scalar functions of the invariants given by (2.16) [558, p. 23]. Thus the Cauchy stress tensor $\mathbf{T}$ and the left Cauchy–Green tensor $\mathbf{B}$ are coaxial, i.e., they have the same eigenvectors.

If the material is incompressible, or more generally, if it satisfies the constraint that $J = \det \mathbf{A}$ is constant, then the Cauchy stress takes the form [415, p. 198–201]

$$\mathbf{T} = -p\,\mathbf{I} + \beta_1 \, \mathbf{B} + \beta_{-1} \, \mathbf{B}^{-1}. \qquad (2.24)$$

When the material is unconstrained, denoting by $\{\alpha_i\}_{i=1,2,3}$ the principal stretches of a given deformation, the corresponding principal components (eigenvalues) of the Cauchy stress tensor are equal to [558, p. 143]

$$T_i = J^{-1} \frac{\partial W}{\partial \alpha_i} \alpha_i = J^{-1} \frac{\partial W}{\partial (\ln \alpha_i)}, \qquad i = 1, 2, 3. \qquad (2.25)$$

Equivalently, by the representation (2.22),

$$T_i = \beta_0 + \beta_1 \alpha_i^2 + \beta_{-1} \alpha_i^{-2}, \qquad i = 1, 2, 3. \qquad (2.26)$$

The principal Cauchy stresses can also be expressed equivalently as follows:

$$T_i = \frac{\partial W}{\partial \iota_1} \frac{\partial \iota_1}{\partial \ln \alpha_i} + \frac{\partial W}{\partial \iota_2} \frac{\partial \iota_2}{\partial \ln \alpha_i} + \frac{\partial W}{\partial \iota_3} \frac{\partial \iota_3}{\partial \ln \alpha_i} = \zeta_0 + \zeta_1 \ln \alpha_i + \zeta_{-1} (\ln \alpha_i)^{-1}, \qquad i = 1, 2, 3,$$
$$(2.27)$$

where

$$\iota_1 = \ln \alpha_1 + \ln \alpha_2 + \ln \alpha_3,$$
$$\iota_2 = \ln \alpha_1 \ln \alpha_2 + \ln \alpha_2 \ln \alpha_3 + \ln \alpha_3 \ln \alpha_1, \qquad (2.28)$$
$$\iota_3 = \ln \alpha_1 \ln \alpha_2 \ln \alpha_3$$

are the principal invariants of the logarithmic stretch tensors $\ln \mathbf{U}$ and $\ln \mathbf{V}$ (with the logarithmic function applied independently on every component of $\mathbf{U}$ and $\mathbf{V}$, respectively), and

$$\zeta_0 = \frac{\partial W}{\partial \iota_1} + \iota_1 \frac{\partial W}{\partial \iota_2}, \qquad \zeta_1 = -\frac{\partial W}{\partial \iota_2}, \qquad \zeta_{-1} = \iota_3 \frac{\partial W}{\partial \iota_3} \qquad (2.29)$$

are scalar functions of the invariants defined by (2.28).

If $J = \alpha_1 \alpha_2 \alpha_3$ is constant, and in particular, if the material is incompressible, i.e., $J = 1$, then the principal Cauchy stress components take the form

$$T_i = -p + \frac{\partial W}{\partial \alpha_i} \alpha_i = -p + \frac{\partial W}{\partial (\ln \alpha_i)}, \qquad i = 1, 2, 3. \qquad (2.30)$$

Equivalently, by the representation (2.24),

$$T_i = -p + \beta_1 \alpha_i^2 + \beta_{-1} \alpha_i^{-2}, \qquad i = 1, 2, 3, \qquad (2.31)$$

where β_1 and β_{-1} are given by (2.23) and p is the arbitrary hydrostatic pressure. In this case, expressing the first two principal invariants as in (2.16) leads to the following equivalent forms of the constitutive coefficients in terms of the principal stretches

$$\beta_1 = \frac{1}{\alpha_1^2 - \alpha_2^2} \left(\frac{\alpha_1^2 + \alpha_3^2}{\alpha_1} \frac{\partial W}{\partial \alpha_1} - \frac{\alpha_2^2 + \alpha_3^2}{\alpha_2} \frac{\partial W}{\partial \alpha_2} \right), \qquad \beta_{-1} = \frac{1}{\alpha_1^2 - \alpha_2^2} \left(\frac{1}{\alpha_1} \frac{\partial W}{\partial \alpha_1} - \frac{1}{\alpha_2} \frac{\partial W}{\partial \alpha_2} \right).$$
$$(2.32)$$

The principal stresses defined by (2.31) are also equal to

$$T_i = -p + \zeta_1 \ln \alpha_i + \zeta_{-1} (\ln \alpha_i)^{-1}, \qquad i = 1, 2, 3, \qquad (2.33)$$

where ζ_1 and ζ_{-1} are given by (2.29).

First Piola–Kirchhoff Stress The first Piola–Kirchhoff stress tensor represents the internal force acting within the deformed solid on an area element that, in its reference state, was one unit of area. This material stress tensor, which is not symmetric in general, takes the form

$$\mathbf{P} = T\mathrm{Cof}\mathbf{A} = J T \mathbf{A}^{-T} = \frac{\partial W}{\partial \mathbf{A}} - p \mathbf{A}^{-T}, \qquad (2.34)$$

where $p = 0$ for unconstrained materials, and $J = \det \mathbf{A} = 1$ for incompressible materials. The transpose tensor $\mathbf{P}^T$ is known as the nominal (engineering) stress [415, pp. 152–153].

When the material is unconstrained, the principal components (eigenvalues) of the first Piola–Kirchhoff stress tensor given by (2.34) are

$$P_i = J T_i \alpha_i^{-1} = \frac{\partial W}{\partial \alpha_i}, \qquad i = 1, 2, 3. \qquad (2.35)$$

Equivalently,

$$P_i = \frac{\partial W}{\partial i_1}\frac{\partial i_1}{\partial \alpha_i} + \frac{\partial W}{\partial i_2}\frac{\partial i_2}{\partial \alpha_i} + \frac{\partial W}{\partial i_3}\frac{\partial i_3}{\partial \alpha_i} = \rho_0 + \rho_1\alpha_i + \rho_{-1}\alpha_i^{-1}, \qquad i = 1, 2, 3,$$

(2.36)

where

$$i_1 = \alpha_1 + \alpha_2 + \alpha_3, \qquad i_2 = \alpha_1\alpha_2 + \alpha_2\alpha_3 + \alpha_3\alpha_1, \qquad i_3 = \alpha_1\alpha_2\alpha_3 \qquad (2.37)$$

are the principal invariants of the stretch tensors $\mathbf{U}$ and $\mathbf{V}$, and

$$\rho_0 = \frac{\partial W}{\partial i_1} + i_1\frac{\partial W}{\partial i_2}, \qquad \rho_1 = -\frac{\partial W}{\partial i_2}, \qquad \rho_{-1} = i_3\frac{\partial W}{\partial i_3} \qquad (2.38)$$

are scalar functions of the invariants given by (2.37).

For materials satisfying the constraint that $J = \alpha_1\alpha_2\alpha_3$ is constant, and in particular, for incompressible materials, with $J = 1$, the principal components of the first Piola–Kirchhoff stress tensor take the form

$$P_i = T_i\alpha_i^{-1} = -p\alpha_i^{-1} + \frac{\partial W}{\partial \alpha_i}, \qquad i = 1, 2, 3, \qquad (2.39)$$

or equivalently,

$$P_i = \rho_0 + \rho_1\alpha_i - p_0\alpha_i^{-1}, \qquad i = 1, 2, 3, \qquad (2.40)$$

where ρ_0 and ρ_1 are given by (2.38) and p_0 is the arbitrary hydrostatic pressure.

Second Piola–Kirchhoff Stress   The second Piola–Kirchhoff stress tensor is equal to

$$\mathbf{S} = \mathbf{A}^{-1}\mathbf{P} = J\mathbf{A}^{-1}\mathbf{T}\mathbf{A}^{-T} = 2\frac{\partial W}{\partial \mathbf{C}} - p\mathbf{C}^{-1}, \qquad (2.41)$$

where $p = 0$ for unconstrained materials and $J = \det \mathbf{A} = 1$ for incompressible materials. This material stress tensor is symmetric and is especially useful in computational approaches [55, 320, 407].

For unconstrained materials, the second Piola–Kirchhoff stress tensor takes the equivalent form

$$\mathbf{S} = 2\left(\frac{\partial W}{\partial I_1}\frac{\partial I_1}{\partial \mathbf{C}} + \frac{\partial W}{\partial I_2}\frac{\partial I_2}{\partial \mathbf{C}} + \frac{\partial W}{\partial I_3}\frac{\partial I_3}{\partial \mathbf{C}}\right) = \gamma_0\mathbf{I} + \gamma_1\mathbf{C} + \gamma_{-1}\mathbf{C}^{-1}, \qquad (2.42)$$

where

$$\gamma_0 = 2\left(\frac{\partial W}{\partial I_1} + I_1\frac{\partial W}{\partial I_2}\right), \qquad \gamma_1 = -2\frac{\partial W}{\partial I_2}, \qquad \gamma_{-1} = 2I_3\frac{\partial W}{\partial I_3} \qquad (2.43)$$

are the scalar functions of the principal invariants given by (2.16). Thus the second Piola–Kirchhoff stress tensor $\mathbf{S}$ and the right Cauchy–Green tensor $\mathbf{C}$ are coaxial.

When $J = \det \mathbf{A}$ is constant, and in particular, when the material is incompressible, i.e., $J = 1$, the second Piola–Kirchhoff stress takes the form

$$\mathbf{S} = \gamma_0\,\mathbf{I} + \gamma_1\,\mathbf{C} - p_0\,\mathbf{C}^{-1}, \tag{2.44}$$

where γ_0 and γ_1 are given by (2.43), and p_0 is the arbitrary hydrostatic pressure.

When the material is unconstrained, the principal components of the second Piola–Kirchhoff stress tensor defined by (2.41) are equal to

$$S_i = \alpha_i^{-1} P_i = 2\frac{\partial W}{\partial \alpha_i^2}, \qquad i = 1, 2, 3, \tag{2.45}$$

or equivalently, by the representation (2.42),

$$S_i = \gamma_0 + \gamma_1 \alpha_i^2 + \gamma_{-1}\alpha_i^{-2}, \qquad i = 1, 2, 3. \tag{2.46}$$

For materials where $J = \alpha_1\alpha_2\alpha_3$ is constant, and in particular, for incompressible materials, the principal second Piola–Kirchhoff stresses are equal to

$$S_i = \alpha_i^{-1} P_i = -p\alpha_i^{-2} + 2\frac{\partial W}{\partial \alpha_i^2}, \qquad i = 1, 2, 3. \tag{2.47}$$

Equivalently, by the representation (2.44),

$$S_i = \gamma_0 + \gamma_1 \alpha_i^2 - p_0 \alpha_i^{-2}, \qquad i = 1, 2, 3, \tag{2.48}$$

where p_0 is the arbitrary hydrostatic pressure.

2.1.4 Adscititious Inequalities

Some empirical (adscititious) conditions are universally accepted as constraints on the constitutive equations in order for the behaviour of hyperelastic materials to be physically plausible [42, 154, 423, 554], [205, p. 291], [558, pp. 153–171]. Such constraints must be chosen carefully, lest they should be unduly restrictive. The following conditions have been selected because they allow for a wide range of interesting and physically realistic material behaviours, as demonstrated in [359–362, 372].

Pressure-Compression (PC) Inequalities For a compressible isotropic material, the fact that the volume of the material is decreased by hydrostatic compression and increased by hydrostatic tension is expressed by the pressure-compression (PC) inequalities stating that each principal stress is a tension or a compression if the corresponding principal stretch is an extension or a contraction [558, p. 155], i.e.,

$$T_i (\alpha_i - 1) > 0, \qquad i = 1, 2, 3. \tag{2.49}$$

Physically, either or both of the following mean versions of the PC conditions are more realistic,

$$T_1 (\alpha_1 - 1) + T_2 (\alpha_2 - 1) + T_3 (\alpha_3 - 1) > 0, \tag{2.50}$$

or

$$T_1 \left(1 - \frac{1}{\alpha_1}\right) + T_2 \left(1 - \frac{1}{\alpha_2}\right) + T_3 \left(1 - \frac{1}{\alpha_3}\right) > 0, \tag{2.51}$$

assuming that not all principal stretches are equal to 1.

Baker–Ericksen (BE) Inequalities These inequalities state that the greater principal Cauchy stress occurs in the direction of the greater principal stretch [35, 345], i.e.,

$$\left(T_i - T_j\right)\left(\alpha_i - \alpha_j\right) > 0 \quad \text{if} \quad \alpha_i \neq \alpha_j, \quad i, j = 1, 2, 3, \tag{2.52}$$

where $\{\alpha_i\}_{i=1,2,3}$ and $\{T_i\}_{i=1,2,3}$ denote the principal stretches and principal Cauchy stresses, respectively. Equivalently, by (2.25),

$$\left(\alpha_i \frac{\partial W}{\partial \alpha_i} - \alpha_j \frac{\partial W}{\partial \alpha_j}\right)(\alpha_i - \alpha_j) > 0 \quad \text{if} \quad \alpha_i \neq \alpha_j, \quad i, j = 1, 2, 3. \tag{2.53}$$

When any two principal stretches are equal, the strict inequality ">" is replaced by "$\geq$".

Theorem 2.1.1 ([345]) *For a homogeneous isotropic hyperelastic material subject to uniaxial tension, with the Cauchy stress tensor taking the form*

$$\mathbf{T} = \text{diag}\,(0, N, 0), \tag{2.54}$$

where $N > 0$, the deformation is a simple extension in the direction of the tensile force, i.e., the left Cauchy–Green tensor has the representation

$$\mathbf{B} = \text{diag}\,(B_{11}, B_{22}, B_{11}), \tag{2.55}$$

where $B_{11} < B_{22}$, if and only if the Baker–Ericksen inequalities (2.52) hold.

Proof For a homogeneous isotropic hyperelastic material, the Cauchy stress is given by (2.22) (or by (2.24) if the material is incompressible). The left Cauchy–Green tensor $\mathbf{B}$ is symmetric and has the general form

$$\mathbf{B} = \begin{bmatrix} B_{11} & B_{12} & B_{13} \\ B_{12} & B_{22} & B_{23} \\ B_{13} & B_{23} & B_{33} \end{bmatrix}.$$

As the Cauchy stress tensor and the left Cauchy–Green tensor are coaxial, we have $\mathbf{TB} = \mathbf{BT}$, from which we deduce that $B_{12} = B_{23} = 0$.

Assuming that the BE inequalities (2.52) hold, since the eigenvalues α_1^2 and α_3^2 of $\mathbf{B}$ are the roots of the characteristic equation

$$\lambda^2 - (B_{11} + B_{33})\lambda + B_{11}B_{33} - B_{13}^2 = 0,$$

while the corresponding principal Cauchy stresses are $T_1 = T_3 = 0$, we obtain $\alpha_1^2 = \alpha_3^2$, i.e., the above quadratic equation has a multiple solution. Hence, $(B_{11} - B_{33})^2 + 4B_{13}^2 = 0$, i.e., $B_{11} = B_{33}$ and $B_{13} = 0$. Thus, the left Cauchy–Green tensor has the form

$$\mathbf{B} = \operatorname{diag}(B_{11}, B_{22}, B_{11}) = \operatorname{diag}\left(\alpha_1^2, \alpha_2^2, \alpha_1^2\right).$$

Finally, since $T_2 - T_1 = N > 0$, the BE inequalities imply $\alpha_2 > \alpha_1$; hence, $B_{22} > B_{11}$.

Conversely, if $B_{22} > B_{11}$, then

$$(T_2 - T_1)(\alpha_2 - \alpha_1) = N\left(\sqrt{B_{22}} - \sqrt{B_{11}}\right) > 0,$$

i.e., the BE inequalities are satisfied. This completes the proof.   □

Generalised Empirical (GE) Inequalities The following conditions on the constitutive coefficients given by (2.23) [360, 361]

$$\beta_0 \leq 0, \qquad \beta_1 > 0 \qquad\qquad (2.56)$$

generalise the more restrictive set of *empirical inequalities*: $\beta_0 \leq 0$, $\beta_1 > 0$, and $\beta_{-1} \leq 0$ [558, p. 158], which imply the BE inequalities, but are not implied by them. For incompressible materials, the condition on β_0 is not needed.

In general, the PC inequalities (2.49), the BE inequalities (2.52), and the GE inequalities (2.56) do not imply each other.

2.2 Nonlinear Elastic Moduli in Isotropic Elasticity

Elastic moduli are defined to quantify the elastic responses of a material during deformation. Such moduli can be inferred approximately from macroscopic shape changes if the applied force is known. For a homogeneous isotropic linearly elastic (Hookean) material, where the force needed to extend or compress it by some distance is proportional to that distance [541, 552], two physical constants, namely, the Young's modulus and the Poisson's ratio, are sufficient to fully characterise its elastic behaviour. These parameters can be derived from a simple tensile or compressive experiment. Then any other linear elastic parameter can be obtained from these two constants [330]. For instance, the shear modulus is directly proportional to the Young's modulus, with the proportionality constant depending on the Poisson's ratio. A similar *universal relation* holds also in nonlinear elasticity, but the moduli are functions of the deformation [362] (see Table 2.1). Universal relations are equations that hold for every material in a specific class [50, 241, 449].

At large strains, the complexity of defining elastic moduli comes from the fact that there are multiple ways of measuring stresses and strains. This gives rise to multiple nonlinear functions that correspond to the same linear parameter at small strains. The choice of these functions depends on how a particular experiment is conducted and how the experimental data are processed [93, 160, 551, 602]. Moreover, although under given forces, many isotropic elastic materials deform uniquely, for nonlinear hyperelastic materials, this is not always the case [368, 369, 398, 484].

For well-known elastic moduli from linear elasticity, such as stretch, bulk, and shear moduli, nonlinear extensions are obtained under homogeneous deformations. These are deformations for which the gradient tensor in Cartesian coordinates is constant. They are *universal* in the sense that they can be obtained in any hyperelastic material from a given class, and *controllable*, i.e., they can be maintained entirely by tractions on the boundary of the body. This is formally stated by the following theorem (see [494, 497] or [415, p. 244–246]).

Theorem 2.2.1 (Ericksen's Theorem [155, 156]) *A deformation of an arbitrary homogeneous isotropic hyperelastic body can be maintained by the application of surface tractions only (without body forces) if and only if it is a homogeneous deformation in the Cartesian coordinates.*

Table 2.1 Universal relation between nonlinear elastic parameters of homogeneous isotropic hyperelastic materials

Universal relation	Equation	Linear elastic limit
$\frac{E(a)}{\tilde{\mu}(a)} = \frac{a^2 - a^{-2v_0(a)}}{\ln a^{1+v_0(a)}}\left(1 + v_0(a) + a v_0'(a)\ln a\right)$	(2.119)	$\frac{\bar{E}}{\bar{\mu}} = \lim_{a\to 1}\frac{E(a)}{\tilde{\mu}(a)} = 2\left(1 + \bar{v}\right)$
$\frac{E(a)}{\tilde{\mu}(a)} = \frac{a^2 - a^{-2\bar{v}}}{\ln a}$	(2.120)	$\frac{\bar{E}}{\bar{\mu}} = \lim_{a\to 1}\frac{E(a)}{\tilde{\mu}(a)} = 2\left(1 + \bar{v}\right)$
$\frac{E(a)}{\hat{\mu}(a)} = \frac{a^2 - a^{-1}}{\ln a}$	(2.121)	$\frac{\bar{E}}{\hat{\mu}} = \lim_{a\to 1}\frac{E(a)}{\hat{\mu}(a)} = 3$

Table 2.2 Nonlinear stretch moduli for homogeneous isotropic hyperelastic materials subject to finite axial stretch (2.54), with stretch parameter $a > 0$. In the small strain limit, these moduli are equal to the Young's modulus from linear elasticity.

Stretch modulus	Equation	Stress tensor required	Linear elastic limit
$E^{\mathrm{incr}}(a) = \frac{\partial T_2}{\partial(\ln a)}$	(2.78)	Cauchy (2.21)	$\overline{E} = \lim_{a \to 1} E^{\mathrm{incr}}(a)$
$E(a) = \frac{T_2}{\ln a - \ln \lambda(a)}\left(1 - \frac{a\lambda'(a)}{\lambda(a)}\right)$	(2.79)	Cauchy (2.21)	$\overline{E} = \lim_{a \to 1} E(a)$
$\widetilde{E}^{\mathrm{incr}}(a) = \frac{\partial P_2}{\partial(a-1)}$	(2.83)	First Piola–Kirchhoff (2.34)	$\overline{\widetilde{E}} = \lim_{a \to 1} \widetilde{E}^{\mathrm{incr}}(a)$
$\widetilde{E}(a) = \frac{P_2}{a - \lambda(a)}\left(1 - \lambda'(a)\right)$	(2.84)	First Piola–Kirchhoff (2.34)	$\overline{\widetilde{E}} = \lim_{a \to 1} \widetilde{E}(a)$
$\overset{\approx}{E}{}^{\mathrm{incr}}(a) = \frac{2\partial S_2}{\partial(a^2-1)}$	(2.88)	Second Piola–Kirchhoff (2.41)	$\overline{\overset{\approx}{E}{}^{\mathrm{incr}}} = \lim_{a \to 1} \overset{\approx}{E}{}^{\mathrm{incr}}(a)$
$\overset{\approx}{E}(a) = \frac{2S_2}{a^2 - \lambda^2(a)}\left(1 - \frac{\lambda(a)\lambda'(a)}{a}\right)$	(2.89)	Second Piola–Kirchhoff (2.41)	$\overline{\overset{\approx}{E}} = \lim_{a \to 1} \overset{\approx}{E}(a)$

Table 2.3 Nonlinear Poisson functions for homogeneous isotropic hyperelastic materials subject to finite axial stretch (2.54), with stretch parameter $a > 0$. In the small strain limit, these functions are equal to the Poisson's ratio from linear elasticity

Poisson function $v_n(a)$	n	Strain tensor required	Linear elastic limit
$v^{(H)}(a) = -\frac{\ln \lambda(a)}{\ln a}$	0	Hencky (2.19)	$\overline{v} = \lim_{a \to 1} v^{(H)}(a)$
$v^{(B)}(a) = \frac{1 - \lambda(a)}{a - 1}$	1	Biot (2.19)	$\overline{v} = \lim_{a \to 1} v^{(B)}(a)$
$v^{(G)}(a) = \frac{1 - \lambda(a)^2}{a^2 - 1}$	2	Green (2.19)	$\overline{v} = \lim_{a \to 1} v^{(G)}(a)$
$v^{(A)}(a) = \frac{\lambda(a)^{-2} - 1}{1 - a^{-2}}$	−2	Almansi (2.19)	$\overline{v} = \lim_{a \to 1} v^{(A)}(a)$

Only ideal deformations of elastic bodies with simple geometries, such as cubes and cylinders, which can be analysed explicitly, are addressed here. The art and challenge of experimental setup are then to get as close as possible to the ideal situations.

Nonlinear parameters obtained under key homogeneous static deformations[6] are summarised in Tables 2.2, 2.3, and 2.4. These parameters represent changes in the material properties as the deformation progresses and can be identified with their linear elastic equivalent when the deformations are small. Such parameters play significant roles in the fundamental understanding and the application of many elastic materials under large elastic strains and also provide a tool for the coupling of elastic responses in multiscale processes where an open challenge is the transfer of meaningful information between scales. Similar parameters can be identified in homogeneous anisotropic elastic materials, where different constitutive parameters may be found in different directions.

In addition, certain nonlinear elastic moduli are derived under particular finite deformations, which are non-homogeneous but controllable for all homogeneous

---

[6] "While simple shear is the most illuminating homogeneous static deformation, other cases are important as well, notably simple extension and uniform dilation."—C. Truesdell [556, p. 115].

Table 2.4 Nonlinear shear moduli for homogeneous isotropic hyperelastic materials subject to simple shear superposed on finite axial stretch (2.98). In the small strain limit, these moduli are equal to the shear modulus from linear elasticity

Shear modulus	Equation	Stress tensor required	Linear elastic limit
$\mu(a, k) = \frac{T_{12}}{ka^2}$	(2.104)	Cauchy (2.21)	$\overline{\mu} = \lim_{a \to 1} \lim_{k \to 0} \mu(a, k)$
$\mu(a, k) = \frac{P_{12}}{ka}$	(2.105)	First Piola–Kirchhoff (2.34)	$\overline{\mu} = \lim_{a \to 1} \lim_{k \to 0} \mu(a, k)$
$\mu(a, k) = \frac{T_1 - T_2}{\alpha_1^2 - \alpha_2^2}$	(2.106)	Cauchy (2.21)	$\overline{\mu} = \lim_{a \to 1} \lim_{k \to 0} \mu(a, k)$
$\widehat{\mu}(k) = \lim_{a \to 1} \mu(a, k)$	(2.112)	Cauchy (2.21) or first Piola–Kirchhoff (2.34)	$\overline{\mu} = \lim_{k \to 0} \widehat{\mu}(k)$
$\widetilde{\mu}(a) = \lim_{k \to 0} \mu(a, k)$	(2.113)	Cauchy (2.21) or first Piola–Kirchhoff (2.34)	$\overline{\mu} = \lim_{a \to 1} \widetilde{\mu}(a)$

isotropic incompressible elastic solids in the absence of body forces [558, pp. 183–186]. Generalisations of these deformations are also possible for specific isotropic compressible materials [88].

Before we present the nonlinear elastic parameters, we briefly review the incremental constitutive parameters for isotropic elastic materials. The incremental Young's modulus, bulk modulus, and Poisson's ratio in isotropic elasticity are treated in detail by Scott [485]. Incremental moduli and their application to finite amplitude wave propagation in elastic solids are further analysed in [476].

2.2.1 Incremental Elastic Moduli

When the strain-energy function W is an analytic function of a strain tensor $\mathbf{e}$, it can be approximated as follows [407, p. 219]:

$$W \approx E_0 + E_{ij}e_{ij} + \frac{1}{2}E_{ijkl}e_{ij}e_{kl}. \tag{2.57}$$

In Eq. (2.57), repeated indices represent summation, E_0 is an arbitrary constant, $\{E_{ij}\}_{i,j=1,2,3}$ are elastic moduli of order 0, and $\{E_{ijkl}\}_{i,j,k,l=1,2,3}$ are elastic moduli of order 1 [415, p. 331]. These moduli measure changes of the stress with the changes of strain. Such changes can be estimated, for example, by the following incremental fourth-order tensors [362]:

- The gradient of the Cauchy stress tensor $\mathbf{T}$ with respect to the logarithmic strain tensor $\ln \mathbf{B}^{1/2}$,

$$\mathbf{E}^{\text{incr}} = \frac{\partial \mathbf{T}}{\partial \left(\ln \mathbf{B}^{1/2}\right)} = \frac{\partial \mathbf{T}}{\partial \left(\ln \mathbf{V}\right)}, \tag{2.58}$$

  with the components

$$E_{ijkl}^{incr} = \frac{\partial T_{ij}}{\partial \left(\ln V_{kl} \right)}, \qquad i, j, k, l = 1, 2, 3. \qquad (2.59)$$

- The gradient of the first Piola–Kirchhoff stress tensor $\mathbf{P}$ with respect to the deformation gradient $\mathbf{A}$, or equivalently, the gradient of $\mathbf{P}$ with respect to the displacement gradient $\mathbf{A} - \mathbf{I}$,

$$\mathbf{E}^{incr} = \frac{\partial \mathbf{P}}{\partial \mathbf{A}} = \frac{\partial \mathbf{P}}{\partial \left(\mathbf{A} - \mathbf{I} \right)}, \qquad (2.60)$$

with the components

$$E_{ijkl}^{incr} = \frac{\partial P_{ij}}{\partial A_{kl}} = \frac{\partial P_{ij}}{\partial \left(A_{kl} - \delta_{kl} \right)}, \qquad i, j, k, l = 1, 2, 3. \qquad (2.61)$$

Then $E_{ijkl}^{incr} > 0$ if the stress component P_{ij} increases as the strain component $A_{kl} - \delta_{kl}$ increases, and $E_{ijkl}^{incr} < 0$ if P_{ij} decreases as $A_{kl} - \delta_{kl}$ increases. The fourth-order tensor (2.60) can be expressed equivalently as [414]

$$\mathbf{E}^{incr} = \frac{\partial^2 W}{\partial \mathbf{A}^2} = \frac{\partial^2 W}{\partial \left(\mathbf{A} - \mathbf{I} \right)^2}. \qquad (2.62)$$

For the unstressed state, we have $\partial W / \partial \left(\mathbf{A} - \mathbf{I} \right) = \mathbf{P} = \mathbf{0}$, and by (2.57), we can write

$$W \approx \frac{1}{2} E_{ijkl}^{incr} \left(A_{ij} - \delta_{ij} \right) \left(A_{kl} - \delta_{kl} \right). \qquad (2.63)$$

- The gradient of the second Piola–Kirchhoff stress tensor $\mathbf{S}$ with respect to the left Cauchy–Green tensor $\mathbf{C}$, or equivalently, half of the gradient of $\mathbf{S}$ with respect to the Green strain tensor $\mathbf{e}^{(G)} = \left(\mathbf{C} - \mathbf{I} \right) / 2$,

$$\mathbf{E}^{incr} = \frac{\partial \mathbf{S}}{\partial \mathbf{C}} = \frac{\partial \mathbf{S}}{\partial \left(\mathbf{C} - \mathbf{I} \right)}, \qquad (2.64)$$

with the components

$$E_{ijkl}^{incr} = \frac{\partial S_{ij}}{\partial C_{kl}} = \frac{\partial S_{ij}}{\partial \left(C_{kl} - \delta_{kl} \right)}, \qquad i, j, k, l = 1, 2, 3. \qquad (2.65)$$

Then $E_{ijkl}^{incr} > 0$ if the stress component S_{ij} increases as the strain component $\left(C_{kl} - \delta_{kl} \right) / 2$ increases, and $E_{ijkl}^{incr} < 0$ if S_{ij} decreases as $\left(C_{kl} - \delta_{kl} \right) / 2$ increases. The fourth-order tensor (2.64) takes the equivalent form

$$\mathbf{E}^{incr} = 2\frac{\partial^2 W}{\partial \mathbf{C}^2}. \tag{2.66}$$

In this case, for the unstressed state, we have $\partial W/\partial \mathbf{e}_G = \mathbf{S} = \mathbf{0}$, and by (2.57), we can write

$$W \approx \frac{1}{8}E^{incr}_{ijkl}\left(C_{ij} - \delta_{ij}\right)\left(C_{kl} - \delta_{kl}\right). \tag{2.67}$$

The incremental elastic moduli (2.58), (2.60), and (2.64) can be calculated for any hyperelastic material for which the strain-energy function W is known, using the definitions for the corresponding stress tensors. When the strain-energy function is not known, assuming that the material is incompressible, these moduli can be approximated from a finite number of experimental measurements where the applied force is prescribed. For compressible materials, suitable body forces also need to be taken into account.

2.2.2 Stretch Moduli

For a hyperelastic body under uniaxial tension (or compression) acting in the second direction, the Cauchy stress takes the form (2.54). In this case, by Theorem 2.1.1, the corresponding deformation is a simple extension (or contraction) in the direction of the positive (or negative) axial force, whereby the ratio between the tensile strain and the strain in the orthogonal direction is greater (or less) than 1, if and only if the Baker–Ericksen (BE) inequalities (2.52) hold [35, 345]. In other words, the deformation corresponding to (2.54) takes the form (see Fig. 2.3)

$$x_1 = \lambda(a)X_1, \qquad x_2 = aX_2, \qquad x_3 = \lambda(a)X_3, \tag{2.68}$$

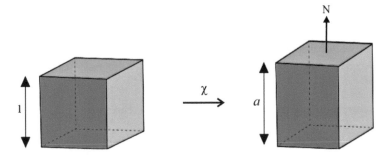

Fig. 2.3 Cuboid (left) deformed by axial stretch, described by Eq. (2.68), under the uniaxial load N (right) [362]

where (X_1, X_2, X_3) and (x_1, x_2, x_3) are the Cartesian coordinates for the reference and current configuration, respectively, a is a positive constant representing the extension (or contraction) ratio, and $\lambda(a)$ is the stretch ratio in the orthogonal direction, if and only if

$$a^2\lambda(a)^2\beta_1 - \beta_{-1} > 0. \tag{2.69}$$

In [46], it was shown that a simple tensile load produces a simple extension, i.e., $a > 1$ in (2.68), if the *empirical inequalities* $\beta_0 \leq 0$, $\beta_1 > 0$, and $\beta_{-1} \leq 0$ hold. However, this result follows from the fact that the BE conditions (2.52) are implied by these empirical inequalities.

In particular, when the deformation described by (2.68) is isochoric, i.e., $a\lambda(a)^2 = 1$, the axial stretch ratios take the form $\alpha_1 = \alpha_3 = 1/\sqrt{a}$ and $\alpha_2 = a$, and the non-zero component of the Cauchy stress is equal to

$$N = (a\beta_1 - \beta_{-1})\left(a - \frac{1}{a^2}\right). \tag{2.70}$$

For this deformation, the BE inequalities (2.52) are equivalent to

$$a\beta_1 - \beta_{-1} > 0, \tag{2.71}$$

i.e., $N > 0$ for $a > 1$, and $N < 0$ for $a < 1$, i.e., axial tension produces elongation in the same direction, and axial compression produces contraction in the same direction.

For the isochoric deformation given by (2.68), with $\lambda(a) = 1/\sqrt{a}$, since $T_{11} = T_{33} = 0$, the pressure-compression (PC) inequality (2.50) becomes

$$N(a - 1) = (a\beta_1 - \beta_{-1})\left(a - \frac{1}{a^2}\right)(a - 1) > 0, \tag{2.72}$$

which is clearly equivalent to the BE inequality (2.71). In this case, if $a > 1$, then the deformation is a uniaxial extension in the second direction, and if $0 < a < 1$, then the deformation is an equibiaxial extension in the orthogonal directions.

The gradient tensor of the homogeneous deformation (2.68) is equal to

$$\mathbf{A} = \mathrm{diag}\,(\lambda(a), a, \lambda(a)), \tag{2.73}$$

and the corresponding left and right Cauchy–Green tensors are equal and take the form

$$\mathbf{B} = \mathbf{C} = \mathrm{diag}\left(\lambda(a)^2, a^2, \lambda(a)^2\right). \tag{2.74}$$

Using (2.74), the strain tensors (2.19) are simply

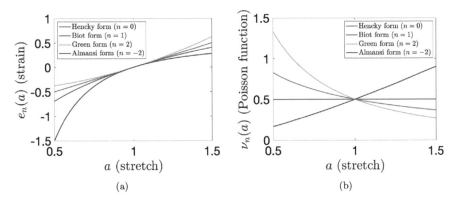

Fig. 2.4 Comparisons of (**a**) different finite axial strains $e_n(a)$ and (**b**) Poisson functions, versus axial stretch ratio a, for an incompressible material. Note that every axial strain increases with increasing axial stretch, but only by using the Hencky (logarithmic) strain, the corresponding Poisson function remains constant and equal to 0.5, capturing the characteristic property that the material volume remains fixed [362]

$$\mathbf{e}_n = \text{diag}\left(e_n(\lambda(a)), e_n(a), e_n(\lambda(a))\right), \tag{2.75}$$

where, for any given stretch ratio $x > 0$, we define the *nonlinear strain* [253]

$$e_n(x) = \begin{cases} (x^n - 1)/n & \text{if } n \neq 0, \\ \ln x & \text{if } n = 0. \end{cases} \tag{2.76}$$

Figure 2.4 (left) shows the values of different strain measures in the second direction as the stretch ratio a varies.

To define a nonlinear version of the Young's modulus that quantifies the nonlinear elastic stress–strain response of an isotropic hyperelastic material under the uniaxial tension (or compression) characterised by (2.54) and (2.68), we introduce the *nonlinear stretch modulus*. This modulus reflects stiffening (or softening) in a material under increasing axial load. That is, when the axial stress increases as the axial deformation increases, there is an increase of the stretch modulus and the material stiffens, and if the axial stress decreases as the axial deformation increases, then there is a corresponding decrease in the stretch modulus as the material softens. As the stretch modulus depends on both a stress and a strain, there are multiple choices based on the particular stress and strain tensors considered. Here, we consider three typical moduli [362]:

- For the Cauchy stress tensor given by (2.54), by subtracting the third from the second principal component defined by (2.27), we obtain

$$\mathcal{T}_2 = T_2 - T_3 = (\ln a - \ln \lambda(a))\left(\zeta_1 - \frac{\zeta_{-1}}{\ln a \ln \lambda(a)}\right). \tag{2.77}$$

It follows that T_2 is proportional to $\ln a - \ln \lambda(a)$, and similarly for incompressible materials, with $\lambda(a) = 1/\sqrt{a}$ if we subtract the third from the second principal component given by (2.33). Applying the general formula for the elastic moduli (2.59), we can define the *incremental stretch modulus* in terms of the logarithmic strain e_0 as follows:

$$E^{\mathrm{incr}}(a) = \frac{\partial T_2}{\partial (\ln a)} = \frac{\partial T_2}{\partial (\ln a - \ln \lambda(a))} \left(1 - \frac{a\lambda'(a)}{\lambda(a)} \right). \qquad (2.78)$$

Alternatively, as T_2 is proportional to $\ln a - \ln \lambda(a)$, we can define a *nonlinear stretch modulus* of the form

$$E(a) = \frac{T_2}{\ln a - \ln \lambda(a)} \left(1 - \frac{a\lambda'(a)}{\lambda(a)} \right). \qquad (2.79)$$

For incompressible materials, where $\lambda(a) = 1/\sqrt{a}$, (2.79) simplifies to

$$E(a) = \frac{T_2(a)}{\ln a}. \qquad (2.80)$$

When $a \to 1$, i.e., for small axial strains, both the incremental modulus defined by (2.78), commonly known as the tangent modulus, and the nonlinear modulus given by (2.79), also known as the secant modulus, converge to the Young's modulus from the linear elastic theory

$$\overline{E} = \lim_{a \to 1} E^{\mathrm{incr}}(a) = \lim_{a \to 1} E(a) = \lim_{a \to 1} \frac{T_2(a)}{\ln a}. \qquad (2.81)$$

- For the corresponding first Piola–Kirchhoff stress tensor, by subtracting the third from the second principal component given by (2.36), we find

$$P_2 = P_2 - P_3 = (a - \lambda(a)) \left(\rho_1 - \frac{\rho_{-1}}{a\lambda(a)} \right). \qquad (2.82)$$

Hence, P_2 is proportional to $a - \lambda(a)$, and similarly for incompressible materials if we subtract the third from the second principal component given by (2.40). Applying the general formula for the elastic moduli (2.61), we define the *incremental stretch modulus* [55, pp. 224]

$$\widetilde{E}^{\mathrm{incr}}(a) = \frac{\partial P_2}{\partial (a - 1)} = \frac{\partial P_2}{\partial (a - \lambda(a))} \left(1 - \lambda'(a) \right). \qquad (2.83)$$

In this case, as P_2 is proportional to $a - \lambda(a)$, we can also define the *nonlinear stretch modulus*

$$\tilde{E}(a) = \frac{P_2}{a - \lambda(a)} \left(1 - \lambda'(a)\right). \tag{2.84}$$

For incompressible materials, (2.84) takes the form

$$\tilde{E}(a) = \frac{P_2}{a^{3/2} - 1} \left(a^{1/2} + \frac{1}{2a}\right). \tag{2.85}$$

When $a \to 1$, both elastic moduli (2.83) and (2.84) converge to the Young's modulus

$$\overline{E} = \lim_{a \to 1} E^{\mathrm{incr}}(a) = \lim_{a \to 1} \tilde{E}(a) = \lim_{a \to 1} \frac{T_2}{\ln a}. \tag{2.86}$$

- For the associated second Piola–Kirchhoff stress tensor, subtracting the third from the second principal component given by (2.46) yields

$$S_2 = S_2 - S_3 = \left(a^2 - \lambda^2(a)\right) \left(\gamma_1 - \frac{\gamma_{-1}}{a^2 \lambda(a)^2}\right), \tag{2.87}$$

i.e., S_2 is proportional to $a^2 - \lambda(a)^2$, and similarly for incompressible materials when we subtract the third from the second principal component given by (2.48). Then, using the formula for the elastic moduli (2.65), we define the following *incremental stretch modulus* in terms of the strain measure $\mathbf{e}_2$:

$$\tilde{E}^{\mathrm{incr}}(a) = \frac{2 \partial S_2}{\partial \left(a^2 - 1\right)} = \frac{2 \partial S_2}{\partial \left(a^2 - \lambda(a)^2\right)} \left(1 - \frac{\lambda(a)\lambda'(a)}{a}\right). \tag{2.88}$$

Alternatively, as S_2 is proportional to $a^2 - \lambda(a)^2$, we can define the *nonlinear stretch modulus*

$$\tilde{\tilde{E}}(a) = \frac{2 S_2}{a^2 - \lambda(a)^2} \left(1 - \frac{\lambda(a)\lambda'(a)}{a}\right). \tag{2.89}$$

If the material is incompressible, then (2.89) becomes

$$\tilde{\tilde{E}}(a) = \frac{2 S_2}{a^3 - 1} \left(a + \frac{1}{2a^2}\right). \tag{2.90}$$

When $a \to 1$, both moduli defined by (2.83) and (2.84), respectively, converge to the Young's modulus

$$\overline{E} = \lim_{a \to 1} E^{\mathrm{incr}}(a) = \lim_{a \to 1} \tilde{\tilde{E}}(a) = \lim_{a \to 1} \frac{T_2}{\ln a}. \tag{2.91}$$

These nonlinear stretch moduli are summarised in Table 2.2. An important observation is that when the strain-energy function is known, the incremental stretch moduli (2.78), (2.83), and (2.88) can be calculated using the definitions of the axial stresses (see Sect. 2.1.3). When the strain-energy function is not known, the nonlinear stretch moduli (2.79), (2.84), and (2.89) can be estimated (approximately) from experimental measurements where the axial force is prescribed. However, while special assumptions regarding the strain-energy function are required in order for the incremental stretch moduli (2.78), (2.83), and (2.88) to be positive, the nonlinear stretch moduli (2.79), (2.84), and (2.89) are always positive given that the PC inequalities (2.50) or (2.51) hold. For anisotropic materials, different stretch moduli may be found as the material is extended or compressed in different directions.

2.2.3 Poisson Functions

Although Poisson's ratio is more often computed for small strains, its usual definition applies also at finite strains [49]. However, whereas in the small strain regime, the Poisson ratio is a constant, at large strains, this ratio is a scalar function of the deformation. Moreover, for a nonlinear elastic material, a Poisson function can be calculated using different definitions of the strain. To obtain the *nonlinear Poisson ratios*, we consider an isotropic elastic material for which uniaxial loading causes a simple tension or compression (2.68). These deformations can be maintained in every homogeneous isotropic hyperelastic body by application of suitable traction [155, 156, 494, 497]. The *Poisson function* is then expressed as the negative quotient of the strain in an orthogonal direction to the strain in the direction of the applied force.

Using the nonlinear strain tensor described by (2.75), we define the following family of nonlinear Poisson functions [362]:

$$\nu_n(a) = -\frac{e_n(\lambda(a))}{e_n(a)}, \tag{2.92}$$

where the strain measure $e_n(\cdot)$ is given by (2.76). In particular, these functions can be specialised to known strain tensors as summarised in Table 2.3. The Poisson ratios $\nu_n(a)$ defined by (2.92) can be calculated directly from experimental measurements, without prior knowledge of the strain-energy function describing the material from which the elastic body undertaking the deformation is made.[7] Under infinitesimal deformations, i.e., when $a \rightarrow 1$, they coincide with the Poisson's ratio from the linear elastic theory, i.e.,

---

[7] A nonlinear Poisson ratio based on logarithmic stretch was defined for the first time by Röntgen [474].

$$\overline{\nu} = \lim_{a \to 1} \nu_n(a) = -\lim_{a \to 1} \lambda'(a), \tag{2.93}$$

where $\lambda'(a) = d\lambda(a)/da$. For an incompressible material, $\lambda(a) = 1/\sqrt{a}$. For this case, different strain values and the corresponding Poisson functions are plotted in Fig. 2.4a and b, respectively.

For many materials, the Poisson's ratio takes values between 0 and 0.5 in the small strain regime [558, p. 154], but apparent Poisson's ratios that are either negative or greater than 0.5 can also be obtained when large deformations occur. Moreover, for anisotropic materials, different Poisson's ratios may be found as the material is extended or compressed in different directions. For example, negative Poisson's ratios were reported for cork under non-radial (axial or transverse) compression [173], while Poisson's ratios with values between 0.6 and 0.8 were measured in some types of wood where the primary strain was extensional in the radial direction and the secondary strain was compressive in the transverse direction [142]. For honeycomb structures with hexagonal cells, an apparent Poisson's ratio equal to 1 was calculated under the small strain assumption in [196]. Poisson's ratios greater than 1 were measured in salamander skin [182], while negative or above 0.5 Poisson's ratios were observed in cat skin [569]. Some liquid crystal elastomers also exhibit negative Poisson's ratios [377].

2.2.4 Bulk Moduli

Volume changes in isotropic hyperelastic models can be quantified also by the *nonlinear bulk modulus*, which is defined under finite triaxial homogeneous deformations, with deformation gradient $\mathbf{A} = \mathrm{diag}(\alpha_1, \alpha_2, \alpha_3)$, as follows [362]:

$$\kappa = \frac{1}{3} \frac{\partial (T_1 + T_2 + T_3)}{\partial (J - 1)}, \tag{2.94}$$

where $\{T_i\}_{i=1,2,3}$ are the axial Cauchy stresses and $J = \det \mathbf{A} = \alpha_1 \alpha_2 \alpha_3$ quantifies the relative change of volume from the reference to the current configuration.

In hydrostatic compression [69, 413], [415, p. 519] (see Fig. 2.5), nonlinear pressure versus volume responses of rubber materials were found. In his case, $T_1 = T_2 = T_3 = -Jp$, where p is the hydrostatic pressure, and the bulk modulus defined by (2.94) takes the form [253]

$$\kappa = -J \frac{\partial p}{\partial J} - p. \tag{2.95}$$

Fig. 2.5 Cube deformed by
hydrostatic compression
[480]

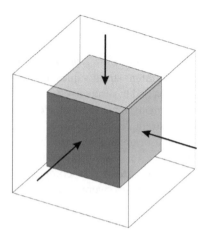

At small strain, the corresponding modulus is equal to

$$\overline{\kappa} = \lim_{J \to 1} \kappa = -J \frac{\partial p}{\partial J}. \tag{2.96}$$

For rubber-like materials, volume changes have also been observed under hydrostatic tension, but in this case the deformation remains small before the elasticity limit is reached and cavitation occurs [415, p. 520]. Experiments that measure volume changes under finite uniaxial tension [415, pp. 516–517], [428] suggest that the bulk modulus remains constant and equal to the classical bulk modulus, i.e., $\kappa = \overline{\kappa}$. When J is close to 1, by setting $T_1 = T_2 = T_3 = T$, the nonlinear bulk modulus given by (2.94) is simply

$$\kappa = \overline{\kappa} = \lim_{J \to 1} \frac{T}{J-1}. \tag{2.97}$$

While cavitation problems are treated later in this book, more experimental evidence is needed regarding finite volume changes in different elastic materials at large strains.

2.2.5 Shear Moduli

In linear elasticity, the Young's modulus, $\overline{E}$, and Poisson's ratio, $\overline{\nu}$, completely characterise a homogeneous isotropic material. In particular, the response of the material under shear is given by the shear modulus $\overline{\mu} = \overline{E}/(2(1 + \overline{\nu}))$. To quantify the shear response of an isotropic hyperelastic material in finite elasticity, we introduce the *nonlinear shear modulus* defined under the following simple shear [465] superposed on axial stretch [454] (see Fig. 2.6),

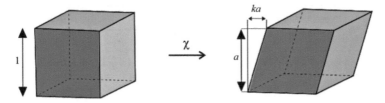

Fig. 2.6 Cuboid (left) deformed by simple shear superposed on axial stretch, described by Eq. (2.98) (right) [362]

$$x_1 = \lambda(a)X_1 + kaX_2, \qquad x_2 = aX_2, \qquad x_3 = \lambda(a)X_3, \qquad (2.98)$$

where (X_1, X_2, X_3) and (x_1, x_2, x_3) are the Cartesian coordinates for the reference and current configuration, as before, and k and a are positive constants representing the shear parameter and axial stretch ratio, respectively. This deformation can be maintained in every homogeneous isotropic hyperelastic body by application of suitable traction [156, 494]. When $a = \lambda(a) = 1$, it reduces to that of simple shear.

For the deformation described by (2.98), the gradient tensor takes the form

$$\mathbf{A} = \begin{bmatrix} \lambda(a) & ka & 0 \\ 0 & a & 0 \\ 0 & 0 & \lambda(a) \end{bmatrix}, \qquad (2.99)$$

and the left Cauchy–Green tensor is equal to

$$\mathbf{B} = \mathbf{A}\mathbf{A}^T = \begin{bmatrix} \lambda(a)^2 + k^2 a^2 & ka^2 & 0 \\ ka^2 & a^2 & 0 \\ 0 & 0 & \lambda(a)^2 \end{bmatrix}. \qquad (2.100)$$

The corresponding principal stretch ratios are given by[8]

---

[8] The principal stretches $\{\alpha_i\}_{i=1,2,3}$ are calculated as the positive square roots of the eigenvalues $\{\alpha_i^2\}_{i=1,2,3}$ of the left Cauchy–Green tensor $\mathbf{B}$ by employing the general formula for nested square roots, $\sqrt{x \pm \sqrt{y}} = \sqrt{\frac{x+\sqrt{x^2-y}}{2}} \pm \sqrt{\frac{x-\sqrt{x^2-y}}{2}}$.

$$\alpha_1 = \frac{\sqrt{\lambda(a)^2 + a^2\left(1+k^2\right) + 2a\lambda(a)}}{2} + \frac{\sqrt{\lambda(a)^2 + a^2\left(1+k^2\right) - 2a\lambda(a)}}{2},$$

$$\alpha_2 = \frac{\sqrt{\lambda(a)^2 + a^2\left(1+k^2\right) + 2a\lambda(a)}}{2} - \frac{\sqrt{\lambda(a)^2 + a^2\left(1+k^2\right) - 2a\lambda(a)}}{2} = \frac{a\lambda(a)}{\alpha_1},$$

$$\alpha_3 = \lambda(a),$$

(2.101)

and the principal invariants (2.16) of the Cauchy–Green tensor are

$$I_1 = \alpha_1^2 + \alpha_2^2 + \alpha_3^2 = k^2 a^2 + a^2 + 2\lambda(a)^2,$$
$$I_2 = \alpha_1^2\alpha_2^2 + \alpha_2^2\alpha_3^2 + \alpha_3^2\alpha_1^2 = k^2 a^2\lambda(a)^2 + 2a^2\lambda(a)^2 + \lambda(a)^4, \qquad (2.102)$$
$$I_3 = \alpha_1^2\alpha_2^2\alpha_3^2 = a^2\lambda(a)^4.$$

By (2.22), the non-zero components of the associated Cauchy stress are among the following components:

$$T_{11} = T_{33} + k^2 a^2 \beta_1,$$

$$T_{12} = ka^2\left(\beta_1 - \frac{\beta_{-1}}{a^2\lambda(a)^2}\right),$$

$$T_{22} = T_{33} + \left(a^2 - \lambda(a)^2\right)\left(\beta_1 - \frac{\beta_{-1}}{a^2\lambda(a)^2}\right) + k^2\frac{\beta_{-1}}{\lambda(a)^2}, \qquad (2.103)$$

$$T_{33} = \beta_0 + \beta_1\lambda(a)^2 + \frac{\beta_{-1}}{\lambda(a)^2}.$$

For incompressible materials, $\lambda(a) = 1/\sqrt{a}$, and an arbitrary pressure is added to the diagonal terms of the Cauchy stress tensor. For compressible and incompressible materials, the principal Cauchy stresses are given by (2.25)–(2.26) and (2.30)–(2.31), respectively.

By (2.103), the shear component of the Cauchy stress tensor, T_{12}, is proportional to ka^2. In this case, a *nonlinear shear modulus* can be defined as follows [362]:

$$\mu(a, k) = \frac{T_{12}}{ka^2} = \beta_1 - \frac{\beta_{-1}}{a^2\lambda(a)^2}. \qquad (2.104)$$

For incompressible materials, the shear component $P_{12} = T_{12}/a$ of the first Piola–Kirchhoff stress tensor (2.34) is proportional to the shear strain ka; hence, the nonlinear shear modulus (2.104) is equal to

$$\mu(a, k) = \frac{P_{12}}{ka} = \frac{T_{12}}{ka^2} = \beta_1 - \frac{\beta_{-1}}{a}. \qquad (2.105)$$

This modulus is independent of the Lagrange multiplier p and can be estimated directly from experimental measurements if the shear force is known.

For both compressible and incompressible materials, by the representations (2.25)–(2.26) and (2.30)–(2.31) of the principal Cauchy stresses, respectively, the nonlinear shear modulus defined by (2.104) can be written equivalently as

$$\mu(a, k) = \frac{T_1 - T_2}{\alpha_1^2 - \alpha_2^2}. \tag{2.106}$$

Hence, the nonlinear shear modulus (2.106) is positive if the BE inequalities (2.52) hold.

If the material is incompressible, then, by (2.39) and (2.47), the nonlinear shear modulus given by (2.106) takes the following equivalent forms in terms of the principal first and second Piola–Kirchhoff stresses, respectively,

$$\mu(a, k) = \frac{P_1\alpha_1 - P_2\alpha_2}{\alpha_1^2 - \alpha_2^2} = \frac{S_1\alpha_1^2 - S_2\alpha_2^2}{\alpha_1^2 - \alpha_2^2}. \tag{2.107}$$

For the cuboid deformed by simple shear superposed on axial stretch (2.98), in the plane of shear (i.e., the plane where X_3 is constant), the unit normal and tangent vectors on the inclined faces are, respectively,

$$\mathbf{n} = \pm \frac{1}{\sqrt{1+k^2}} \begin{bmatrix} 1 \\ -k \\ 0 \end{bmatrix}, \qquad \mathbf{t} = \pm \frac{1}{\sqrt{1+k^2}} \begin{bmatrix} k \\ 1 \\ 0 \end{bmatrix}. \tag{2.108}$$

Then, for the Cauchy stress and left Cauchy–Green tensor, the normal components are, respectively,

$$T_n = \mathbf{n}^T \mathbf{Tn} = T_{33} - \frac{k^2\lambda(a)^2}{1+k^2}\left(\beta_1 - \frac{\beta_{-1}}{a^2\lambda(a)^2}\right) + \beta_{-1}\frac{k^2}{\lambda(a)^2}, \qquad B_n = \mathbf{n}^T \mathbf{Bn} = \frac{\lambda(a)^2}{1+k^2}, \tag{2.109}$$

and the corresponding tangent components are

$$T_t = \mathbf{t}^T \mathbf{Tn} = \frac{k\lambda(a)^2}{1+k^2}\left(\beta_1 - \frac{\beta_{-1}}{a^2\lambda(a)^2}\right), \qquad B_t = \mathbf{t}^T \mathbf{Bn} = \frac{k\lambda(a)^2}{1+k^2}. \tag{2.110}$$

Hence, (2.104) is also equal to

$$\mu(a, k) = \frac{T_t}{B_t} = \beta_1 - \frac{\beta_{-1}}{a^2\lambda(a)^2}. \tag{2.111}$$

When $a \to 1$, i.e., for simple shear superposed on infinitesimal axial stretch, in both compressible and incompressible materials, the nonlinear shear modulus given by (2.104) converges to the nonlinear shear modulus for simple shear [558, pp. 174–175],

$$\widehat{\mu}(k) = \lim_{a \to 1} \mu(a, k) = \widehat{\beta}_1 - \widehat{\beta}_{-1}, \tag{2.112}$$

where $\widehat{\beta}_1 = \lim_{a \to 1} \beta_1$ and $\widehat{\beta}_{-1} = \lim_{a \to 1} \beta_{-1}$.

When $k \to 0$, i.e., for infinitesimal simple shear superposed on finite axial stretch, the nonlinear shear modulus given by (2.104) converges to

$$\widetilde{\mu}(a) = \lim_{k \to 0} \mu(a, k) = \widetilde{\beta}_1 - \frac{\widetilde{\beta}_{-1}}{a^2 \lambda(a)^2}, \tag{2.113}$$

where $\widetilde{\beta}_1 = \lim_{k \to 0} \beta_1$ and $\widetilde{\beta}_{-1} = \lim_{k \to 0} \beta_{-1}$. For incompressible materials,

$$\widetilde{\mu}(a) = \widetilde{\beta}_1 - \frac{\widetilde{\beta}_{-1}}{a}. \tag{2.114}$$

These nonlinear shear moduli are summarised in Table 2.4. When $a \to 1$ and $k \to 0$, these moduli converge to the linear shear modulus of the infinitesimal theory [558, p. 179], i.e.,

$$\overline{\mu} = \lim_{a \to 1} \lim_{k \to 0} \mu(a, k) = \lim_{k \to 0} \widehat{\mu}(k) = \lim_{a \to 1} \widetilde{\mu}(a) = \overline{\beta}_1 - \overline{\beta}_{-1}, \tag{2.115}$$

where $\overline{\beta}_1 = \lim_{a \to 1} \lim_{k \to 0} \beta_1$ and $\overline{\beta}_{-1} = \lim_{a \to 1} \lim_{k \to 0} \beta_{-1}$.

Note that, for simple shear (i.e., when $a = 1$), the shear modulus (2.104) coincides with the shear modulus defined by Moon and Truesdell [384], and also with the generalised shear modulus defined in [558, pp. 174–175]. However, for simple shear superposed on axial stretch (i.e., when $a \neq 1$), the shear modulus (2.104) differs by a factor a^2 from the shear modulus in [384]. Nevertheless, for the nonlinear shear modulus defined here, the equivalent form (2.106) is valid for any $a > 0$, including $a = 1$ as in the simple shear case [558, p. 175].

Table 2.5 provides explicit forms of the nonlinear shear modulus $\mu(a, k)$ given by (2.106), its limit in the case of small shear superposed on finite axial stretch, $\widetilde{\mu}(a) = \lim_{k \to 0} \mu(a, k)$ defined by (2.114), and its linear elastic limit $\overline{\mu} = \lim_{a \to 1} \lim_{k \to 0} \mu(a, k) = \lim_{a \to 1} \widetilde{\mu}(a)$ given by (2.115) for some popular incompressible isotropic hyperelastic models. For each model, the nonlinear shear modulus under simple shear, $\widehat{\mu}(k) = \lim_{a \to 1} \mu(a, k)$ defined by (2.112), can also be obtained. Note that although some materials have the same linear shear modulus (e.g., $\overline{\mu} = c_1$ for neo-Hookean, Yeoh, Fung and Gent models; $\overline{\mu} = c_1 + c_2$ for Mooney–Rivlin, Carroll, Gent–Thomas, and Gent–Gent models; $\overline{\mu} = \sum_{j=1}^{n} c_j$ for Ogden and Lopez-Pamies models), the nonlinear shear moduli are specific to each model, distinguishing them by their elastic responses at large strains.

Table 2.5 Explicit forms of the shear moduli $\mu(a, k)$ of (2.106), $\widetilde{\mu}(a) = \lim_{k \to 0} \mu(a, k)$ of (2.114), and $\overline{\mu} = \lim_{a \to 1} \widetilde{\mu}(a)$ of (2.115) for selected incompressible isotropic hyperelastic models. For the shear moduli of these incompressible materials, the principal stretches are given by (2.101) with $\lambda(a) = 1/\sqrt{a}$

Hyperelastic model	Strain-energy function $W(\alpha_1, \alpha_2, \alpha_3)$	Shear moduli
neo-Hookean [549]	$\frac{c_1}{2}\left(\alpha_1^2 + \alpha_2^2 + \alpha_3^2 - 3\right)$ c_1 independent of deformation	$\mu(a, k) = c_1$ $\widetilde{\mu}(a) = c_1$ $\overline{\mu} = c_1$
Mooney–Rivlin [385, 465]	$\frac{c_1}{2}\left(\alpha_1^2 + \alpha_2^2 + \alpha_3^2 - 3\right)$ $+ \frac{c_2}{2}\left(\alpha_1^{-2} + \alpha_2^{-2} + \alpha_3^{-2} - 3\right)$ c_1, c_2 independent of deformation	$\mu(a, k) = c_1 + \frac{c_2}{a}$ $\widetilde{\mu}(a) = c_1 + \frac{c_2}{a}$ $\overline{\mu} = c_1 + c_2$
Ogden [412]	$\sum_{j=1}^{n} \frac{c_j}{2p_j^2}\left(\alpha_1^{2p_j} + \alpha_2^{2p_j} + \alpha_3^{2p_j} - 3\right)$ c_j, p_j independent of deformation	$\mu(a, k) = \sum_{j=1}^{n} \frac{c_j}{p_j} \frac{\alpha_1^{2p_j} - \alpha_2^{2p_j}}{\alpha_1^2 - \alpha_2^2}$ $\widetilde{\mu}(a) = \sum_{j=1}^{n} \frac{c_j}{p_j} \frac{a^{1-p_j}\left(1 - a^{3p_j}\right)}{1 - a^3}$ $\overline{\mu} = \sum_{j=1}^{n} c_j$
Lopez-Pamies [327]	$\sum_{j=1}^{n} \frac{3c_j}{2p_j}\left[\left(\frac{\alpha_1^2 + \alpha_2^2 + \alpha_3^2}{3}\right)^{p_j} - 1\right]$ c_j, p_j independent of deformation	$\mu(a, k) =$ $\sum_{j=1}^{n} c_j \left(\frac{k^2a^2 + a^2 + 2/a}{3}\right)^{p_j - 1}$ $\widetilde{\mu}(a) = \sum_{j=1}^{n} c_j \left(\frac{a^2 + 2/a}{3}\right)^{p_j - 1}$ $\overline{\mu} = \sum_{j=1}^{n} c_j$
Arruda–Boyce [30]	$\sum_{j=1}^{n} \frac{c_j \alpha}{2j}\left[\left(\frac{\alpha_1^2 + \alpha_2^2 + \alpha_3^2}{\alpha}\right)^{j} - \left(\frac{3}{\alpha}\right)^{j}\right]$ c_j, α independent of deformation	$\mu(a, k) =$ $\sum_{j=1}^{n} c_j \left(\frac{k^2a^2 + a^2 + 2/a}{\alpha}\right)^{j - 1}$ $\widetilde{\mu}(a) = \sum_{j=1}^{n} c_j \left(\frac{a^2 + 2/a}{\alpha}\right)^{j - 1}$ $\overline{\mu} = \sum_{j=1}^{n} c_j \left(\frac{3}{\alpha}\right)^{j - 1}$
Yeoh [615, 616]	$\sum_{j=1}^{n} \frac{c_j}{2j}\left(\alpha_1^2 + \alpha_2^2 + \alpha_3^2 - 3\right)^{j}$ c_j independent of deformation	$\mu(a, k) =$ $\sum_{j=1}^{n} c_j \left(k^2a^2 + a^2 + 2/a - 3\right)^{j - 1}$ $\widetilde{\mu}(a) =$ $\sum_{j=1}^{n} c_j \left(a^2 + 2/a - 3\right)^{j - 1}$ $\overline{\mu} = c_1$
Carroll [87]	$\frac{c_1}{2}\left(\alpha_1^2 + \alpha_2^2 + \alpha_3^2 - 3\right)$ $+ \sqrt{3}c_2\left(\sqrt{\alpha_1^{-2} + \alpha_2^{-2} + \alpha_3^{-2}} - \sqrt{3}\right)$ c_1, c_2 independent of deformation	$\mu(a, k) = c_1 + \frac{c_2\sqrt{3}}{\sqrt{k^2a^3 + 2a^3 + 1}}$ $\widetilde{\mu}(a) = c_1 + \frac{c_2\sqrt{3}}{\sqrt{2a^3 + 1}}$ $\overline{\mu} = c_1 + c_2$

(continued)

Table 2.5 (continued)

Hyperelastic model	Strain-energy function $W(\alpha_1, \alpha_2, \alpha_3)$	Shear moduli
Dobrynin–Carrillo	$\frac{c_1}{6}(\alpha_1^2 + \alpha_2^2 + \alpha_3^2)$ $+ c_1\left(\frac{1}{\gamma} - \frac{\alpha_1^2+\alpha_2^2+\alpha_3^2}{3}\right)^{-1}$	$\mu(a,k) = \frac{c_1}{3} +$ $\frac{2c_1}{3}\left[1 - \frac{\gamma}{3}(k^2a^2 + a^2 + 2/a)\right]^{-2}$
[143]	c_1, γ independent of deformation	$\tilde{\mu}(a) = \frac{c_1}{3}$ $+ \frac{2c_1}{3}\left[1 - \frac{\gamma}{3}(a^2 + 2/a)\right]^{-2}$
		$\bar{\mu} = \frac{c_1}{3}\left[1 - 2(1 - \gamma)^{-2}\right]$
Fung	$\frac{c_1}{2\alpha}\left[e^{\alpha(\alpha_1^2+\alpha_2^2+\alpha_3^2-3)} - 1\right]$	$\mu(a,k) = c_1 e^{\alpha(k^2a^2+a^2+2/a-3)}$
[186]	c_1, α independent of deformation	$\tilde{\mu}(a) = c_1 e^{\alpha(a^2+2/a-3)}$
		$\bar{\mu} = c_1$
Anssari-Benam and Bucchi	$c_1 N\left[\frac{1}{6N}(\alpha_1^2 + \alpha_2^2 + \alpha_3^2 - 3)\right.$ $\left. - \ln\frac{\alpha_1^2+\alpha_2^2+\alpha_3^2-3N}{3(1-N)}\right]$	$\mu(a,k) = \frac{c_1(k^2a^2+a^2+2/a-9N)}{3(k^2a^2+a^2+2/a-3N)}$
[18]	c_1, N independent of deformation	$\tilde{\mu}(a) = \frac{c_1(a^2+2/a-9N)}{3(a^2+2/a-3N)}$
		$\bar{\mu} = \frac{c_1(1-3N)}{3(1-N)}$
Gent	$-\frac{c_1}{2\beta}\ln\left[1 - \beta(\alpha_1^2 + \alpha_2^2 + \alpha_3^2 - 3)\right]$	$\mu(a,k) = \frac{c_1}{1-\beta(k^2a^2+a^2+2/a-3)}$
[191]	c_1, β independent of deformation	$\tilde{\mu}(a) = \frac{c_1}{1-\beta(a^2+2/a-3)}$
		$\bar{\mu} = c_1$
Gent–Thomas	$\frac{c_1}{2}(\alpha_1^2 + \alpha_2^2 + \alpha_3^2 - 3)$ $+ \frac{3c_2}{2}\ln\frac{\alpha_1^{-2}+\alpha_2^{-2}+\alpha_3^{-2}}{3}$	$\mu(a,k) = c_1 + \frac{3c_2}{k^2a^2+2a^2+1/a}$
[194]	c_1, c_2 independent of deformation	$\tilde{\mu}(a) = c_1 + \frac{3c_2}{2a^2+1/a}$
		$\bar{\mu} = c_1 + c_2$
Gent–Gent	$-\frac{c_1}{2\beta}\ln\left[1 - \beta(\alpha_1^2 + \alpha_2^2 + \alpha_3^2 - 3)\right]$ $+ \frac{3c_2}{2}\ln\frac{\alpha_1^{-2}+\alpha_2^{-2}+\alpha_3^{-2}}{3}$	$\mu(a,k) =$ $\frac{c_1}{1-\beta(k^2a^2+a^2+2/a-3)} + \frac{3c_2}{k^2a^2+2a^2+1/a}$
[417, 450]	c_1, c_2, β independent of deformation	$\tilde{\mu}(a) =$ $\frac{c_1}{1-\beta(a^2+2/a-3)} + \frac{3c_2}{2a^2+1/a}$
		$\bar{\mu} = c_1 + c_2$

2.2.6 Universal Relations

Various universal relations can be established between the nonlinear shear and stretch moduli and Poisson functions defined above. Taking a unit cube of unconstrained material subject to simple shear superposed on finite axial stretch (2.98), if $T_{33} = 0$ in (2.103), then the normal force is equal to

$$N(a,k) = T_{22} = \left(a^2 - \lambda(a)^2\right)\left(\beta_1 - \frac{\beta_{-1}}{a^2\lambda(a)^2}\right) + k^2\frac{\beta_{-1}}{\lambda(a)^2}. \tag{2.116}$$

In the limit of infinitesimal shear, this normal force becomes

$$\tilde{N}(a) = \lim_{k \to 0} N(a, k) = \left(a^2 - \lambda(a)^2\right)\left(\tilde{\beta}_1 - \frac{\tilde{\beta}_{-1}}{a^2 \lambda(a)^2}\right), \tag{2.117}$$

and by (2.113),

$$\frac{\tilde{N}(a)}{\tilde{\mu}(a)} = \lim_{k \to 0} \frac{N(a, k)}{\mu(a, k)} = a^2 - \lambda(a)^2. \tag{2.118}$$

Therefore, as the axial stretch ratio $a > 1$ increases, the magnitude of the normal force $\tilde{N}$ relative to the shear modulus $\tilde{\mu}$ increases. This is a *universal relation*, i.e., it holds independently of β_1 and β_{-1}. Recalling that, under infinitesimal simple shear, the normal force is zero [360, 361], the following universal relation holds between the nonlinear shear modulus in the small shear limit (2.113) and the nonlinear stretch modulus (2.79) for the axial stretch ratio a,

$$\frac{E(a)}{\tilde{\mu}(a)} = \frac{\tilde{N}(a)}{\tilde{\mu}(a)} \frac{1}{\ln a - \ln \lambda(a)} \left(1 - \frac{a \lambda'(a)}{\lambda(a)}\right) = \frac{a^2 - a^{-2v_0(a)}}{\ln a^{1+v_0(a)}} \left(1 + v_0(a) + a v_0'(a) \ln a\right). \tag{2.119}$$

In particular, if the Poisson's ratio defined by $v^{(H)} = v_0$ is constant, $v_0 = \bar{v}$, then $\lambda(a) = a^{-v_0} = a^{-\bar{v}}$, and the universal relation (2.119) becomes

$$\frac{E(a)}{\tilde{\mu}(a)} = \frac{\tilde{N}(a)}{\tilde{\mu}(a) \ln a} = \frac{a^2 - a^{-2\bar{v}}}{\ln a}. \tag{2.120}$$

Specifically, for incompressible materials, $\bar{v} = 1/2$, and therefore,

$$\frac{E(a)}{\tilde{\mu}(a)} = \frac{\tilde{N}(a)}{\tilde{\mu}(a) \ln a} = \frac{a^2 - a^{-1}}{\ln a}. \tag{2.121}$$

For the general case, in the linear elastic limit, i.e., when $a \to 1$, from (2.119), the classical relation between the Young's modulus and the linear shear modulus is recovered, i.e.,

$$\frac{\overline{E}}{\overline{\mu}} = \lim_{a \to 1} \frac{E(a)}{\tilde{\mu}(a)} = 2(1 + \bar{v}). \tag{2.122}$$

Hence, for incompressible materials, $\overline{E} = 3\overline{\mu}$. The universal relations (2.119) and (2.120) and their linear elastic limits are summarised in Table 2.1.

2.2.7 Torsion Moduli

A right circular cylinder of incompressible isotropic hyperelastic material is deformed by the simple torsion superposed on axial stretch [558, pp. 189–191]

Fig. 2.7 Circular cylinder
(left) deformed by simple
torsion superposed on axial
stretch, described by
Eq. (2.123) (right) [362]

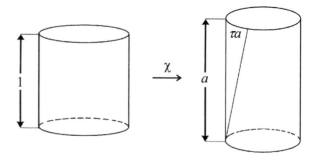

(see Fig. 2.7)

$$r = \frac{R}{\sqrt{a}}, \qquad \theta = \Theta + \tau a Z, \qquad z = a Z, \qquad (2.123)$$

where (R, Θ, Z) and (r, θ, z) are the cylindrical coordinates for the reference and
the current configurations, respectively.

For this deformation, the gradient tensor is equal to

$$\mathbf{A} = \begin{bmatrix} \partial r/\partial R & 0 & 0 \\ 0 & (r/R)\partial\theta/\partial\Theta & r\partial\theta/\partial Z \\ 0 & 0 & \partial z/\partial Z \end{bmatrix} = \begin{bmatrix} 1/\sqrt{a} & 0 & 0 \\ 0 & 1/\sqrt{a} & r\tau a \\ 0 & 0 & a \end{bmatrix}, \qquad (2.124)$$

where a and τ are positive constants representing the axial stretch ratio and the
torsion parameter, respectively, and the left Cauchy–Green tensor is

$$\mathbf{B} = \mathbf{A}\mathbf{A}^T = \begin{bmatrix} 1/a & 0 & 0 \\ 0 & 1/a + r^2\tau^2 a^2 & r\tau a^2 \\ 0 & r\tau a^2 & a^2 \end{bmatrix}. \qquad (2.125)$$

By (2.24), the non-zero components of the Cauchy stress tensor are among the
following components:

$$T_{rr} = -p + \frac{\beta_1}{a} + \beta_{-1}a,$$

$$T_{\theta\theta} = T_{rr} + \beta_1 r^2\tau^2 a^2,$$

$$T_{\theta z} = r\tau a^2 \left(\beta_1 - \frac{\beta_{-1}}{a} \right), \qquad (2.126)$$

$$T_{zz} = T_{rr} + \left(a^2 - \frac{1}{a} \right)\left(\beta_1 - \frac{\beta_{-1}}{a} \right) + r^2\tau^2 a^2 \frac{\beta_{-1}}{a},$$

where p depends only on r.

The classical torsion modulus is measured as the ratio between the torque and the twist. For the deformation (2.123), if $B_{rr} < 1$ and $T_{rr} = -p_0 \le 0$ at the external surface $r = r_0$, then at this surface, the torque is equal to [558, pp. 190–191]

$$T(a, \tau) = 2\pi \int_0^{r_0} T_{\theta z} r^2 dr = 2\pi \tau a^2 \int_0^{r_0} \left(\beta_1 - \frac{\beta_{-1}}{a} \right) r^3 dr. \qquad (2.127)$$

The resultant normal force is [558, p. 191]

$$\begin{aligned}
N(a, \tau) =& 2\pi \int_0^{r_0} T_{zz} r dr \\
=& 2\pi \int_0^{r_0} (T_{zz} - T_{rr}) r dr + 2\pi \int_0^{r_0} T_{rr} r dr \\
=& -\pi p_0 r_0^2 + 2\pi \left(a^2 - \frac{1}{a} \right) \int_0^{r_0} \left(\beta_1 - \frac{\beta_{-1}}{a} \right) r dr \\
& -\pi \tau^2 a^2 \int_0^{r_0} \left(\beta_1 - \frac{2\beta_{-1}}{a} \right) r^3 dr.
\end{aligned} \qquad (2.128)$$

As the torque is proportional to the twist, we define the nonlinear torsion modulus as the ratio between the torque T given by (2.127) and the amount of twist τa,

$$\mu_T(a, \tau) = \frac{T}{\tau a} = 2\pi a \int_0^{r_0} \left(\beta_1 - \frac{\beta_{-1}}{a} \right) r^3 dr = \frac{2\pi}{a} \int_0^{R_0} \left(\beta_1 - \frac{\beta_{-1}}{a} \right) R^3 dR. \qquad (2.129)$$

Note that this modulus increases as the radius R_0 of the (undeformed) cylinder increases. When $a \to 1$, i.e., for simple torsion superposed on infinitesimal axial stretch, the modulus defined by (2.129) converges to the torsion modulus for simple torsion [558, p. 192],

$$\widehat{\mu}_T(\tau) = \lim_{a \to 1} \frac{T}{\tau} = \frac{\pi r_0^4}{2} \left(\widehat{\beta}_1 - \widehat{\beta}_{-1} \right) = \frac{\pi R_0^4}{2} \left(\widehat{\beta}_1 - \widehat{\beta}_{-1} \right), \qquad (2.130)$$

where $\widehat{\beta}_1 = \lim_{a \to 1} \beta_1$ and $\widehat{\beta}_{-1} = \lim_{a \to 1} \beta_{-1}$. When $\tau \to 0$, i.e., for infinitesimal torsion superposed on finite axial stretch, the modulus given by (2.129) converges to

$$\widetilde{\mu}_T(a) = \lim_{\tau \to 0} \mu_T(a, \tau) = \lim_{\tau \to 0} \frac{T}{\tau a} = \frac{\pi a r_0^4}{2} \left(\widetilde{\beta}_1 - \frac{\widetilde{\beta}_{-1}}{a} \right) = \frac{\pi R_0^4}{2a} \left(\widetilde{\beta}_1 - \frac{\widetilde{\beta}_{-1}}{a} \right), \qquad (2.131)$$

where $\widetilde{\beta}_1 = \lim_{\tau \to 0} \beta_1$ and $\widetilde{\beta}_{-1} = \lim_{\tau \to 0} \beta_{-1}$. In [466], it was noted for the first time that, when $p_0 = 0$, if $\tau \to 0$, then

$$\frac{Nr_0^2}{\widetilde{\mu}_T(a)} = 2\left(a - \frac{1}{a^2}\right) \tag{2.132}$$

is independent of the material parameters. This universal relation was proven experimentally in [470]. When $\tau \to 0$ and $a \to 1$, the above nonlinear torsion moduli converge to the linear elastic modulus

$$\overline{\mu}_T = \lim_{a \to 1} \lim_{\tau \to 0} \mu_T(\tau, a) = \lim_{a \to 1} \widehat{\mu}_T(\tau) = \lim_{\tau \to 0} \widetilde{\mu}_T(a) = \frac{\pi r_0^4}{2}\left(\overline{\beta}_1 - \overline{\beta}_{-1}\right) = \frac{\pi R_0^4}{2}\left(\overline{\beta}_1 - \overline{\beta}_{-1}\right),$$
$$\tag{2.133}$$

where $\overline{\beta}_1 = \lim_{a \to 1} \lim_{\tau \to 0} \beta_1$ and $\overline{\beta}_{-1} = \lim_{a \to 1} \lim_{\tau \to 0} \beta_{-1}$.

2.2.8 Poynting Moduli

The elastic *Poynting effect* [443, 444] naturally captures the coupling between normal and shear deformations in a cube and between axial and torsion deformations in a right circular cylinder. This phenomenon can be observed when an elastic cube is sheared between two plates and a stress appears in the direction normal to the sheared faces, or when an axial stress develops in a cylinder subjected to finite torsion, while its length is kept fixed[9] (see Fig. 2.8). If the material tends to elongate in the direction perpendicular to the shear stress, then the positive Poynting effect occurs, and when the material tends to contract, the negative Poynting effect is obtained.[10] The simple shear deformation of homogeneous isotropic hyperelastic cubes is analysed theoretically in [21, 136, 137, 258, 360, 361, 384, 467, 528, 538, 553, 631]. The effects of strain stiffening and limiting chain extensibility on the extension and torsion of incompressible homogeneous isotropic hyperelastic cylinders, where there is a critical axial stretch at which Poynting-type effects reverse, are investigated in [20, 296]. Experimental observations of the Poynting effect are reported in [38, 279, 376, 528]. Torsional instabilities of stretched rubber cylinders were identified experimentally in [192].

For an incompressible cube that is free on its outer surface and subject to simple shear, the axial stretch is proportional to the square of the shear, i.e.,

---

[9] "The association of a thrust with torsion of a circular cylinder and the lengthening of a cylinder as a result of the application of a torque are called *Poynting effects*. Sometimes this name is also applied to *any* effect that arises from the requirement that in simple shear of an elastic material the normal components of the stress are other than a hydrostatic pressure or torsion. However, it seems to me that the term *normal stress effect* is more appropriate for general use since somewhat analogous effects can occur in inelastic solids and in non-Newtonian fluids."— R.S. Rivlin, Autobiographical Postscript [469, pp. xxiii–xxiv].

[10] The *Poynting effect* is closely related to the *Weissenberg effect* [599, 600], which is the normal stress phenomenon consisting in the climbing of a non-Newtonian fluid around a spinning rod.

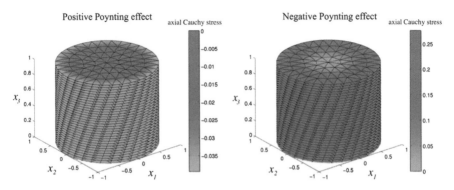

Fig. 2.8 Finite element simulations of Poynting effects due to simple torsion of circular cylinder of Mooney–Rivlin material, with colour bar showing axial Cauchy stresses: compressive for positive Poynting effect (left); tensile for negative Poynting effect (right) [361]

$$|a - 1| = |a(k) - 1| = \mu_P k^2 a^2. \tag{2.134}$$

In (2.134), the parameter μ_P is a positive constant and represents the *Poynting modulus* in shear. If $a > 1$, then the classical (positive) Poynting effect occurs, and when $a < 1$, the negative Poynting effect is observed. To estimate the value of μ_P, assuming $T_{33} = 0$ in (2.103), since $\lambda(a) = 1/\sqrt{a}$, the normal force is equal to

$$N(a, k) = T_{22} = \left(a^2 - \frac{1}{a}\right)\left(\beta_1 - \frac{\beta_{-1}}{a}\right) + k^2 a^2 \frac{\beta_{-1}}{a}. \tag{2.135}$$

Then, taking $N(a, k) = 0$ in (2.135) provides an equation for the axial stretch ratio a corresponding to the amount of shear ka. By (2.134) and (2.135), noting that $a \to 1$ as $k \to 0$, we obtain

$$\lim_{k \to 0} \frac{|a^2 - 1/a|}{k^2 a^2} = \lim_{k \to 0} \left|\frac{\tilde{\beta}_{-1}/a}{\tilde{\beta}_1 - \tilde{\beta}_{-1}/a}\right| = \frac{|\overline{\beta}_{-1}|}{\overline{\beta}_1 - \overline{\beta}_{-1}}, \tag{2.136}$$

and by (2.134) and (2.136),

$$\mu_P = \lim_{k \to 0} \frac{|a - 1|}{k^2 a^2} = \frac{1}{3} \lim_{k \to 0} \frac{|a^3 - 1|}{k^2 a^2} = \frac{1}{3} \lim_{k \to 0} \frac{|a^2 - 1/a|}{k^2 a^2} = \frac{|\overline{\beta}_{-1}|}{3\left(\overline{\beta}_1 - \overline{\beta}_{-1}\right)}. \tag{2.137}$$

When $\beta_{-1} = 0$, the Poynting modulus vanishes, meaning that the Poynting effect is not observed, if $\beta_{-1} < 0$, then the positive Poynting effect occurs, and if $\beta_{-1} > 0$, then the negative Poynting effect is obtained. As shown in [360, 361], positive or negative Poynting effects are possible if the GE inequalities (2.56) hold.

For an incompressible cylinder under torsion, the axial stretch is proportional to the square of the amount of twist [38, 279, 361, 467, 553], i.e.,

$$|a - 1| = |a(\tau) - 1| = \mu_P \tau^2 a^2, \tag{2.138}$$

where the positive constant μ_P is identified as the *Poynting modulus* in torsion [558, p. 193]. To find the value of this modulus, we note that setting $N(a, \tau) = -p_0 \pi r_0^2$ in (2.128) provides an equation for the axial stretch ratio a corresponding to the amount of twist τa. In this case, by (2.128), $a \to 1$ as $\tau \to 0$, and

$$\lim_{\tau \to 0} \frac{|a^2 - 1/a|}{\tau^2 a^2} = \lim_{\tau \to 0} \left| \frac{\int_0^{r_0} \left(\tilde{\beta}_1 - 2\tilde{\beta}_{-1}/a \right) r^3 dr}{2 \int_0^{r_0} \left(\tilde{\beta}_1 - \tilde{\beta}_{-1}/a \right) r dr} \right| = \frac{R_0^2}{4} \left| 1 - \frac{\overline{\beta}_{-1}}{\overline{\beta}_1 - \overline{\beta}_{-1}} \right|. \tag{2.139}$$

Then, by (2.138), as $a = a(\tau) \to 1$ as $\tau \to 0$, we obtain

$$\mu_P = \lim_{\tau \to 0} \frac{|a - 1|}{\tau^2 a^2} = \frac{1}{3} \lim_{\tau \to 0} \frac{|a^3 - 1|}{\tau^2 a^2} = \frac{1}{3} \lim_{\tau \to 0} \frac{|a^2 - 1/a|}{\tau^2 a^2} = \frac{R_0^2}{12} \left| 1 - \frac{\overline{\beta}_{-1}}{\overline{\beta}_1 - \overline{\beta}_{-1}} \right|, \tag{2.140}$$

where the last equality follows from (2.139). Hence, the Poynting modulus given by (2.140) increases as the radius R_0 of the (undeformed) cylinder increases.

2.3   Nonlinear Elastic Moduli in Anisotropic Elasticity

In general, for a homogeneous anisotropic hyperelastic material with two preferred directions, the strain-energy function requires seven independent invariants [6, 7], including the principal invariants, I_1, I_2, I_3, of the Cauchy–Green tensors $\mathbf{B}$ and $\mathbf{C}$, the pseudo-invariants given by

$$I_4 = (\mathbf{C}\mathbf{M}_1) \cdot \mathbf{M}_1, \qquad I_6 = (\mathbf{C}\mathbf{M}_2) \cdot \mathbf{M}_2, \tag{2.141}$$

and two other pseudo-invariants defined by

$$I_5 = \left(\mathbf{C}^2 \mathbf{M}_1 \right) \cdot \mathbf{M}_1, \qquad I_7 = \left(\mathbf{C}^2 \mathbf{M}_2 \right) \cdot \mathbf{M}_2. \tag{2.142}$$

For anisotropic materials, with different elastic properties in different directions, various constitutive formulations are discussed in [395] (see also [261]).

Here, we confine our attention to a simple material model with two families of extensible fibres embedded in a neo-Hookean material. The corresponding strain-energy function is

$$W(I_1, I_4, I_6) = \frac{\mu_1}{2} (I_1 - 3) + \frac{\mu_4}{4} (I_4 - 1)^2 + \frac{\mu_6}{4} (I_6 - 1)^2, \tag{2.143}$$

where μ_1, μ_4, and μ_6 are positive constants. To derive the associated nonlinear shear moduli, we consider a cuboid sample of the material described by (2.143), such that the edges of the cuboid are aligned with the Cartesian axes in the reference configuration. The two preferred directions are defined by

$$\mathbf{M}_1 = \begin{bmatrix} M_{11} \\ M_{12} \\ M_{13} \end{bmatrix} = \begin{bmatrix} \cos \Phi \\ \sin \Phi \\ 0 \end{bmatrix} \quad \text{and} \quad \mathbf{M}_2 = \begin{bmatrix} M_{21} \\ M_{22} \\ M_{23} \end{bmatrix} = \begin{bmatrix} -\cos \Psi \\ \sin \Psi \\ 0 \end{bmatrix},$$
(2.144)

where $\Phi, \Psi \in [0, \pi/2]$. The stretch ratios α_4 and α_6 in these two directions, respectively, satisfy

$$I_4 = \alpha_4^2, \qquad I_6 = \alpha_6^2,$$
(2.145)

where I_4 and I_6 are given by (2.141).

For the deformed material, the Cauchy stress tensor takes the form

$$\mathbf{T} = -p\mathbf{I} + \beta_1 \mathbf{B} + \beta_4 \mathbf{AM}_1 \otimes \mathbf{AM}_1 + \beta_6 \mathbf{AM}_2 \otimes \mathbf{AM}_2,$$
(2.146)

where $\mathbf{A}$ is the deformation gradient, $\mathbf{B} = \mathbf{AA}^T$ is the left Cauchy–Green tensor, p is the Lagrange multiplier for the incompressibility constraint $I_3 = 1$, and the constitutive coefficients are as follows:

$$\beta_1 = \frac{2}{\sqrt{I_3}} \frac{\partial W}{\partial I_1}, \qquad \beta_4 = \frac{2}{\sqrt{I_3}} \frac{\partial W}{\partial I_4}, \qquad \beta_6 = \frac{2}{\sqrt{I_3}} \frac{\partial W}{\partial I_6}.$$
(2.147)

First, we deform the cuboid by simple shear, such that

$$x_1 = X_1 + kX_2, \qquad x_2 = X_2, \qquad x_3 = X_3,$$
(2.148)

where (X_1, X_2, X_3) and (x_1, x_2, x_3) are the Cartesian coordinates for the reference and current configurations, respectively. For the deformation (2.148), the gradient tensor is equal to

$$\mathbf{A} = \begin{bmatrix} 1 & k & 0 \\ 0 & 1 & 0 \\ 0 & 0 & 1 \end{bmatrix},$$
(2.149)

where $k > 0$ is the shear parameter. The associated left and right Cauchy–Green deformation tensors are, respectively,

$$\mathbf{B} = \mathbf{AA}^T = \begin{bmatrix} 1+k^2 & k & 0 \\ k & 1 & 0 \\ 0 & 0 & 1 \end{bmatrix}, \qquad \mathbf{C} = \mathbf{A}^T \mathbf{A} = \begin{bmatrix} 1 & k & 0 \\ k & 1+k^2 & 0 \\ 0 & 0 & 1 \end{bmatrix}.$$
(2.150)

The two preferred directions, defined by (2.144), deform into the following directions, respectively,

$$\mathbf{m}_1 = \mathbf{A}\mathbf{M}_1 = \begin{bmatrix} \cos\Phi + k\sin\Phi \\ \sin\Phi \\ 0 \end{bmatrix}, \qquad \mathbf{m}_2 = \mathbf{A}\mathbf{M}_2 = \begin{bmatrix} -\cos\Psi + k\sin\Psi \\ \sin\Psi \\ 0 \end{bmatrix}.$$

$$(2.151)$$

Then, the non-zero components of the associated Cauchy stress tensor given by (2.146) are

$$T_{11} = -p + \beta_1 \left(1 + k^2\right) + \beta_4 \left(\cos\Phi + k\sin\Phi\right)^2 + \beta_6 \left(\cos\Psi - k\sin\Psi\right)^2,$$

$$T_{12} = \beta_1 k + \beta_4 \sin\Phi \left(\cos\Phi + k\sin\Phi\right) - \beta_6 \sin\Psi \left(\cos\Psi - k\sin\Psi\right),$$

$$T_{22} = -p + \beta_1 + \beta_4 \sin^2\Phi + \beta_6 \sin^2\Psi,$$

$$T_{33} = -p + \beta_1,$$

$$(2.152)$$

where, by (2.147),

$$\beta_1 = \mu_1, \qquad \beta_4 = \mu_4 k\sin\Phi \left(k\sin\Phi + 2\cos\Phi\right), \qquad \beta_6 = \mu_6 k\sin\Psi \left(k\sin\Psi - 2\cos\Psi\right).$$

$$(2.153)$$

In this case, a nonlinear shear modulus can be defined as the ratio of the shear stress to the shear strain under large strain [373], i.e.,

$$\widehat{\mu}_{12}(k) = \frac{T_{12}}{k}.$$

$$(2.154)$$

By (2.152) and (2.153), this shear modulus is equal to

$$\widehat{\mu}_{12}(k) = \mu_1 + \mu_4 \sin^2\Phi \left(2\cos^2\Phi + 3k\sin\Phi\cos\Phi + k^2\sin^2\Phi\right)$$

$$+ \mu_6 \sin^2\Psi \left(2\cos^2\Psi - 3k\sin\Psi\cos\Psi + k^2\sin^2\Psi\right).$$

$$(2.155)$$

Under infinitesimal shear, such that $k \to 0$, the modulus (2.155) converges to the linear elastic limit

$$\overline{\mu}_{12} = \lim_{k\to 0} \widehat{\mu}_{12}(k) = \mu_1 + 2\mu_4 \sin^2\Phi\cos^2\Phi + 2\mu_6 \sin^2\Psi\cos^2\Psi. \qquad (2.156)$$

Next, the original (undeformed) cuboid is subjected to the simple shear deformation

$$x_1 = X_1 + kX_3, \qquad x_2 = X_2, \qquad x_3 = X_3. \qquad (2.157)$$

By similar arguments as those detailed above, the following nonlinear shear modulus can be defined

$$\widehat{\mu}_{13}(k) = \frac{T_{13}}{k}. \tag{2.158}$$

Hence, by (2.152) and (2.153),

$$\widehat{\mu}_{13}(k) = \mu_1. \tag{2.159}$$

As μ_1 is constant, the shear modulus described by (2.159) coincides with its linear elastic limit

$$\overline{\mu}_{13} = \lim_{k \to 0} \widetilde{\mu}_{13}(k) = \mu_1. \tag{2.160}$$

Finally, assuming that the original cuboid deforms by the simple shear

$$x_1 = X_1, \qquad x_2 = X_2 + kX_3, \qquad x_3 = X_3, \tag{2.161}$$

the following nonlinear shear modulus is defined,

$$\widetilde{\mu}_{23}(k) = \frac{T_{23}}{k}. \tag{2.162}$$

In this case also,

$$\widetilde{\mu}_{23}(k) = \mu_1, \tag{2.163}$$

and its linear elastic limit is

$$\overline{\mu}_{23} = \lim_{k \to 0} \widetilde{\mu}_{23}(k) = \mu_1. \tag{2.164}$$

The shear moduli (2.156), (2.160), and (2.164), obtained under infinitesimal shear, are equivalent to those described by equation (6.1) of [396].

The effects of fibre reorientation on simple shear deformations are also considered in [257]. Experimental results on the Poynting effect in fibre-reinforced materials are presented in [29]. Positive and negative Poynting effects in anisotropic soft materials are treated in [259].

2.4   Quasi-Equilibrated Motion

For large strain dynamic deformations of an elastic solid, Cauchy's laws of motion (balance laws of linear and angular momentum) are governed by the following Eulerian field equations [558, p. 40]:

$$\rho\ddot{\mathbf{x}} - \operatorname{div} \mathbf{T} = \rho\mathbf{b}, \tag{2.165}$$

$$\mathbf{T} = \mathbf{T}^T, \tag{2.166}$$

where ρ is the material density (mass per unit volume, which is assumed constant), $\mathbf{x} = \chi(\mathbf{X}, t)$ is the motion of the elastic solid, with velocity $\dot{\mathbf{x}} = \partial\chi(\mathbf{X}, t)/\partial t$ and acceleration $\ddot{\mathbf{x}} = \partial^2\chi(\mathbf{X}, t)/\partial t^2$, $\mathbf{b} = \mathbf{b}(\mathbf{x}, t)$ is the body force (force per unit volume), and $\mathbf{T} = \mathbf{T}(\mathbf{x}, t)$ is the Cauchy stress tensor (which is symmetric).

To obtain possible dynamical solutions, one can solve Cauchy's equation for particular motions or generalise known static solutions to dynamical forms using the following notion of *quasi-equilibrated motion* (*motus quasi aequilibratus*[11]) introduced by Truesdell [555] and reviewed extensively in [558, Sec 61].

Definition 2.4.1 ([558, p. 208]) A quasi-equilibrated motion, $\mathbf{x} = \chi(\mathbf{X}, t)$, is the motion of an incompressible homogeneous elastic solid subject to a given body force, $\mathbf{b} = \mathbf{b}(\mathbf{x}, t)$, whereby, for each value of t, $\mathbf{x} = \chi(\mathbf{X}, t)$ defines a static deformation that satisfies the equilibrium conditions under the body force $\mathbf{b} = \mathbf{b}(\mathbf{x}, t)$.

Theorem 2.4.2 ([558, p. 208]) *A quasi-equilibrated motion,* $\mathbf{x} = \chi(\mathbf{X}, t)$, *of an incompressible homogeneous elastic solid subject to a given body force,* $\mathbf{b} = \mathbf{b}(\mathbf{x}, t)$, *is dynamically possible, subject to the same body force, if and only if the motion is circulation preserving with a single-valued acceleration potential ξ, i.e.,*

$$\ddot{\mathbf{x}} = -\mathrm{grad}\,\xi(\mathbf{x}, t). \tag{2.167}$$

The condition (2.167) *is satisfied if*

$$\mathrm{curl}\,\ddot{\mathbf{x}} = \mathbf{0}. \tag{2.168}$$

In this case, the stress field is determined by the present configuration alone and takes the form

$$\mathbf{T} = -\rho\xi\mathbf{I} + \mathbf{T}^{(0)}, \tag{2.169}$$

where $\mathbf{T}^{(0)}$ *is the Cauchy stress for the equilibrium state at time t and* $\mathbf{I}$ *is the identity tensor. Therefore, the shear stresses are the same as those of the equilibrium state at time t.*

Proof The Cauchy stress $\mathbf{T}^{(0)}$ for the equilibrium state under the body force $\mathbf{b} = \mathbf{b}(\mathbf{x}, t)$ at time t satisfies

---

[11] "Denominetur motus talis, qualis omni momento temporis t praebet configurationem capacem aequilibrii corporis iisdem viribus massalibus sollicitati, 'motus quasi aequilibratus'. Generatim motus quasi aequilibratus non congruet legibus dynamicis et proinde motus verus corporis fieri non potest, manentibus iisdem viribus masalibus."—C. Truesdell [555].

$$- \operatorname{div} \mathbf{T}^{(0)} = \rho \mathbf{b}. \tag{2.170}$$

If (2.167) holds, with ξ a single-valued function, then, substitution of (2.167) and (2.170) in (2.165) gives

$$- \rho \operatorname{grad} \xi = \operatorname{div} \left(\mathbf{T} - \mathbf{T}^{(0)} \right), \tag{2.171}$$

at any time instant t. From (2.171), it follows that the Cauchy stress takes the form (2.169). Hence, the motion is quasi-equilibrated according to Definition 2.4.1.

Conversely, if the motion $\mathbf{x} = \chi(\mathbf{X}, t)$ is quasi-equilibrated under the body force $\mathbf{b} = \mathbf{b}(\mathbf{x}, t)$, then, by Definition 2.4.1, at any fixed time instant t, the Cauchy stress takes the form (2.169). Substituting (2.169) in (2.165) gives

$$\rho \ddot{\mathbf{x}} = -\rho \operatorname{grad} \xi + \operatorname{div} \mathbf{T}^{(0)} + \rho \mathbf{b}. \tag{2.172}$$

Then, (2.167) follows from (2.170) and (2.172).   □

An immediate consequence of the above theorem is that a quasi-equilibrated motion is dynamically possible under a given body force in all elastic materials if, at every time instant, it can be reduced to a static equilibrium state that is possible in all those materials under the specified load. Quasi-equilibrated motions of isotropic materials subject to surface tractions alone are obtained by taking the arbitrary constant in those deformations to be arbitrary functions of time. Under this type of motion, a body can be brought instantly to rest by applying a suitable pressure impulse on its boundary. Examples include the homogeneous motions that are possible in all homogeneous incompressible materials, and certain non-homogeneous motions, some of which will be treated in Chaps. 4 and 5 (see also [358, 374]). For compressible materials, Theorem 2.4.2 on quasi-equilibrated dynamics is not applicable [558, p. 209].

Chapter 3
Are Elastic Materials Like Gambling Machines?

Now as human knowledge comes by the senses in such a way that the existence of things external is only inferred from the harmonious (not similar) testimony of the different senses, understanding, acting by the laws of right reason, will assign to different truths (or facts, or testimonies, or what shall I call them) different degrees of probability.

—J.C. Maxwell [84, p. 143]

In this chapter, stochastic hyperelastic models are defined, and a numerical procedure is described for the calibration of such models to experimental datasets. The procedure is mathematically tractable and builds directly on knowledge from deterministic finite elasticity. To illustrate its efficiency, a rubber-like material under uniaxial stretch is considered. For this material, several models listed in Table 2.5 are suitable candidates. To capture the data variability, it is sufficient then to treat the model coefficients as random variables independent of the other parameters, which can remain deterministic. Finally, Bayes' theorem [47] is applied to select the model that is most likely to reproduce the data. These calibration and selection procedures can be extended to anisotropic materials [95].

For an elastic material at large strains, standard experimental tests are conducted mostly under uniaxial or biaxial loads [54, 100, 109, 183, 187, 287, 299, 436], [376, 455, 457, 471] and less frequently under simple or pure shear and torsional loading [279, 287, 376, 406, 456], while combined shear and axial, or torsion and axial, experiments are still few [268, 356, 436].

When the geometries and boundary conditions of the deforming body are more complex, or application-specific, inverse finite element modelling can be employed [55, 320, 407, 610]. This involves the simulation of experiments whereby the material parameters are altered until the force-displacement responses in the simulations match those derived experimentally [116, 183, 235]. For many applications, this can be computationally intensive and prone to noise, especially if complex geometries and a very fine mesh are involved. In addition, as the modelling errors and the computational ones are undistinguishable, the model verification and validation

© Springer Nature Switzerland AG 2022
L. A. Mihai, *Stochastic Elasticity*, Interdisciplinary Applied Mathematics 55,
https://doi.org/10.1007/978-3-031-06692-4_3

processes become prohibitive [33, 410, 411]. In this case, the choice of one set of output parameters versus another remains unclear [133].

In practice, constitutive models containing fewer terms and constant coefficients, which can be altered more easily or related directly to the linear elastic moduli, are usually preferred even if their approximation of the experimental data is not the best [53, 92, 109, 134, 138, 236, 341, 355, 521]. This is further underpinned by the fact that, for more complex models, no particular physical interpretation can be attributed to every individual constituent, which may increase the risk of overfitting [237, 238].

The approach adopted here is to regard the individual constants appearing in a hyperelastic model as (non-unique) contributors to general nonlinear parameters that are explicit functions of the deformation, and which can be estimated directly from experiments. In particular, the nonlinear shear modulus defined in the previous chapter is useful. For the experimental setup, the challenge is to get as close as possible to the ideal situations that can be analysed mathematically. From the mathematical modelling point of view, stochastic representations accounting for data dispersion are needed to improve assessment and predictions [152, 195, 270, 292, 408, 420, 509, 529].

3.1   Stochastic Hyperelastic Models

The stochastic hyperelastic models defined here rely on the following fundamental assumptions:

(A1) *Material objectivity*, stating that material properties are independent of a superimposed rigid-body transformation (which involves a change of position after deformation).

(A2) *Material isotropy*, requiring that the strain-energy function is unaffected by a rigid-body transformation prior to deformation.

(A3) *Baker–Ericksen (BE) inequalities* (2.52), by which the greater principal Cauchy stress occurs in the direction of the greater principal stretch.

(A4) For any given finite deformation, at any point in the material, the shear modulus, μ, *and its inverse*, $1/\mu$, *are second-order random variables*, i.e., they have finite mean value and finite variance.

Conditions (A1)–(A3) are directly inherited from the finite elasticity theory (see Chap. 2). However, while (A1) and (A2) are necessary for all homogeneous isotropic hyperelastic materials, (A3) is chosen because BE inequalities hold for many different constitutive laws showing interesting and seemingly opposite mechanical responses. Noteworthy examples are the positive or negative Poynting effects mentioned in the previous chapter, and the supercritical or subcritical bifurcation when an elastic sphere cavitates, which will be discussed in the next chapter. In addition, condition (A4) places random variables at the foundation of stochastic hyperelastic models [371, 519–521]. A random variable is usually described in terms of its *mean value* and its *variance*, which contains information about the range of values about the mean value [72, 89, 220, 270, 347, 402, 403]. In practice, elastic moduli can take different values, corresponding to possible outcomes of experimental tests, and different probability distributions can be fitted to the same datasets consisting of the mean values and standard deviations of these moduli [325]. In other words:

> The mean and the variance are unambiguously determined by the distribution, but a distribution is, of course, not determined by its mean and variance: A number of different distributions have the same mean and the same variance.—Richard von Mises [576, p. 212]

To find suitable probability distributions, we employ the principle of maximum entropy [280, 281] stating that:

> The probability distribution which best represents the current state of knowledge is the one with largest entropy (or uncertainty) in the context of precisely stated prior data (or testable information).—E.T. Jaynes [282]

Approaches for the explicit derivation of probability distributions for the elastic parameters of stochastic homogeneous isotropic hyperelastic models calibrated to experimental data for rubber-like material and soft tissues were proposed in [371, 521]. Capturing the variability of Rivlin and Saunders' original data [470] (see also [550, p. 224], or [558, p. 181]), probability distributions for the random shear modulus under relatively small strains were obtained in [357].

3.1.1 Model Calibration

A calibration procedure for stochastic homogeneous isotropic incompressible hyperelastic models was developed in [371] to capture the intriguing behaviour of brain tissue when tested under combined shear and axial stretch [77, 436]. There are wide variations in the conditions of measurement of brain tissue, particularly those reported by different authors, and the data provided are far from abundant [38, 92, 109, 134, 138, 236, 268, 287, 341, 436, 455–457]. This precludes conclusive quantitative comparisons between independent results [93] and makes mathematical modelling of this tissue extremely challenging. Nevertheless, it can reasonably be inferred that when brain tissue is subject to large strains, the nonlinear shear modulus increases strongly and almost linearly as axial compression increases, while increasing only moderately as axial tension increases, regardless of the stress–strain response under simple shear [77, 355, 356, 436, 446]. A similar behaviour has been observed in other soft tissues with large lipid content, such as liver and adipose tissues [356, 431, 565]. Although biological tissues have a viscoelastic mechanical behaviour, hyperelastic modelling is useful as a starting point for the development of more complex models. A hyperelastic constitutive model has a unique stress–strain relationship, which is independent of strain rate, whereas for viscoelastic materials, the stress–strain response changes with the strain rate. Nevertheless, for some soft tissues where the shape of the stress–strain curve is almost invariant with respect to strain rate, at fixed strain rate, the shear modulus may be captured by a nonlinear hyperelastic model. The nonlinear shear modulus can be useful in quantifying results from such experiments.

We restrict our attention to stochastic hyperelastic strain-energy functions derived from experimental measurements for the nonlinear shear modulus $\widetilde{\mu}(a)$ defined by (2.113) under small shear superposed on finite axial stretch.

The following general notation convention is used: a quantity with an overbar denotes a value appearing in the theory of linear elasticity (e.g., $\overline{\mu}$); an underlined quantity denotes the mean value of that quantity (e.g., $\underline{\widetilde{\mu}}$, $\underline{\overline{\mu}}$). Given the dataset for the nonlinear shear modulus $\widetilde{\mu}(a)$ consisting of the mean values $\{\underline{\widetilde{\mu}}_s\}_{s=1,\cdots,m}$ and standard deviations $\{d_s\}_{s=1,\cdots,m}$ at the prescribed stretches $\{a_s\}_{s=1,\cdots,m}$, respectively, the following two-step procedure is employed [371]:

Step 1 First, similarly to a deterministic approach [355, 356, 417], the optimal mean values $\{\underline{C}_p\}_{p=1,\cdots,n}$ of the random constant coefficients $\{C_p\}_{p=1,\cdots,n}$ are obtained by minimising the residual between the mean nonlinear shear modulus $\underline{\widetilde{\mu}}$ and the mean data values $\{\underline{\widetilde{\mu}}_s\}_{s=1,\cdots,m}$ at the prescribed stretches $\{a_s\}_{s=1,\cdots,m}$,

$$\mathcal{R}_{\text{mean}} = \sqrt{\sum_{s=1}^{m} \left(\underline{\widetilde{\mu}}(a_s) - \underline{\widetilde{\mu}}_s \right)^2}. \tag{3.1}$$

If other model parameters (e.g., exponents) are not fixed a priori, these parameters are also identified in the same process.

For example, for each model listed in Table 2.5, by (2.113), the mean value of the nonlinear shear modulus and its linear elastic limit (2.115) take the respective forms

$$\underline{\widetilde{\mu}}(a) = \sum_{p=1}^{n} \underline{C}_p g_p(a), \tag{3.2}$$

$$\underline{\overline{\mu}} = \lim_{a \to 1} \underline{\widetilde{\mu}}(a) = \sum_{p=1}^{n} \underline{C}_p g_p(1). \tag{3.3}$$

Step 2 Second, based on the mean values derived at the first step, the goal is to identify the probability distributions followed by the random model parameters. For the nonlinear shear modulus $\widetilde{\mu}$, the variance is equal to

$$\text{Var}[\widetilde{\mu}(a)] = \sum_{p=1}^{n} \text{Var}[C_p] g_p(a)^2 + 2 \sum_{p_1=1}^{n} \left(\sum_{p_2=p_1+1}^{n} \text{Cov}[C_{p_1}, C_{p_2}] g_{p_1}(a) g_{p_2}(a) \right),$$

$$\tag{3.4}$$

where $\mathrm{Var}[C_p]$ is the variance of C_p, and $\mathrm{Cov}[C_{p_1}, C_{p_2}]$ is the covariance of C_{p_1} and C_{p_2}. The standard deviation of the nonlinear shear modulus is the square root of the variance,

$$\|\widetilde{\mu}(a)\| = \sqrt{\mathrm{Var}[\widetilde{\mu}(a)]}, \tag{3.5}$$

and, similarly, for every random constant coefficient, C_p, $p = 1, \cdots, n$, the standard deviation is $\|C_p\| = \sqrt{\mathrm{Var}[C_p]}$. To find $\|C_p\|$, we need to minimise the residual

$$\mathcal{R}_{\mathrm{std}} = \sqrt{\sum_{s=1}^{m} (\|\widetilde{\mu}(a_s)\| - d_s)^2}, \tag{3.6}$$

between the standard deviation (3.5) and the associated data $\{d_s\}_{s=1,\cdots,m}$ at the prescribed stretches $\{a_s\}_{s=1,\cdots,m}$. Before we do so, we fix the value of the stretch parameter to a particular value $a_0 > 0$ that is used for calibration. The corresponding random shear modulus (2.113) is equal to

$$\widetilde{\mu}(a_0) = \sum_{p=1}^{n} C_p g_p(a_0). \tag{3.7}$$

In particular, when $a_0 = 1$, $\widetilde{\mu}(a_0)$ is simply $\overline{\mu}$. Assuming $C_p > b$, for all $p = 1, \cdots, n$, where $b > -\infty$ is fixed, such that the mean values of the random coefficients, determined at Step 1, are bounded away from b, i.e., $\underline{C}_p > b$, $p = 1, \cdots, n$, we define the auxiliary random variables

$$R_p(a_0) = g_p(a_0) \left(C_p - b\right) \left(\widetilde{\mu}(a_0) - b \sum_{p=1}^{n} g_p(a_0)\right)^{-1}, \qquad p = 1, \cdots, n. \tag{3.8}$$

Note that $R_p(a_0) \in [0, 1]$, and by (3.7),

$$\sum_{p=1}^{n} R_p(a_0) = \left(\widetilde{\mu}(a_0) - b \sum_{p=1}^{n} g_p(a_0)\right)^{-1} \left(\sum_{p=1}^{n} C_p g_p(a_0) - b \sum_{p=1}^{n} g_p(a_0)\right) = 1. \tag{3.9}$$

Then, by (3.8), the random coefficients take the form

$$C_p = \frac{R_p(a_0)}{g_p(a_0)} \left(\widetilde{\mu}(a_0) - b \sum_{p=1}^{n} g_p(a_0)\right) + b, \qquad p = 1, \cdots, n. \tag{3.10}$$

Next, for the random nonlinear shear modulus $\widetilde{\mu}(a_0)$, defined by (3.7), we set the following mathematical expectations [225, 226, 519–521]:

$$E\left[\widetilde{\mu}(a_0)\right] = \underline{\widetilde{\mu}}(a_0), \qquad \underline{\widetilde{\mu}}(a_0) > 0, \tag{3.11}$$

$$E\left[\log\,\widetilde{\mu}(a_0)\right] = \omega, \qquad |\omega| < \infty. \tag{3.12}$$

The constraint (3.11) specifies the positive mean value $\underline{\widetilde{\mu}}(a_0)$ for $\widetilde{\mu}(a_0)$, while (3.12) implies that $1/\widetilde{\mu}(a_0)$ is a second-order random variables (i.e., it has finite mean value and finite variance) [507], [510, p. 270]. Then, by the principle of maximum entropy [280–282], $\widetilde{\mu}(a_0)$ follows a Gamma distribution (the maximum entropy distribution) [506, 507] (see Appendix B.6), $\Gamma\,(\rho_1(a_0),\rho_2(a_0))$ (see Table A.2), with shape and scale parameters $\rho_1(a_0) > 0$ and $\rho_2(a_0) > 0$, respectively, satisfying

$$\underline{\widetilde{\mu}}(a_0) = \rho_1(a_0)\rho_2(a_0), \qquad \mathrm{Var}[\widetilde{\mu}(a_0)] = \rho_1(a_0)\rho_2(a_0)^2. \tag{3.13}$$

The word "hyperparameters" is typically used for ρ_1 and ρ_2 to distinguish them from μ and other material constants [510, p. 8]. Here, the mean value $\underline{\widetilde{\mu}}(a_0)$ found at Step 1.

For the random vector $(R_1(a_0), \cdots, R_n(a_0))$, setting the constraints [519, 521]

$$E[\log\,R_p(a_0)] = \omega_p, \qquad |\omega_p| < \infty, \qquad p = 1, \cdots, n-1, \tag{3.14}$$

$$E\left[\log\left(1 - \sum_{p=1}^{n-1} R_p(a_0)\right)\right] = \omega_n, \qquad |\omega_n| < \infty, \tag{3.15}$$

this vector follows a Dirichlet distribution [288, 305], $D\,(\xi_1(a_0), \cdots, \xi_n(a_0))$ (see Table A.2). Then, every random variable $R_p(a_0)$, $p = 1, \cdots, n$, is characterised by a standard Beta distribution [1, 288], $B(\xi_p(a_0), \psi_p(a_0))$ (see Table A.2), with $\xi_p(a_0) > 0$ and $\psi_p(a_0) = \sum_{q=1,q\neq p}^{n} \xi_q(a_0) > 0$ satisfying

$$\underline{R}_p(a_0) = \frac{\xi_p(a_0)}{\sum_{q=1}^{n} \xi_q(a_0)}, \qquad \mathrm{Var}\left[R_p(a_0)\right] = \frac{R_p(a_0)^2 \psi_p(a_0)}{\xi_p(a_0)\left(\xi_p(a_0) + \psi_p(a_0) + 1\right)}, \tag{3.16}$$

where $\underline{R}_p(a_0)$ is the mean value and $\mathrm{Var}\left[R_p(a_0)\right]$ is the variance of $R_p(a_0)$ (with the standard deviation $\|R_p(a_0)\| = \sqrt{\mathrm{Var}\left[R_p(a_0)\right]}$). By (3.8), the mean value is equal to

$$\underline{R}_p(a_0) = g_p(a_0)\left(\underline{C}_p - b\right)\left(\underline{\widetilde{\mu}}(a_0) - b\sum_{p=1}^{n} g_p(a_0)\right)^{-1} \tag{3.17}$$

and is calculated from the mean values obtained at Step 1. The optimal hyperparameter vectors $(\rho_1(a_0), \rho_2(a_0))$ and $(\xi_1(a_0), \cdots, \xi_n(a_0))$ are identified by minimising the residual for the standard deviation (3.6) and taking into account relations (3.10), (3.13), and (3.16).

The variance of each random coefficient C_p, $p = 1, \cdots, n$, is then computed from (3.10).

3.1.2   Multiple-Term Models

It is instructive to consider the particular case when the calibrated modulus is the shear modulus at small strain, taking the form $\overline{\mu} = \sum_{p=1}^{n} C_p$, with $C_p > 0$, $p = 1, \cdots, n$ (see Table 2.5 for relevant examples of strain-energy functions). For this modulus, the expectations (3.11)–(3.12) take the form

$$E[\overline{\mu}] = \overline{\mu}, \qquad \overline{\mu} > 0, \tag{3.18}$$

$$E[\log \overline{\mu}] = \omega, \qquad |\omega| < \infty. \tag{3.19}$$

Thus, there is a Gamma distribution, $\Gamma(\rho_1, \rho_2)$ (see Table A.2), with $\rho_1 > 0$ and $\rho_2 > 0$ satisfying

$$\overline{\mu} = \rho_1 \rho_2, \qquad \text{Var}[\overline{\mu}] = \rho_1 \rho_2^2, \tag{3.20}$$

where $\overline{\mu}$ is the mean value and $\text{Var}[\overline{\mu}]$ is the variance of $\overline{\mu}$ (the corresponding standard deviation is $\|\overline{\mu}\| = \sqrt{\text{Var}[\overline{\mu}]}$). In this case, by (3.10), the individual random coefficients C_p, $p = 1, \cdots, n$, are equal to

$$C_p = \overline{\mu} R_p, \tag{3.21}$$

where $R_p \in [0, 1]$, such that $\sum_{p=1}^{n} R_p = 1$. For the auxiliary random variables, the mathematical expectations (3.14)–(3.15) are

$$E[\log R_p] = \omega_p, \qquad |\omega_p| < \infty, \qquad p = 1, \cdots, n - 1, \tag{3.22}$$

$$E\left[\log \left(1 - \sum_{p=1}^{n-1} R_p \right) \right] = \omega_n, \qquad |\omega_n| < \infty. \tag{3.23}$$

Hence, each random variable R_p follows a Beta distribution, $B(\xi_p, \psi_p)$ (see Table A.2), with $\xi_p > 0$ and $\psi_p = \sum_{q=1, q \neq p}^{n} \xi_q > 0$ satisfying

$$R_p = \frac{\xi_p}{\xi_p + \psi_p}, \qquad \text{Var}[R_p] = \frac{R_p^2 \psi_p}{\xi_p (\xi_p + \psi_p + 1)}. \tag{3.24}$$

Then, for every random coefficient C_p, $p = 1, \cdots, n$, the mean value is equal to

$$\underline{C}_p = \underline{\mu}\underline{R}_p = \frac{\rho_1\rho_2\xi_p}{\xi_p + \psi_p}, \tag{3.25}$$

and the variance takes the explicit form

$$\begin{aligned}
\text{Var}\left[C_p\right] &= \overline{\mu}^2\text{Var}\left[R_p\right] + \underline{R}_p^2\text{Var}[\overline{\mu}] + \text{Var}[\overline{\mu}]\text{Var}\left[R_p\right] \\
&= \frac{\rho_1\rho_2^2\xi_p\left(\xi_p^2 + \xi_p + \psi_p + \xi_p\psi_p + \psi_p\rho_1\right)}{\left(\xi_p + \psi_p\right)^2\left(\xi_p + \psi_p + 1\right)}.
\end{aligned} \tag{3.26}$$

3.1.3   One-Term Models

For any one-term model (see Table 2.5), there is only one random coefficient to be determined, namely C_1. For such a model, at any stretch $a = a_0$, the random shear modulus (2.113) is equal to

$$\tilde{\mu}(a_0) = C_1 g_1(a_0) = \overline{\mu}g_1(a_0), \tag{3.27}$$

where $\overline{\mu}$ is the linear shear modulus given by (2.115). In this case, the two-step calibration procedure simplifies as follows:

Step 1 First, determine the mean value coefficient $\underline{C}_1$, and any other unknown constant parameter appearing in the expression of the strain-energy function, by minimising the residual function for the mean values (3.1). For the one-term models listed in Table 2.5, the mean value of the random shear modulus (3.27) is equal to

$$\underline{\tilde{\mu}}(a) = \underline{C}_1 g_1(a), \tag{3.28}$$

and its linear elastic limit is

$$\underline{\overline{\mu}} = \underline{C}_1. \tag{3.29}$$

Step 2 The variance (3.4) of the shear modulus defined by (3.28) is

$$\text{Var}[\tilde{\mu}(a)] = \text{Var}[C_1]g_1(a)^2, \tag{3.30}$$

and the corresponding standard deviation, given by (3.5), is

$$\|\tilde{\mu}(a)\| = \|C_1\|g_1(a). \tag{3.31}$$

By (3.27), the mathematical constraints (3.11)–(3.12) for $\widetilde{\mu}(a_0)$ are equivalent to (3.18)–(3.19). Hence, $\overline{\mu}$, defined by (3.29), follows a Gamma distribution, $\Gamma(\rho_1, \rho_2)$ (see Table A.2), with $\rho_1 > 0$ and $\rho_2 > 0$ satisfying

$$\overline{\mu} = \underline{C}_1 = \rho_1 \rho_2, \qquad \|\overline{\mu}\| = \|C_1\| = \underline{C}_1/\sqrt{\rho_1}, \tag{3.32}$$

with $\underline{C}_1$ determined at Step 1. After the optimal value of $\|C_1\|$ is found by minimising the residual (3.6) for the standard deviation, the hyperparameters (ρ_1, ρ_2) are derived from (3.32).

3.2 Stochastic Modelling of Rubber-Like Materials

In [167], experimental measurements that capture the inherent variation between the constitutive behaviour of different material samples are recorded for silicone rubber under simple uniaxial tensile tests. The small datasets from two sample batches are illustrated in Fig. 3.1 where the arithmetic mean values, which are typically used for the calibration of deterministic models, are marked by red lines. In this figure, in addition to the applied force versus maximum vertical displacement observed experimentally, the computed values of the following nonlinear quantities of interest defined in Chap. 2 are represented:

- The first Piola–Kirchhoff tensile stress is equal to the applied tensile force $\mathcal{F}$ divided by the cross-sectional area $\mathcal{A} = 40\,\text{mm}^2$ in the reference configuration,

$$P = \frac{\mathcal{F}}{\mathcal{A}}. \tag{3.33}$$

- The nonlinear stretch modulus is computed by the formula (2.80),

$$E(a) = \frac{T}{\ln a} = \frac{aP}{\ln a}, \tag{3.34}$$

  where a is the stretch ratio and $T = aP$ is the Cauchy stress, with P the first Piola–Kirchhoff stress given by (3.33).
- The nonlinear shear modulus is calculated by the universal formula (2.121)

$$\widetilde{\mu}(a) = \frac{E \ln a}{a^2 - a^{-1}} = \frac{aP}{a^2 - a^{-1}}, \tag{3.35}$$

  where E is the stretch modulus given by (3.34).

Generally, in nonlinear continuum mechanics, several different models can be found to reproduce well a given set of experimental data. However, they may provide different predictions when employed to solve further problems. Ideally, material constitutive models should be calibrated and validated on multiaxial test data

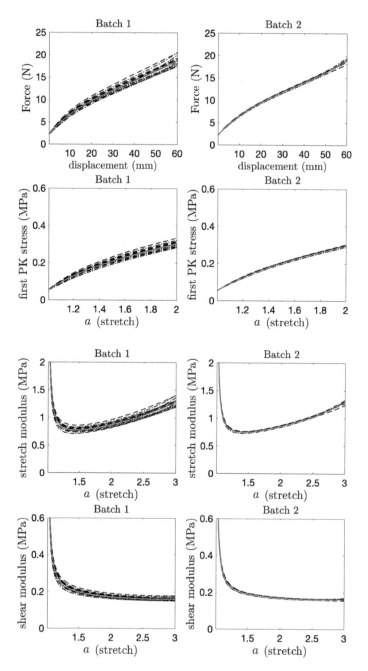

Fig. 3.1 Experimental data for applied force versus maximum vertical displacement in tensile loading of the two batches of silicone specimens, together with the first Piola–Kirchhoff (PK) tensile stress given by Eq. (3.33), and the nonlinear stretch and shear moduli defined by Eqs. (3.34) and (3.35), respectively. The red lines indicate the arithmetic mean data values [167]

[355, 356, 376, 513]. In this case also, the stochastic calibration method described in Sect. 3.1.1 (see also [371]) can be employed to obtain suitable models. Nevertheless, the datasets from uniaxial tests are sufficient here to illustrate the technique.

3.2.1 Hypothesis Testing

Before attempting to construct material models from the collected data, standard statistical tests [172, 175] are applied to verify whether the entire dataset can be treated as one, or the dataset of each batch must be modelled separately (see also the review article [232]). First, an unpaired Student's t-test [426] is used to compare the shear modulus data at small strain for the two batches. As this test provides a p-value of approximately 10^{-12}, i.e., less than the standard 0.05 significance level, the null hypothesis that the moduli of the two batches come from the same distribution has to be rejected. The significant differences between the data corresponding to the two batches are illustrated in Fig. 3.2. A χ^2 (chi-square) goodness-of-fit test [425] is also applied to check the null hypothesis that the shear modulus data of each batch come from a Gamma distribution. In this case, the p-values for batch 1 and 2 fits are 0.1 and 0.07, respectively. We conclude that the hypothesis that the shear moduli are Gamma-distributed cannot be rejected.

Figure 3.3 illustrates the Gamma and normal (Gaussian) probability distributions fitted to the shear modulus data at small strain. Notably, the represented Gamma distributions are similar to normal distributions (see also [167, 357]). This is due to the fact that, when $\rho_1 \rightarrow \infty$, the Gamma probability distribution is approximated by a normal distribution (a detailed proof of this result can be found in Appendix B.4). The parameters of the fitted distributions are recorded in Table 3.1 (note that ρ_1 is very large compared to ρ_2). However, shear moduli cannot be characterised by the

Fig. 3.2 Box plots of the shear modulus at small strain data of the two batches. On each box, purple dots represent data, the central mark indicates the median, and the bottom and top edges of the box indicate the 25th and 75th percentiles, respectively. The whiskers extend to the most extreme data points [167]

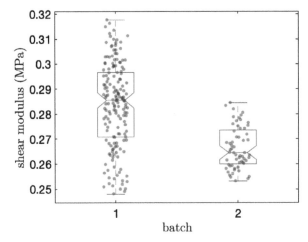

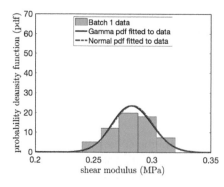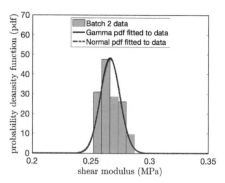

Fig. 3.3 Probability distributions derived from the data values for the random shear modulus at small strain. The parameters for these distributions are recorded in Table 3.1 [167]

Table 3.1 Parameters of the probability distributions derived from the data values for the random shear modulus at small strain (see Fig. 3.3)

Probability density function (pdf)	$\overline{\mu}$	$\|\overline{\mu}\|$	ρ_1	ρ_2
Gamma pdf fitted to batch 1 data	0.2837	0.0171	275.4403	0.0010
Normal pdf fitted to batch 1 data	0.2837	0.0170	-	–
Gamma pdf fitted to batch 2 data	0.2663	0.0083	1029.4047	0.0003
Normal pdf fitted to batch 2 data	0.2663	0.0084	-	–

normal distribution since this distribution is defined on the entire real line, whereas shear moduli are positive.

3.2.2   Parameter Estimation

We apply the stochastic method described in Sect. 3.1.1 to three different models listed in Table 2.5, which we calibrate to the given experimental data. The model parameters are recorded in Table 3.2. As shown by the load-deformation curves plotted in Figs. 3.4 and 3.5, all three models agree well with the experimental measurements. The values of the respective shear moduli are also comparable to those recorded in Table 3.1.

3.2.3   Bayesian Model Selection

The calibrated models can further be incorporated into Bayesian approaches for model selection or updates [472]. To select the best performing model, we employ *Bayes' theorem* [47, 78, 349, 622]. This theorem states that the posterior probability $P(M|D)$ of the model M, given the data values D, is

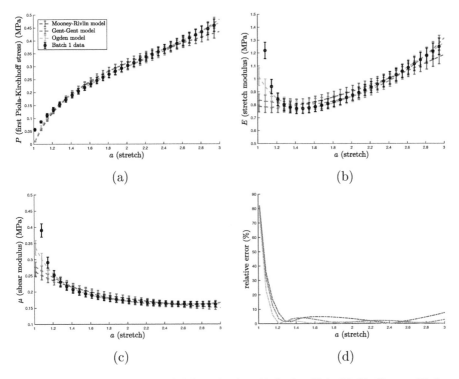

Fig. 3.4 Models calibrated to batch 1 data, showing: (**a**) the first Piola–Kirchhoff stress, (**b**) the nonlinear stretch modulus, (**c**) the nonlinear shear modulus, and (**d**) the relative error [167]

$$P(M|D) = \frac{P(M)P(D|M)}{P(D)}, \tag{3.36}$$

where $P(M)$ is the prior probability of the model M before the data values D were taken into account, $P(D|M)$ denotes the likelihood, or the probability of obtaining the data values D from the model M, and $P(D)$ is the normalisation value, known as the marginal likelihood.

Bayes' theorem also provides a methodology for estimating the odds for a model $M^{(i)}$ to another model $M^{(j)}$ in light of the data D, i.e.,

$$O_{ij} = \frac{P(M^{(i)}|D)}{P(M^{(j)}|D)} = \frac{P(M^{(i)})P(D|M^{(i)})}{P(M^{(j)})P(D|M^{(j)})} = \frac{P(M^{(i)})}{P(M^{(j)})}B_{ij}, \tag{3.37}$$

where

$$B_{ij} = \frac{P(D|M^{(i)})}{P(D|M^{(j)})} \tag{3.38}$$

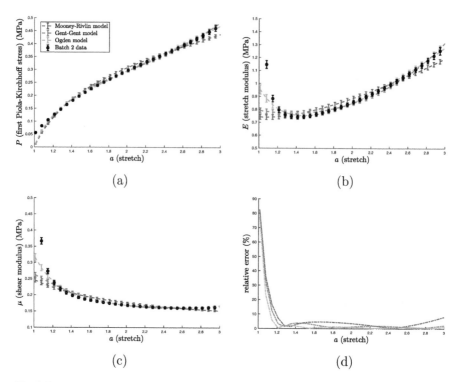

Fig. 3.5 Models calibrated to batch 2 data, showing: (**a**) the first Piola–Kirchhoff stress, (**b**) the nonlinear stretch modulus, (**c**) the nonlinear shear modulus, and (**d**) the relative error [167]

is the Bayes factor. According to (3.37), the posterior odds O_{ij} for the model $M^{(i)}$ against the model $M^{(j)}$, given the data D, are equal to the prior odds multiplied by the Bayes factor. In particular, if the models have equal prior probabilities, $P(M^{(i)}) = P(M^{(j)})$, i.e., there is no prior favourite, then, by (3.37), the posterior odds are equal to the Bayes factor, i.e., $O_{ij} = B_{ij}$. If the Bayes factor is equal to 1, then *Occam's razor* [283–285, 539] implies that a larger prior probability must be assigned to the simpler model than to the more complex one for reasons of parsimony.

Maintaining a general framework, we assume $P(D|M)$ to be an arbitrary probability that is symmetric about the mean value $D = 0$ and decreasing in the absolute value of D (see also the imaginative cartoon by D.M. Titterington (1982) [542]). In this case, the Bayes factor B_{ij} satisfies the inequality [56]

$$B_{ij} \geq \frac{\|D^{(j)}\| + \sqrt{2 \ln \left(\|D^{(j)}\| + 1.2\right)}}{e^{\|D^{(i)}\|^2/2}} \sqrt{\frac{2}{\pi}}, \qquad (3.39)$$

Table 3.2 Parameters of stochastic constitutive models given in Table 2.5 calibrated to the data, and the corresponding random nonlinear shear modulus $\mu = \mu(a)$ at $a = 1.15$

Stochastic model	Calibrated parameters (mean value ± standard deviation)		Shear modulus (MPa) (mean value ± standard deviation)	
	Batch 1	Batch 2	Batch 1	Batch 2
Mooney–Rivlin	$c_1 = 0.0936 \pm 0.0030$	$c_1 = 0.1029 \pm 0.0001$	$\mu = 0.2411 \pm 0.0130$	$\mu = 0.2277 \pm 0.0056$
	$c_2 = 0.1696 \pm 0.0115$	$c_2 = 0.1435 \pm 0.0063$		
Gent–Gent	$c_1 = 0.0971 \pm 0.0042$	$c_1 = 0.1007 \pm 0.0011$	$\mu = 0.2532 \pm 0.0136$	$\mu = 0.2397 \pm 0.0056$
	$c_2 = 0.1826 \pm 0.0110$	$c_2 = 0.1625 \pm 0.0053$		
	$\beta = 0.0434$	$\beta = 0.0421$		
Ogden $p_1 = 1, p_2 = -1, p_3 = -2$	$c_1 = -0.0645 \pm 0.0143$	$c_1 = -0.0437 \pm 0.0111$	$\mu = 0.2719 \pm 0.0720$	$\mu = 0.2563 \pm 0.0272$
	$c_2 = -0.0764 \pm 0.0155$	$c_2 = -0.0844 \pm 0.0112$		
	$c_3 = 0.4861 \pm 0.0534$	$c_3 = 0.4505 \pm 0.0345$		

where $\|D^{(i)}\|$ and $\|D^{(j)}\|$ designate the standard deviation that the predicted quantity of interest computed with the models $M^{(i)}$ and $M^{(j)}$, respectively, deviates from the observed data value D. The formula for calculating $\|D^{(i)}\|$ is as follows:

$$\|D^{(i)}\| = \frac{|Q^{(i)} - \underline{D}|}{\|D\|}, \tag{3.40}$$

where, for the quantity of interest, $Q^{(i)}$ is the expected value computed with the model $M^{(i)}$, and $\underline{D}$ and $\|D\|$ represent the experimentally observed mean value and standard deviation, respectively. The expression on the right-hand side of (3.39) provides an explicit lower bound on the Bayes factor B_{ij}. Taking equal prior probabilities, the lower bound on the Bayes factor given by (3.39) represents a lower bound on the posterior odds. Such lower bound is an estimate of the amount of evidence against the model $M^{(i)}$, i.e., the maximum support for the model $M^{(j)}$ provided by the data.

We apply this to select the best performing model among the calibrated models. For example, at $a = 1.15$, for each model recorded in Table 3.2, the standard deviations that the mean shear modulus $\tilde{\mu}$ deviates from the experimental mean data value 0.2909 is calculated, given that the experimental standard deviation is 0.0170. By formula (3.40),

$$\|D^{(1)}\| = \frac{|0.2411 - 0.2909|}{0.0170} = 2.9294 \qquad \text{for the Mooney–Rivlin model,}$$

$$\|D^{(2)}\| = \frac{|0.2532 - 0.2909|}{0.0170} = 2.2176 \qquad \text{for the Gent–Gent model,}$$

$$\|D^{(3)}\| = \frac{|0.2719 - 0.2909|}{0.0170} = 1.1176 \qquad \text{for the Ogden model.}$$

Assuming no prior favourite model (equal prior probabilities), by (3.39), it follows that the Bayes factors (or the odds) for the Ogden model against each of the other two models satisfy, respectively, $B_{31} \geq 1.9713$ and $B_{32} \geq 1.6174$. Next, taking $P(D|M^{(1)}) = 1 - P(D|M^{(3)})$, by the lower bound on the Bayes factor B_{31}, the likelihood of obtaining the data with the Ogden model is $P(D|M^{(3)}) \geq 0.6634$. Similarly, assuming equal prior probabilities and taking $P(D|M^{(2)}) = 1 - P(D|M^{(3)})$, the lower bound on B_{32} implies $P(D|M^{(3)}) \geq 0.6179$. Therefore, the data at $a = 1.15$ are more likely to be reproduced by the Ogden model than by the other two models.

In Fig. 3.6, the lower bounds on the Bayes factor B_{ij} are represented for each model $M^{(i)}$ against another model $M^{(j)}$ at different stretch ratios. By these bounds, the Gent–Gent model is generally more likely to represent the data than the Mooney–Rivlin model, and the Ogden model is more likely than both the Mooney–Rivlin and the Gent–Gent models. These results are consistent with those of Figs. 3.4d and 3.5d where the *relative errors* $|\tilde{\mu}_{model} - \tilde{\mu}_{data}|/\tilde{\mu}_{data}$, with $\tilde{\mu}_{model}$ and $\tilde{\mu}_{data}$ the mean value of the shear modulus for the model and for the experimental data, respectively, are shown.

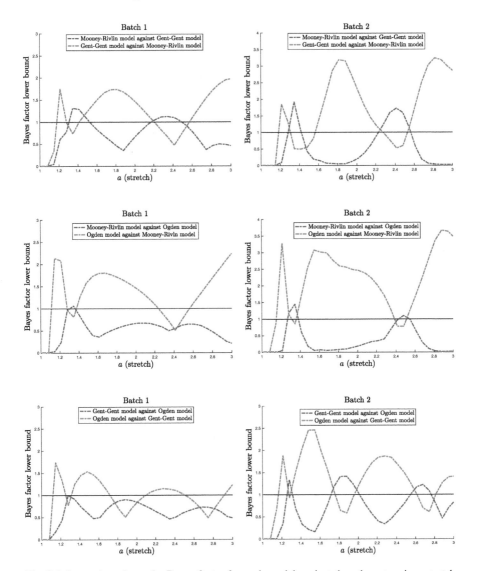

Fig. 3.6 Lower bounds on the Bayes factor for each model against the other at various stretch ratios [167]

Chapter 4
Elastic Instabilities

> *In particular, the conditions for instability should be predictable from this kind of analysis, just as they are in physics, meteorology, and engineering.*
>
> —E.T. Jaynes [282]

Stochastic elastic modeling raises many questions about the behaviour of materials. A very important one is the problem of stability. In this chapter, we examine within the stochastic elasticity framework the necking instability of a prismatic body under tensile load, the Rivlin cube, which changes shape under equitriaxial surface traction, and the cavitation of an elastic sphere where a void nucleates at its centre when a uniform tensile surface load is imposed. The aim is to derive condition for the onset of instability when the material properties are homogeneous but sampled from specific probability density functions. In each case, a homogeneous isotropic incompressible hyperelastic body is considered, characterised by a stochastic strain-energy function $W(\lambda_1, \lambda_2, \lambda_3)$ that satisfies assumptions (A1)–(A4) introduced in Chap. 3. The body is subjected to a prescribed *dead-load traction*, i.e., the magnitude and direction of the applied force are maintained independently of how the body deforms. The corresponding model parameters are defined as (spatially independent) random variables that follow non-Gaussian probability distributions, as described in Chap. 3. These are prior probability distributions that can be made more precise if experimental data are provided. Intuitively, such a stochastic body can be regarded as an ensemble (or population) of bodies that are equal in size and have the same geometry, and each body in the ensemble is made from a single homogeneous isotropic hyperelastic material with the elastic parameters not known with certainty but distributed according to known probability density functions. For every individual body, the finite elasticity theory applies. Then, under the Baker–Ericksen (BE) conditions (2.52), guaranteed by assumption (A3), the shear modulus is always positive, but the individual model coefficients may be either positive or negative, allowing for some interesting nonlinear elastic effects to be captured [360–362].

© Springer Nature Switzerland AG 2022
L. A. Mihai, *Stochastic Elasticity*, Interdisciplinary Applied Mathematics 55,
https://doi.org/10.1007/978-3-031-06692-4_4

The particular choice of numerical values in our calculations is for illustrative purposes only. To compare the stochastic results with the deterministic ones, we sample from distributions where the parameters have mean values corresponding to the deterministic system. Thus, the mean value of the distribution is guaranteed to converge to the expected value. In contrast to the deterministic problem where a single critical value strictly separates the stable and unstable cases, for the stochastic problem, we obtain probabilistic intervals where the stable and unstable states compete in the sense that both have a quantifiable chance to be observed with a given probability. Moreover, for the stochastic solution, although values are initially close, with a tight distribution around the mean value, as deformation progresses, their variance changes non-uniformly around the mean value, suggesting that the average value may be less significant from the physical point of view if fluctuations become large.

We also examine the elastic responses under dead loading of a hyperelastic cylindrical tube of stochastic anisotropic material where the model parameters are random variables. In this case, the dead loading consists of simultaneous internal pressure, axial tension, and torque, and the resulting universal deformation is a combined inflation, extension, and torsion from the reference circular cylindrical configuration to a deformed, also circular cylindrical, state. For the anisotropic material, two non-orthogonal preferred directions are assumed, corresponding to the (mean) directions of two families of aligned fibres embedded in an isotropic matrix material. In the deterministic elastic case, the tube may undergo "inversion" in the deformation, such that the radius first decreases and then increases, or "perversion" whereby the torsion chirality changes from right-handed to left-handed [210, 351]. The possible existence of such responses depends on the material constitutive model. For the stochastic problem, we derive the probability distribution of the deformations and find that, due to the probabilistic nature of the material parameters, the different states always compete. In particular, at a critical load, the radius may decrease or increase with a given probability, and similarly, right-handed or left-handed torsion may occur with a given probability.

4.1   Necking

When homogeneous isotropic incompressible hyperelastic materials for which the BE inequalities (2.52) hold are subject to uniaxial tension, described by (2.54), the deformation is a simple extension in the direction of the positive axial force (see Theorem 2.1.1). For this deformation, *necking instability* is said to occur if there is a maximum tensile load or, in other words, if there exists a critical extension ratio, such that the force required to extend the material beyond this critical value changes from increasing to decreasing [22, 24, 32, 108, 157, 185, 254, 421, 581]. The relation between the onset of necking and the maximum load was originally analysed for ductile materials by Considére [112]. For a class of hyperelastic materials where the load–displacement curve does not possess a maximum, in [500], it was proved that the homogeneous deformation is the only absolute minimiser of the elastic energy. In particular, incompressible neo-Hookean or Mooney–Rivlin hyperelastic models do not exhibit necking. For a stochastic hyperelastic body, we consider the following question: *What is the probability that necking is observed under a given tensile dead load?*

A simple extension takes the form

$$x_1 = \frac{1}{\sqrt{\lambda}} X_1, \quad x_2 = \lambda X_2, \quad x_3 = \frac{1}{\sqrt{\lambda}} X_3, \tag{4.1}$$

where (X_1, X_2, X_3) and (x_1, x_2, x_3) are the Cartesian coordinates for the reference and current configuration, respectively, and $\lambda > 1$ is the extension ratio. For this deformation, the principal stretches are $\alpha_1 = \alpha_3 = 1/\sqrt{\lambda}$ and $\alpha_2 = \lambda$.

We denote

$$\mathcal{W}(\lambda) = W(\lambda^{-1/2}, \lambda, \lambda^{-1/2}) \tag{4.2}$$

and calculate the first derivative of the above function with respect to λ,

$$\frac{d\mathcal{W}}{d\lambda} = -\lambda^{-3/2} \frac{\partial W}{\partial \alpha_1} + \frac{\partial W}{\partial \alpha_2} = -P_1 \lambda^{-3/2} + P_2, \tag{4.3}$$

where P_1 and P_2 are the principal components of the first Piola–Kirchhoff stress tensor $\mathbf{P}$, defined by (2.39), associated with the principal stretches $\alpha_1 = \lambda^{-1/2}$ and $\alpha_2 = \lambda$, respectively.

Under uniaxial tension, $P_1 = 0$, and by (4.3), $P_2 = d\mathcal{W}/d\lambda$. In this case, necking occurs if there exists a critical stretch ratio $\lambda_* > 1$, such that P_2 is not everywhere increasing as $\lambda > \lambda_*$ increases. In other words, the second derivative $d^2\mathcal{W}/d\lambda^2$ changes sign at $\lambda = \lambda_*$.

To illustrate this, we consider the following Ogden-type strain-energy function (see Table 2.5),

$$W(\alpha_1, \alpha_2, \alpha_3) = \frac{\mu_1}{2} \left(\alpha_1^2 + \alpha_2^2 + \alpha_3^2 - 3 \right) + 2\mu_2 \left(\alpha_1^{-1} + \alpha_2^{-1} + \alpha_3^{-1} - 3 \right), \tag{4.4}$$

where $\mu = \mu_1 + \mu_2$ is the random shear modulus at small strain, with $\mu_1 > 0$ and $\mu_2 > 0$ random variables.

By (4.2), we have

$$\mathcal{W}(\lambda) = \frac{\mu_1}{2} \left(2\lambda^{-1} + \lambda^2 - 3 \right) + 2\mu_2 \left(2\alpha_1^{1/2} + \lambda^{-1} - 3 \right), \tag{4.5}$$

and assuming $P_1 = 0$, by (4.3), we obtain

$$P_2 = \frac{d\mathcal{W}}{d\lambda} = \mu_1 \left(-\lambda^{-2} + \lambda \right) + 2\mu_2 \left(\lambda^{-1/2} - \lambda^{-2} \right). \tag{4.6}$$

Equivalently,

$$P_2 = \mu_1 \lambda + 2\mu_2 \lambda^{-1/2} - (\mu_1 + 2\mu_2) \lambda^{-2}. \tag{4.7}$$

To verify the monotonicity of the tensile stress when $\lambda > 1$, we differentiate P_2 given by (4.7) with respect to λ,

$$\frac{dP_2}{d\lambda} = \frac{d^2 \mathcal{W}}{d\lambda^2} = 2(\mu_1 + 2\mu_2) \lambda^{-3} - \mu_2 \lambda^{-3/2} + \mu_1. \tag{4.8}$$

Noting that the above function of λ is a quadratic function in $\lambda^{-3/2}$, we deduce that $dP_2/d\lambda$ takes negative values, i.e., P_2 decreases, if μ_1 and μ_2 satisfy

$$8 \left(\frac{\mu_1}{\mu_2} \right)^2 + 16 \frac{\mu_1}{\mu_2} - 1 < 0. \tag{4.9}$$

As $\mu_1 > 0$ and $\mu_2 > 0$, the above inequality holds when

$$\frac{\mu}{\mu_1} = 1 + \frac{\mu_2}{\mu_1} > \frac{3\sqrt{2}}{3\sqrt{2} - 4}. \tag{4.10}$$

The behaviour of the tensile load P_2 as a function of the extension ratio λ is illustrated in Fig. 4.1 for different deterministic values of the parameter ratio μ_1/μ.

Depending on the material parameters, if the shear modulus, μ, follows a Gamma distribution, then the probability of onset of necking is

$$P \left(\mu > \mu_1 \frac{3\sqrt{2}}{3\sqrt{2} - 4} \right) = 1 - \int_0^{\mu_1 \frac{3\sqrt{2}}{3\sqrt{2} - 4}} g(u; \rho_1, \rho_2) du \tag{4.11}$$

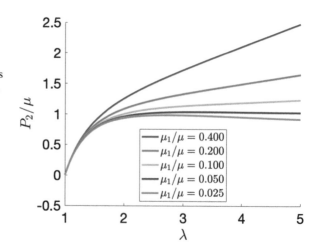

Fig. 4.1 The normalised tensile load P_2, defined by (4.7), as a function of the extension ratio $\lambda > 1$, for different deterministic values of the parameter ratio μ_1/μ. For sufficiently small values of this ratio, the tensile load changes from increasing to decreasing

P_2/μ

λ

— $\mu_1/\mu = 0.400$
— $\mu_1/\mu = 0.200$
— $\mu_1/\mu = 0.100$
— $\mu_1/\mu = 0.050$
— $\mu_1/\mu = 0.025$

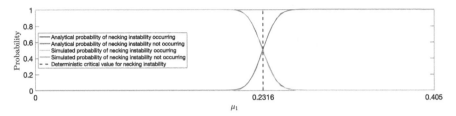

Fig. 4.2 Probability distributions of necking instability occurring or not for a material characterised by the strain-energy function given by (4.4) with random shear modulus, $\mu > 0$, following
the Gamma distribution with $\rho_1 = 405$ and $\rho_2 = 0.01$. The vertical black line separates the
expected regions based only on the mean value of the shear modulus, $\underline{\mu} = \rho_1 \rho_2$

Fig. 4.3 Probability
distribution of the normalised
tensile load P_2 defined by
(4.7) when the random shear
modulus, $\mu > 0$, is drawn
from a Gamma distribution
with $\rho_1 = 405$, $\rho_2 = 0.01$,
and $R_1 = \mu_1/\mu$ is drawn
from a Beta distribution with
$\xi_1 = 64$, $\xi_2 = 1000$. The
black lines correspond to the
expected values based only
on the mean values of the
parameters

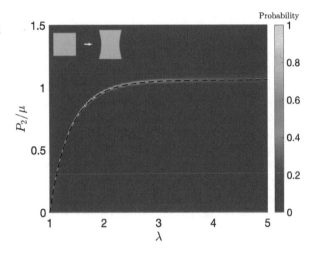

and that of necking not occurring is

$$P\left(\mu < \mu_1 \frac{3\sqrt{2}}{3\sqrt{2}-4}\right) = \int_0^{\mu_1 \frac{3\sqrt{2}}{3\sqrt{2}-4}} g(u; \rho_1, \rho_2)\mathrm{d}u. \qquad (4.12)$$

For example, when the shear modulus, μ, follows a Gamma probability density
function, with hyperparameters $\rho_1 = 405$ and $\rho_2 = 0.01$, the probability
distributions given by (4.11) and (4.12) are represented in Fig. 4.2. Numerically,
the interval $(0, \underline{\mu}/10)$, where $\underline{\mu} = \rho_1\rho_2$, was divided into 100 steps, then for each
value of $\mu_1 \in (0, \underline{\mu})$, 10^3 random values of μ were generated from the specified
Gamma distribution, then compared with the inequalities involving μ_1.

The stochastic tensile load is illustrated in Fig. 4.3, where μ follows a Gamma
distribution with $\rho_1 = 405$ and $\rho_2 = 0.01$, and $R_1 = \mu_1/\mu$ follows a Beta
distribution with $\xi_1 = 64$ and $\xi_2 = 1000$. In this case, the mean value is
$\underline{\mu} = \rho_1\rho_2 = 4.05$ for μ and $\underline{R_1} = \xi_1/(\xi_1 + \xi_2) = 0.0602$ for R_1. Thus, the

mean value of μ_1 is $\underline{\mu}_1 = \underline{R}_1 \cdot \underline{\mu} = 0.2436$, and necking instability is not expected. However, there is about 15% chance that necking may also occur (see also Fig. 4.2).

Note that μ and R_1 are independent random variables, depending on parameters (ρ_1, ρ_2) and (ξ_1, ξ_2), respectively, which are derived by fitting distributions to given data. However, μ_1 and μ_2 are codependent variables as they both require (μ, R_1) (see Chap. 3).

4.2   Equilibria of an Elastic Cube Under Equitriaxial Surface Loads

The classic problem of the *Rivlin cube* has played a central role in the development of new concepts in nonlinear elasticity [275]. This problem, which was first posed by Rivlin [464], consists in obtaining the equilibria of a homogeneous isotropic incompressible hyperelastic cube on which equal triaxial dead loads are applied to each face. Under equitriaxial surface dead loads, a homogeneous deformation is equivalent to a homogeneous triaxial stretch [43]. The questions are: *What are the possible equilibrium states?* and How does their stability depend on the material constitutive law? [43, 48, 252, 468].

For a cube of (incompressible) neo-Hookean material, Rivlin found that (neutrally) stable non-trivial triaxial deformations with two equal stretches are possible if the equitriaxial tractions are sufficiently large [48, 252, 464, 468]. In addition to these so-called *plate-like* and *rod-like* equilibrium states (see Fig. 4.4), for a similar cube of (incompressible) Mooney–Rivlin material, it was shown in [43] that neutrally stable equilibria with three unequal stretches are also possible.

Extensions to the case of a cube of compressible Mooney-type material were studied in [535], while the case of anisotropic materials was analysed in [511]. For an isotropic cube, when the three pairs of equal and opposite forces differ from each

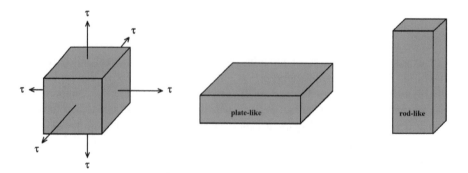

Fig. 4.4 Schematic of a Rivlin cube under normal dead-load tractions, $\tau > 0$, uniformly distributed on all faces in the reference configuration, showing the reference state (left) and the plate-like (middle) or rod-like (right) deformations [372]

other by a small amount [464], or if two pairs are the same but different from the third by a small amount [481], neutrally stable homogeneous triaxial deformations with three different stretches can occur as the reference state becomes unstable. Further discussions on stability analysis under small perturbations can be found in [202].

The aim here is to present the Rivlin cube problem in the context of stochastic elasticity, as formulated originally in [372]. Similarly to the deterministic case, the cube is subject to equitriaxial dead-load tractions, uniformly distributed on all faces in the reference configuration, and deforms by a homogeneous triaxial stretch. In addition to finding all possible homogeneous triaxial deformations and identifying the stable equilibrium states, for the stochastic cube, the question is: *What is the probability distribution of stable triaxial deformation under a prescribed dead load, given the distributions of the material parameters?*

4.2.1   The Rivlin Cube

We consider a cube of hyperelastic material described by a stochastic Mooney–Rivlin strain-energy function of the form (see Table 2.5),

$$W(\alpha_1, \alpha_2, \alpha_3) = \frac{\mu_1}{2} \left(\alpha_1^2 + \alpha_2^2 + \alpha_3^2 - 3 \right) + \frac{\mu_2}{2} \left(\alpha_1^{-2} + \alpha_2^{-2} + \alpha_3^{-2} - 3 \right),$$

(4.13)

where $\mu = \mu_1 + \mu_2$ is the random shear modulus at small strain, with μ_1 and μ_2 random variables.

When the cube is subject to constant normal tension or compression, τ, uniformly distributed on all faces in the reference configuration, in the absence of body forces, assuming that the resulting deformation is a homogeneous triaxial stretch, the deformation gradient tensor, measured from a reference configuration, is equal to

$$\mathbf{A} = \mathrm{diag}(\alpha_1, \alpha_2, \alpha_3),$$

(4.14)

with the constants $\alpha_i > 0, i = 1, 2, 3$, satisfying the incompressibility condition,

$$J = \det \mathbf{A} = \alpha_1 \alpha_2 \alpha_3 = 1.$$

(4.15)

Under the deformation (4.14), the principal components of the Cauchy stress tensor $\mathbf{T}(\mathbf{x})$, with $\mathbf{x} = (x_1, x_2, x_3)^T$ the Cartesian coordinates in the current configuration, are

$$T_i = -p + \alpha_i \frac{\partial W}{\partial \alpha_i} = -p + \mu_1 \alpha_i^2 - \frac{\mu_2}{\alpha_i^2}, \qquad i = 1, 2, 3,$$

(4.16)

where p is the Lagrange multiplier for the incompressibility constraint (4.15). The corresponding first Piola–Kirchhoff stress tensor $\mathbf{P}(\mathbf{X}) = J\mathbf{T}\mathbf{A}^{-T}$, with $\mathbf{X} = (X_1, X_2, X_3)^T$ the Cartesian coordinates in the reference configuration, has the principal components

$$P_i = \frac{T_i}{\alpha_i} = -\frac{p}{\alpha_i} + \mu_1 \alpha_i - \frac{\mu_2}{\alpha_i^3}, \qquad i = 1, 2, 3. \tag{4.17}$$

In the absence of body forces, the elastic equilibrium equations in the reference configuration are

$$\frac{\partial P_i}{\partial X_i} = 0, \qquad i = 1, 2, 3. \tag{4.18}$$

Equivalently, in the current configuration,

$$\frac{\partial T_i}{\partial x_i} = 0, \qquad i = 1, 2, 3. \tag{4.19}$$

Thus, by (4.16), the Lagrange multiplier p satisfies

$$\frac{\partial p}{\partial x_i} = 0, \qquad i = 1, 2, 3, \tag{4.20}$$

i.e., p is constant. The equitriaxial surface tractions, τ, are constant in the reference configuration (dead loading), i.e.,

$$P_i = \tau, \qquad i = 1, 2, 3, \tag{4.21}$$

or equivalently, by (4.17),

$$\mu_1 \alpha_i^2 - \frac{\mu_2}{\alpha_i^2} - \tau \alpha_i - p = 0, \qquad i = 1, 2, 3. \tag{4.22}$$

When an elastic body of homogeneous isotropic incompressible hyperelastic material deforms by $\mathbf{x}(\mathbf{X})$ under the sole action of the equitriaxial dead-load traction τ, the *total free energy* is equal to

$$E(\mathbf{x}) = \int_V [W(\mathbf{A}) - \tau \mathrm{tr}(\mathbf{A})]\, dV, \tag{4.23}$$

where $W(\mathbf{A})$ is the strain-energy density function of the material, expressed in terms of the gradient tensor $\nabla \mathbf{x} = \mathbf{A}$. As we are interested in the minimisers of this energy for an incompressible material, we only consider deformation gradients that satisfy the incompressibility constraint $\det \mathbf{A} = 1$. This incompressibility constraint can be easily enforced by considering the minimisers of the following unconstrained form:

$$E_o(\mathbf{x}) = \int_V [W(\mathbf{A}) - \tau \mathrm{tr}(\mathbf{A}) - p\,(\det \mathbf{A} - 1)]\,dV, \tag{4.24}$$

where p is a Lagrange multiplier that is interpreted as the hydrostatic pressure. In the case of a homogeneous triaxial stretch $\mathbf{x}(\mathbf{X})$, with gradient tensor $\nabla \mathbf{x} = \mathrm{diag}(\alpha_1, \alpha_2, \alpha_3)$, we define the function

$$\Psi(\alpha_1, \alpha_2, \alpha_3; \tau) = W(\alpha_1, \alpha_2, \alpha_3) - \tau\,(\alpha_1 + \alpha_2 + \alpha_3), \tag{4.25}$$

where $W(\alpha_1, \alpha_2, \alpha_3)$ is the strain-energy density function for the isotropic elastic material expressed in terms of the principal stretches $\{\alpha_i\}_{i=1,2,3}$. In this case, the total free energy, given by (4.23), takes the equivalent form

$$E(\mathbf{x}) = \int_V \Psi(\alpha_1, \alpha_2, \alpha_3; \tau)\,dV, \tag{4.26}$$

and we are interested in the minimisers of this energy when the deformation gradients satisfy the incompressibility condition (4.15). Alternatively, we introduce the incompressibility constraint by defining

$$\Psi_o(\alpha_1, \alpha_2, \alpha_3; \tau) = W(\alpha_1, \alpha_2, \alpha_3) - \tau\,(\alpha_1 + \alpha_2 + \alpha_3) - p\,(\alpha_1\alpha_2\alpha_3 - 1), \tag{4.27}$$

the corresponding total free energy given by (4.24) is equal to

$$E_o(\mathbf{x}) = \int_V \Psi_o(\alpha_1, \alpha_2, \alpha_3; \tau)\,dV. \tag{4.28}$$

Definition 4.2.1 We say that the deformation $\mathbf{x} = \mathbf{x}(\mathbf{X})$, with gradient $\nabla \mathbf{x}$, such that $\det(\nabla \mathbf{x}) = 1$, is stable if it is a local minimum of the total free energy, i.e., if the following inequality holds:

$$E(\mathbf{x}) < E(\mathbf{y}), \tag{4.29}$$

for all $\mathbf{y} = \mathbf{y}(\mathbf{X})$ that are continuous, piecewise differentiable deformation mappings, with gradients $\nabla \mathbf{y}$ satisfying $\det(\nabla \mathbf{y}) = 1$. When relation (4.29) holds with "$\leq$" instead of the strict inequality "$<$," the deformation is neutrally stable. Otherwise, the deformation is unstable.

The following result is central to the stability analysis of the Rivlin cube problem (see also Theorem 2.2 of [43]).

Theorem 4.2.1 *When a homogeneous isotropic incompressible hyperelastic body, with the strain-energy density function $W(\alpha_1, \alpha_2, \alpha_3)$, deforms by a homogeneous triaxial stretch under the equitriaxial dead-load traction τ, the following statements hold:*

(i) *If $\tau < 0$ (compressive loading), then only the trivial (undeformed, reference) configuration, with $\alpha_1 = \alpha_2 = \alpha_3 = 1$, is possible, and this state is unstable.*

(ii) *If $\tau = 0$ (no loading), then the reference state is neutrally stable.*

(iii) *If $\tau > 0$ (tensile loading), then the reference state is stable when it is a strict local minimum of $\Psi(\cdot, \tau)$, given by (4.25), neutrally stable when it is a non-strict local minimum of $\Psi(\cdot, \tau)$, and unstable otherwise.*

(iv) *If $\tau > 0$ (tensile loading), then any non-trivial local minimum of $\Psi(\cdot, \tau)$ is neutrally stable.*

Proof

(i) The fact that there are no pure homogeneous equilibrium configurations other than the reference state when $\tau < 0$ follows from the BE inequalities (2.52). For the reference configuration, the gradient deformation tensor is the identity tensor, $\mathbf{I} = \mathrm{diag}(1, 1, 1)$. To prove that it is not stable in the sense of Definition 4.2.1, we define the function

$$H(\mathbf{A}) = W(\mathbf{A}) - \tau \mathrm{tr}(\mathbf{A}) \tag{4.30}$$

and take a skew-symmetric matrix $\mathbf{A}$ and the exponential mapping

$$e^{\epsilon \mathbf{A}} = \sum_{n=0}^{\infty} \frac{(\epsilon \mathbf{A})^n}{n!}, \tag{4.31}$$

where $0 < \epsilon \ll 1$. Since $\mathbf{A}^T = -\mathbf{A}$, it follows that $\mathrm{tr}(\mathbf{A}) = 0$ and $e^{\epsilon \mathbf{A}}$ is proper orthogonal (i.e., $e^{-\epsilon \mathbf{A}} = \left(e^{\epsilon \mathbf{A}}\right)^T$ and $\det e^{\epsilon \mathbf{A}} = 1$). We consider the following Taylor expansion about the identity tensor in powers of ϵ:

$$H(e^{\epsilon \mathbf{A}}) = W(\mathbf{I}) - \tau \, \mathrm{tr}\left(\mathbf{I} + \epsilon \mathbf{A} + \frac{\epsilon^2}{2}\mathbf{A}^2 + \mathcal{O}(\epsilon^3)\right) = H(\mathbf{I}) + \frac{\epsilon^2}{2}\tau \, \mathrm{tr}\left(\mathbf{A}\mathbf{A}^T\right) + \mathcal{O}(\epsilon^3).$$

As $\tau < 0$ and $\mathrm{tr}\left(\mathbf{A}\mathbf{A}^T\right) > 0$, the above identity implies $H(e^{\epsilon \mathbf{A}}) < H(\mathbf{I}) + \mathcal{O}(\epsilon^3)$, showing that $\mathbf{I}$ is not a local minimum for the total free energy (4.23). Hence, the reference state is unstable.

(ii) When $\tau = 0$, we consider a deformation $\mathbf{y}(\mathbf{X})$, such that $\nabla \mathbf{y} = \mathbf{Q}\mathrm{diag}(\alpha_1, \alpha_2, \alpha_3)\mathbf{R}$, where $\mathbf{Q} = \mathbf{Q}(\mathbf{X})$ and $\mathbf{R} = \mathbf{R}(\mathbf{X})$ are orthonormal tensors, satisfying $(\mathbf{RQ})_{ii} \leq 1$, $i = 1, 2, 3$, and $\alpha_i = \alpha_i(\mathbf{X})$, $i = 1, 2, 3$, are the principal stretches of $\nabla \mathbf{y}$, satisfying $\sup \sum_{i=1}^{3} |\alpha_i - 1| < \epsilon \ll 1$ and $\alpha_1\alpha_2\alpha_3 = 1$. Then, the total free energy (4.23) of the elastic body deformed by $\mathbf{y}(\mathbf{X})$ is equal to

$$E(\mathbf{y}) = \int_V W(\alpha_1, \alpha_2, \alpha_3) dV = E(\mathbf{X}) + \int_V [\Psi(\alpha_1, \alpha_2, \alpha_3; \tau) - \Psi(1, 1, 1; \tau)] dV.$$

From the above expression, we deduce that, if the reference state, $\mathbf{X}$, with the gradient tensor $\nabla \mathbf{X} = \mathbf{I} = \mathrm{diag}(1, 1, 1)$, is a strict or non-strict local minimum of $\Psi(\cdot, \tau)$, then $E(\mathbf{y}) \geq E(\mathbf{X})$. Next, taking $\mathbf{R}$ proper orthogonal implies $E(\mathbf{R}\mathbf{X}) = E(\mathbf{X})$, and hence the reference state is neutrally stable.

(iii) When $\tau > 0$, for the reference state, $\mathbf{X}$, with gradient tensor $\nabla \mathbf{X} = \mathbf{I}$, we consider a deformation $\mathbf{y}(\mathbf{X})$ as in (ii). In this case, the total free energy (4.23) of the elastic body deformed by $\mathbf{y}(\mathbf{X})$ is equal to

$$E(\mathbf{y}) = \int_V \left[W(\alpha_1, \alpha_2, \alpha_3) - \tau\, \mathrm{tr}\,(\mathbf{Q}\mathrm{diag}(\alpha_1, \alpha_2, \alpha_3)\mathbf{R}) \right] dV$$

$$= E(\mathbf{X}) + \int_V [\Psi(\alpha_1, \alpha_2, \alpha_3; \tau) - \Psi(1, 1, 1; \tau)]\, dV + \tau \int_V \sum_{i=1}^{3} [1 - (\mathbf{R}\mathbf{Q})_{ii}]\, \alpha_i\, dV.$$

From the above expression, we deduce that stability or neutral stability of the reference state occurs as follows:

 – if $\tau > 0$ and the reference state is a strict local minimum of $\Psi(\cdot, \tau)$, as $(\mathbf{R}\mathbf{Q})_{ii} \leq 1$, $i = 1, 2, 3$, then $E(\mathbf{y}) > E(\mathbf{X})$, i.e., the reference state is stable;
 – if $\tau > 0$ and the reference state is a non-strict local minimum of $\Psi(\cdot, \tau)$, then $E(\mathbf{y}) \geq E(\mathbf{X})$, i.e., the reference state is neutrally stable.

(iv) When $\tau > 0$, for a non-trivial homogeneous triaxial stretch $\mathbf{x}(\mathbf{X})$, with gradient tensor $\nabla \mathbf{x} = \mathrm{diag}(\alpha_1, \alpha_2, \alpha_3)$, we consider a deformation $\mathbf{y}(\mathbf{X})$ as in (iii), and obtain

$$E(\mathbf{y}) = E(\mathbf{x}) + \int_V [\Psi(\alpha_1, \alpha_2, \alpha_3; \tau) - \Psi(\alpha_1, \alpha_2, \alpha_3; \tau)]\, dV + \tau \int_V \sum_{i=1}^{3} [1 - (\mathbf{R}\mathbf{Q})_{ii}]\, \alpha_i\, dV.$$

In this case, if the triaxial stretch $\mathbf{x}(\mathbf{X})$ is a (strict or non-strict) local minimum of $\Psi(\cdot, \tau)$, then setting $\nabla \mathbf{y} = \mathbf{R}^T \mathrm{diag}(\alpha_1, \alpha_2, \alpha_3)\mathbf{R}$, with $\mathbf{R}$ proper orthogonal, implies $E(\mathbf{y}) = E(\mathbf{x})$, hence the triaxial stretch $\mathbf{x}(\mathbf{X})$ is neutrally stable. This concludes the proof.

According to the above theorem, in the absence of loading, i.e., when $\tau = 0$, the reference state is neutrally stable, i.e., the trivial deformation is a minimum for the elastic energy of a stochastic Mooney–Rivlin cube. Furthermore, under compressive dead load, $\tau < 0$, the reference state is unstable, and there are no other homogeneous triaxial states for the cube. Thus, when $\tau \leq 0$, there is no uncertainty to be resolved about the reference state or its expected stability, and we only need to consider the case when the surface dead load is tensile, i.e., when $\tau > 0$. In this case, we derive the probability distribution of the homogeneous triaxial deformations for which the stochastic cube is in stable equilibrium. We do this by specifying the magnitude of the dead load and the probability distributions of the hyperelastic material parameters.

4.2.2 The Stochastic neo-Hookean Cube

If $\mu_2 = \mu - \mu_1 = 0$ and $\mu = \mu_1 > 0$, then the Mooney–Rivlin model (4.13) reduces to a neo-Hookean form [549]. In this case, at $\tau = 2\mu$, the trivial solution becomes unstable and there is a bifurcation into a plate-like deformation (one stretch less than 1 and two equal stretches greater than 1) and a rod-like deformation (one stretch greater than 1 and two equal stretches less than 1) [43, 468]. By rotation, there are six such non-trivial deformations.

When the non-trivial homogeneous triaxial stretches occur, $0 < \lambda \neq 1$ is a solution of the equation

$$\lambda^{3/2} - \frac{\tau}{\mu}\lambda^{1/2} + 1 = 0. \tag{4.32}$$

We label the axes of Cartesian coordinates, such that the rod-like solution is obtained by taking $\alpha_1 = \lambda > 1$ and $\alpha_2 = \alpha_3 = 1/\sqrt{\lambda} < 1$. Then, the plate-like solution is obtained by taking $\alpha_1 = \lambda < 1$ and $\alpha_2 = \alpha_3 = 1/\sqrt{\lambda} > 1$. These non-trivial deformations are neutrally stable if $\lambda < \lambda_* = \tau/(3\mu)$ and unstable if $\lambda > \lambda_* = \tau/(3\mu)$. The corresponding dead loads take the form

$$\tau = \mu\left(\lambda + \frac{1}{\sqrt{\lambda}}\right) > 0. \tag{4.33}$$

Note that these loads are always positive. When $\tau = \tau_* = 3\mu/2^{2/3}$, there is only one non-trivial solution to the cubic equation (4.32), namely $\lambda = \lambda_* = \tau_*/(3\mu) = 1/2^{2/3}$, and the corresponding plate-like deformation is neutrally stable. The behaviour of the neo-Hookean cube under equitriaxial tensile tractions is illustrated in Fig. 4.5.

For the stochastic neo-Hookean cube, first, we look at the probability distributions for the number of equilibria as a function of τ, given that μ follows a Gamma probability density function, $g(\mu, \rho_1, \rho_2)$. Under a dead-load traction $\tau > 0$, the following three cases are possible:

- $\tau < 3\mu/2^{2/3}$, in which case the trivial reference state is the unique stable state.
- $0 < 2\mu < \tau$, in which case there are two non-trivial solutions, with $\lambda \neq 1$ satisfying equation (4.33).
- $3\mu/2^{2/3} < \tau < 2\mu$, in which case there are three possible equilibria, namely both the previous cases coexist.

Depending on the applied dead load, the probability of having one, two, or three possible equilibria is, respectively,

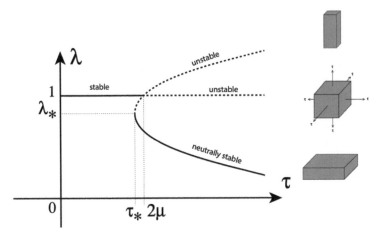

Fig. 4.5 Bifurcation diagram showing possible homogeneous equilibria of a neo-Hookean cube under normal dead-load tractions, $\tau > 0$, uniformly distributed on all faces in the reference configuration [372]

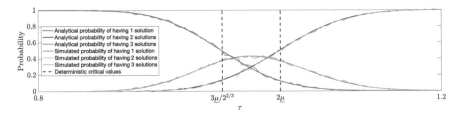

Fig. 4.6 Probability distributions of different numbers of possible equilibria for a stochastic neo-Hookean cube with random shear modulus, $\mu > 0$, following the Gamma distribution with $\rho_1 = 400$ and $\rho_2 = 0.0013$. Dark coloured lines represent analytically derived solutions, given by Eqs. (4.34)–(4.36), whereas the lighter versions represent stochastically generated data. The vertical black lines separate the expected regions of equilibria based only on the mean value of the shear modulus, $\underline{\mu} = \rho_1\rho_2$ [372]

$$P(\mu > 2^{2/3}\tau/3) = 1 - \int_0^{2^{2/3}\tau/3} g(u; \rho_1, \rho_2)du, \qquad (4.34)$$

$$P(0 < \mu < \tau/2) = \int_0^{\tau/2} g(u; \rho_1, \rho_2)du, \qquad (4.35)$$

$$P(\tau/2 < \mu < 2^{2/3}\tau) = \int_{\tau/2}^{2^{2/3}\tau/3} g(u; \rho_1, \rho_2)du. \qquad (4.36)$$

To illustrate these, we assume that the shear modulus, μ, follows a Gamma probability density function, with hyperparameters $\rho_1 = 400$ and $\rho_2 = 0.0013$. The resulting probability distributions given by Eqs. (4.34)–(4.36) are represented in Fig. 4.6 and compared with the distribution generated from stochastic simulations.

Numerically, the dead-load interval of $(0.8, 1.2)$ was divided into 100 steps, and then for each value of τ, 10^3 random values of μ were numerically generated from a specified Gamma distribution and then compared with the inequalities defining the three intervals for values of τ.

Figure 4.6 shows that, for a cube with the mechanical properties defined by the average of the Gamma distribution, $\underline{\mu}$, and dead load $\tau = 2\underline{\mu} = 2\rho_1\rho_2 = 1.04$, in the deterministic case, this is a bifurcation point where the cube transitions from having three equilibria (one trivial and two non-trivial) to having only two non-trivial equilibria. However, in the stochastic case, there is approximately 10% chance of a randomly chosen cube presenting only the single trivial state (red solid and dashed lines), 50% chance of a randomly chosen cube presenting two non-trivial sates (blue solid and dashed lines), and 40% chance of randomly choosing a cube that will present all three equilibria (green solid and dashed lines). In order to ensure that only the trivial equilibrium exists or, alternatively, that only non-trivial equilibria exist, the values of the dead load must be chosen beyond the expected bifurcation points.

As in the deterministic case, although multiple equilibria may exist, not all of them are stable. For the deterministic problem, the stable branches are noted on the bifurcation diagram of Fig. 4.5. Based on this information, one can derive simulated probability distributions of the stretches seen in the stochastic case. Specifically, during the stochastic trials, instead of coarsely categorising how many equilibria are possible, we use the generated value of μ and the given value of τ to calculate the observed stable stretch values. There are two possible cases: first, there is a single stable solution branch for the given value of τ, either the trivial or the non-trivial configuration with smallest λ; alternatively, when there are multiple stable branches, as one of the branches is always the trivial reference state, the reference state is chosen. This choice is based on the fact that the dead load is applied to the reference state and, unless perturbed in some way, the cube tends to remain in this state. The stable equilibria are illustrated in the stochastic bifurcation diagram shown in Fig. 4.7 where the shear modulus μ satisfies a Gamma distribution with $\rho_1 = 400$ and $\rho_2 = 0.0013$ (compare with Fig. 4.5).

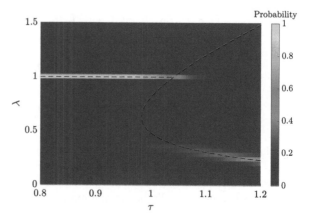

Fig. 4.7 Stochastic bifurcation diagram for a stochastic neo-Hookean cube with hyperparameters $\rho_1 = 400$ and $\rho_2 = 0.0013$. The probability distribution of stretches follows the stable branches of the bifurcation diagram. The black lines correspond to the expected values based only on the mean value of the shear modulus, $\underline{\mu} = \rho_1\rho_2$ (compare with Fig. 4.5) [372]

4.2.3   The Stochastic Mooney–Rivlin Cube

When $0 < \mu = \mu_1 + \mu_2 \neq \mu_1$, under the kinematic assumption (A3), stating that the BE inequalities (2.52) hold, two cases are distinguished, namely, with $\mu_2 > 0$ and $\mu_2 < 0$, respectively.[1]

Case 1   If $0 < \mu_2 = \mu - \mu_1 < \mu$, then, under sufficiently large tensile loading, the reference state is unstable, and there is a bifurcation into rod-like and plate-like deformations [43]. By rotation, there are six non-trivial deformations, with the corresponding dead load taking the form

$$\tau = \left(\mu_1 + \frac{\mu_2}{\lambda}\right)\left(\lambda + \frac{1}{\sqrt{\lambda}}\right) > 0. \tag{4.37}$$

Case 1.1   When $\mu_2 = \mu - \mu_1 \geq \mu_1/3 > 0$, neutrally stable rod-like configurations are possible, such that $\lambda > \lambda_*$, for some $\lambda_* \geq 1$, while all other configurations are unstable (see Fig. 4.8a). If $\mu_2 = \mu_1/3 > 0$, then $\lambda_* = 1$ and rod-like configurations are neutrally stable, while plate-like ones are unstable.

Case 1.2   When $0 < \mu_2 = \mu - \mu_1 < \mu_1/3$, under sufficiently small dead loads, plate-like states are neutrally stable, while rod-like ones are unstable. When the dead load is increased, a secondary bifurcation into a neutrally stable deformation with three unequal stretches, $\alpha_1 \neq \alpha_2 \neq \alpha_3 \neq \lambda_1$, that links the rod-like branch to the plate-like one is possible, after which rod-like deformations are neutrally stable (see Fig. 4.8b). Indeed, in the case of three unequal stretches, by (4.17) and (4.21),

$$\tau = \left(\mu_1 + \mu_2 \alpha_1^2\right)(\alpha_2 + \alpha_3) = \left(\mu_1 + \mu_2 \alpha_3^2\right)(\alpha_1 + \alpha_2). \tag{4.38}$$

Eliminating τ from the above identities implies

$$\mu_1 = \mu_2 \left(\alpha_1 \alpha_2 + \alpha_2 \alpha_3 + \alpha_3 \alpha_1\right). \tag{4.39}$$

Then, by (4.15) and (4.39), the dead load takes the form

$$\tau = \frac{\mu_2}{\alpha_3}\left(\frac{\mu_1}{\mu_2} + \alpha_3^2\right)\left(\frac{\mu_1}{\mu_2} - \frac{1}{\alpha_3}\right) \tag{4.40}$$

and has a maximum, $\tau^{**} > 0$, where it intersects the rod-like branch, with $\lambda = \lambda^{**} > 1$ and a minimum, $\tau_{**} > 0$, where it intersects the plate-like branch, with

---

[1] For rubber, some negative values of μ_2 were obtained in [298] from experimental data reported in [429].

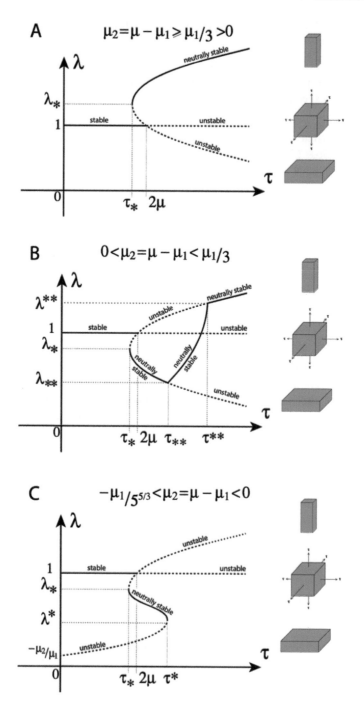

Fig. 4.8 Bifurcation diagrams showing possible homogeneous equilibria of Mooney–Rivlin cubes under normal dead-load tractions [372]

$\lambda = \lambda_{**} < 1$. The points of intersection can be found by setting $\alpha_3 = 1/\sqrt{\lambda}$ in (4.40).

For $\mu_2 > 0$, we set $b = 0$ in (3.8) and define the Beta-distributed random variable

$$R_1 = \frac{\mu_1}{\mu}. \tag{4.41}$$

Due to the stochastic nature of μ_1 and μ_2, one cannot guarantee a priori the sign of $\mu_2 - \mu_1/3$. However, we can calculate the probability of a cube presenting a given sign. Namely, $\mu_2 - \mu_1/3$ is negative if and only if $R_1 > 3/4$, and positive otherwise. The corresponding probabilities are

$$P(\mu_2 < \mu_1/3) = 1 - \int_0^{3/4} h(r; \xi_1, \xi_2) dr, \tag{4.42}$$

$$P(\mu_2 > \mu_1/3) = \int_0^{3/4} h(r; \xi_1, \xi_2) dr. \tag{4.43}$$

For example, if μ follows a Gamma distribution with $\rho_1 = 240$ and $\rho_2 = 0.01$, and R_1, given by (4.41), satisfies a Beta distribution with $\xi_1 = 200$ and $\xi_2 = 100$, then the expected values are $\underline{\mu} = \rho_1 \rho_2 = 2.4$ and $\underline{\mu_1} = \mu \xi_1/(\xi_1 + \xi_2) = 1.6$. Hence, $\underline{\mu_2} = \underline{\mu} - \underline{\mu_1} = 0.8 > \underline{\mu_1}/3$, and the stochastic system tends to follow the stochastic bifurcation diagram shown in Fig. 4.9a (corresponding to Fig. 4.8a). Alternatively, if $\xi_1 = 400$ and $\xi_2 = 100$, then the expected values are $\underline{\mu} = \rho_1 \rho_2 = 2.4$ and $\underline{\mu_1} = \mu \xi_1/(\xi_1 + \xi_2) = 1.92$. Hence, $\underline{\mu_2} = \underline{\mu} - \underline{\mu_1} = 0.4 < \underline{\mu_1}/3$, and the stochastic system follows the stochastic bifurcation diagram of Fig. 4.9b (corresponding to Fig. 4.8b).

Case 2 When $-\mu_1 < \mu_2 = \mu - \mu_1 < 0$, combining (4.16) and (2.52) gives

$$\mu_1 + \mu_2 \alpha_k^2 > 0 \quad \text{if} \quad \alpha_i \neq \alpha_j, \ i \neq k \neq j, \quad i, j, k = 1, 2, 3. \tag{4.44}$$

The above inequalities are satisfied if $\mu_1 > 0$ and $\mu_2 \geq 0$. When $0 > \mu_2 > -\mu_1$, these inequalities are equivalent to

$$0 < \alpha_k^2 < -\frac{\mu_1}{\mu_2} \quad \text{if} \quad \alpha_i \neq \alpha_j, \ i \neq k \neq j, \quad i, j, k = 1, 2, 3. \tag{4.45}$$

Assuming the deformation with three unequal stretches, $\alpha_1 \neq \alpha_2 \neq \alpha_3 \neq \lambda_1$, by (4.40) and (4.44), $\tau > 0$. However, as $\mu_1 > 0$ and $\alpha_i > 0$, $i = 1, 2, 3$, (4.39) implies that three unequal stretches are impossible when $\mu_2 < 0$. Therefore, only nontrivial deformations with two equal stretches (rod-like or plate-like) require further consideration. In this case, setting $\alpha_k = 1/\sqrt{\lambda}$ in (4.45) implies $\lambda > -\mu_2/\mu_1$, and the dead load, τ, takes the form given by (4.37). As $0 < -\mu_2/\mu_1 < 1$, both $\lambda > 1$ (rod-like states) and $\lambda < 1$ (plate-like states) are possible. By direct calculations, we obtain the following:

Fig. 4.9 Stochastic
bifurcation diagram for
stochastic Mooney–Rivlin
cubes with hyperparameters:
(**a**) $\rho_1 = 240$, $\rho_2 = 0.01$,
$\xi_1 = 200$, $\xi_2 = 100$; (**b**)
$\rho_1 = 240$, $\rho_2 = 0.01$,
$\xi_1 = 400$, $\xi_2 = 100$; and (**c**)
$\rho_1 = 721$, $\rho_2 = 0.01$,
$\xi_1 = 10000$, $\xi_2 = 500$. The
probability distribution of
stretches follows the stable
branches of the bifurcation
diagram. The black lines
correspond to the expected
values based only on mean
parameter values (compare
with Fig. 4.8) [372]

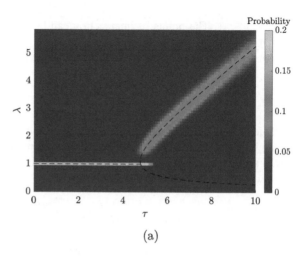

(a)

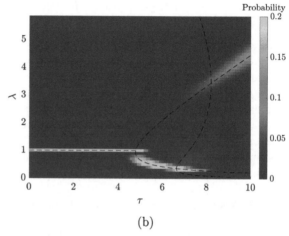

(b)

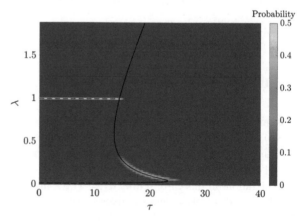

Case 2.1 When $-\mu_1/5^{5/3} < \mu_2 < 0$, there is a maximum dead load, $\tau^* > 0$, attained for some λ^*, satisfying $-\mu_2/\mu_1 < \lambda^* < -5\mu_2/\mu_1 < 1/5^{2/3}$, and a minimum dead load, $\tau_* > 0$, attained for some, λ_*, such that $-5\mu_2/\mu_1 < \lambda_* < 1$. In this case, neutrally stable plate-like solutions are possible under dead loads satisfying $\tau_* < \tau < \tau^*$ (see Fig. 4.8c). For example, taking $\mu_1 = 2.484$ and $\mu_2 = -0.148$, as reported in [298], we obtain $\mu_2/\mu_1 = -0.0596 > -1/5^{5/3} \approx -0.0684$, and thus such possibilities are not unrealistic. When $\tau^* > \tau_0 = 2\mu$, neutrally stable plate-like deformations are obtained after the reference state becomes unstable. This situation occurs if

$$\inf_{0<\lambda<1} \frac{\lambda^{5/2} - 2\lambda^{3/2} + \lambda}{\lambda^{3/2} - 1} \approx -0.045 < \frac{\mu_2}{\mu_1} < 0, \qquad (4.46)$$

where "inf" denotes infimum.

Case 2.2 When $\mu_2/\mu_1 \leq -1/5^{5/3} < 0$, the non-trivial homogeneous states are unstable.

Thus, for $\mu_2 < 0$, equilibria only exist in the small finite region $-\mu_1/5^{5/3} < \mu_2 < 0$. In this case, setting $b = -\mu_1/5^{5/3}$ in (3.8), we define the Beta-distributed random variable

$$R_1 = \frac{\mu_1(1 + 1/5^{5/3})}{\mu + 2\mu_1/5^{5/3}}. \qquad (4.47)$$

The corresponding stochastic simulations are illustrated in Fig. 4.9c, where μ follows a Gamma distribution with $\rho_1 = 721$ and $\rho_2 = 0.01$, and R_1, given by equation (4.47), follows a Beta distribution with $\xi_1 = 10000$ and $\xi_2 = 500$ (compare with Fig. 4.8c).

In all the above cases, the probability of the trivial reference state being a stable equilibrium is exactly 1 when $0 < \tau < 2\mu$ or, equivalently, when $\mu > \tau/2 > 0$. Thus, if we specify a dead-load traction, τ, then the probability that a cube chosen at random will present the trivial stable state is equal to

$$P(\mu > \tau/2) = 1 - \int_0^{\tau/2} g(u; \rho_1, \rho_2)du. \qquad (4.48)$$

However, if $\tau > 2\mu$, then the stable state will have to be calculated based on randomly generated values of μ and μ_1 and the given value of τ.

In summary, for stochastic neo-Hookean or Mooney–Rivlin cubes under uniform tensile dead loads, the probabilities of stable equilibria are obtained, given that the model parameters are generated from known probability density functions. In the deterministic case, which is based on mean parameter values, there are single-valued critical loads that strictly separate the cases where either the trivial or a non-trivial stable configuration occurs. In the stochastic case, there are probabilistic load intervals, containing the deterministic (expected) critical values, where there is a quantifiable chance for both the trivial and non-trivial equilibrium states may be found. This is reflected by the corresponding stochastic bifurcation diagrams where the probability distributions are more diffuse near the expected critical values.

4.3 Cavitation of Spheres Under Radially Symmetric Tensile Loads

Cavitation in solids is the formation of a void within a solid under tensile loads, by analogy to the similar phenomenon observed in fluids. For rubber-like materials, following early studies of damage under loads [79, 617], this phenomenon was first reported in [193] where experiments showed that rubber cylinders ruptured under relatively small tensile dead loads by opening an internal cavity. The nonlinear elastic analysis in [39] provided the first systematic theory for the formation of a spherical cavity at the centre of a sphere of isotropic incompressible hyperelastic material under a radially symmetric tensile load. Mathematically, cavitation was treated as a bifurcation from the trivial state at a critical value of the surface traction or displacement, at which the trivial solution became unstable. This paved the way for numerous applied and theoretical studies devoted to this inherently nonlinear mechanical effect, which is not captured by the linear elasticity theory. Spheres of particular homogeneous isotropic incompressible materials were discussed in [97], homogeneous anisotropic spheres with transverse isotropy about the radial direction were examined in [25, 353, 438], concentric homogeneous spheres of different hyperelastic material were analysed in [264, 295, 439, 498],; non-spherical cavities were investigated in [276], cavities with non-zero pressure were presented in [499], cavitation in an elastic membrane was studied in [524], the homogenisation problem of nonlinear elastic materials was treated in [326, 328], growth-induced cavitation in nonlinearly elastic solids was explored in [209, 350, 427], and cavitation in pseudo-elastic models [141, 145, 263, 267, 316, 416, 462], associated with the so-called Mullins effect [390–394, 400, 543, 544] of stress-softening in filled rubber [68, 317, 323, 333], was considered in [148, 149]. Important results focusing on cavitation in rubber-like materials are reviewed in [170, 190, 265]. Recent experimental studies regarding the onset, healing, and growth of cavities in elastomers can be found in [273, 441, 442, 452]. Finite element calculations are presented in [294].

In addition to the well-known stable cavitation post-bifurcation at the critical dead load, such that the cavity radius monotonically increases with the applied load, it was shown in [359] that unstable (snap) cavitation is also possible for some homogeneous isotropic incompressible hyperelastic materials where Backer–Ericksen inequalities hold. For the stochastic hyperelastic sphere, the question is: *What is the probability distribution of stable radially symmetric deformation under a given surface dead load?*

We consider a sphere of stochastic incompressible hyperelastic material described by the strain-energy function

$$W(\alpha_1, \alpha_2, \alpha_3) = \frac{\mu_1}{2m^2}\left(\alpha_1^{2m} + \alpha_2^{2m} + \alpha_3^{2m} - 3\right) + \frac{\mu_2}{2n^2}\left(\alpha_1^{2n} + \alpha_2^{2n} + \alpha_3^{2n} - 3\right),$$
$$(4.49)$$

where m and n are deterministic constants, and μ_1 and μ_2 are random variables.

The sphere is subjected to a radially symmetric deformation caused by the sole action of a given radial tensile dead load. As for the deterministic elastic sphere [39], we obtain conditions on the constitutive law, such that, setting the internal pressure equal to zero, where the radius tends to zero, the required external dead load is finite, and therefore cavitation occurs. We further analyse the stability of the cavitated solution and distinguish between supercritical cavitation, where the cavity radius monotonically increases as the dead load increases, and subcritical (snap) cavitation, with a sudden jump to a finite internal radius immediately after initiation.

For the stochastic sphere, the radially symmetric deformation takes the form

$$r = g(R), \qquad \theta = \Theta, \qquad \phi = \Phi, \qquad (4.50)$$

where (R, Θ, Φ) and (r, θ, ϕ) are the spherical polar coordinates in the reference and current configuration, respectively, such that $0 \leq R \leq B$, and $g(R) \geq 0$ is to be determined. The corresponding deformation gradient is equal to $\mathbf{A} = \mathrm{diag}\,(\alpha_1, \alpha_2, \alpha_3)$, with

$$\alpha_1 = \frac{dg}{dR} = \lambda^{-2}, \qquad \alpha_2 = \alpha_3 = \frac{g(R)}{R} = \lambda, \qquad (4.51)$$

where α_1 and $\alpha_2 = \alpha_3$ are the radial and hoop stretches, respectively, and dg/dR denotes the derivative of g with respect to R. By (4.51),

$$g^2 \frac{dg}{dR} = R^2, \qquad (4.52)$$

hence,

$$g(R) = \left(R^3 + c^3\right)^{1/3}, \qquad (4.53)$$

where $c \geq 0$ is a constant to be calculated. If $c > 0$, then $g(R) \to c > 0$ as $R \to 0_+$, and a spherical cavity of radius c forms at the centre of the sphere, from zero initial radius (see Fig. 4.10), otherwise the sphere remains undeformed.

Assuming that the deformation (4.50) is due to a prescribed radial tensile dead load, applied uniformly on the sphere surface in the reference configuration, in the absence of body forces, the radial equation of equilibrium is

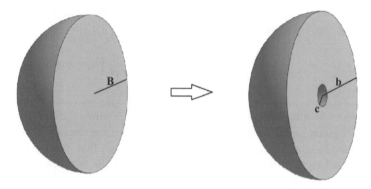

Fig. 4.10 Schematic of cross section of a sphere, showing the reference state, with outer radius B (left), and the deformed state, with cavity radius c and outer radius b (right), respectively

$$\frac{dP_{11}}{dR} + \frac{2}{R}(P_{11} - P_{22}) = 0, \tag{4.54}$$

or equivalently,

$$\frac{dP_{11}}{d\lambda}\lambda^{-2} + 2\frac{P_{11} - P_{22}}{1 - \lambda^3} = 0, \tag{4.55}$$

where $\mathbf{P} = (P_{ij})_{i,j=1,2,3}$ is the first Piola–Kirchhoff stress tensor. For an incompressible material,

$$P_{11} = \frac{\partial W}{\partial \alpha_1} - \frac{p}{\alpha_1}, \qquad P_{22} = \frac{\partial W}{\partial \alpha_2} - \frac{p}{\alpha_2}.$$

Denoting

$$\mathcal{W}(\lambda) = W(\lambda^{-2}, \lambda, \lambda), \tag{4.56}$$

where $\lambda = r/R = g(R)/R = (1 + c^3/R^3)^{1/3} > 1$, we obtain

$$\frac{d\mathcal{W}}{d\lambda} = -\frac{2}{\lambda^3}\frac{\partial W}{\partial \alpha_1} + 2\frac{\partial W}{\partial \alpha_2} = -\frac{2P_{11}}{\lambda^3} + 2P_{22}. \tag{4.57}$$

Then, setting the internal pressure (at $R \to 0_+$) equal to zero, by (4.55) and (4.57), the external tension (at $R = B$) is equal to

$$T = \frac{P_{11}}{\lambda^2}\Big|_{\lambda=\alpha_b} = \int_{\alpha_b}^{\alpha_c} \frac{d\mathcal{W}}{d\lambda}\frac{d\lambda}{\lambda^3 - 1}, \tag{4.58}$$

and the applied dead load, in the reference configuration, is

$$P = T\alpha_b^2 = \alpha_b^2 \int_{\alpha_b}^{\alpha_c} \frac{dW}{d\lambda} \frac{d\lambda}{\lambda^3 - 1}, \tag{4.59}$$

where α_c and α_b represent the stretches at the centre and outer surface, respectively. The value of the required dead load, P_0, for the onset of cavitation (bifurcation from the trivial solution) is obtained by taking $\alpha_c \to \infty$ and $\alpha_b = (1 + c^3/B^3)^{1/3} \to 1$ as $c \to 0_+$ in (4.59), i.e.,

$$P_0 = \int_1^\infty \frac{dW}{d\lambda} \frac{d\lambda}{\lambda^3 - 1}. \tag{4.60}$$

The BE inequalities (2.53) imply

$$\frac{dW}{d\lambda} \frac{1}{\lambda^3 - 1} > 0, \tag{4.61}$$

and hence, $P_0 > 0$. Then, if the critical dead load given by (4.60) is finite, cavitation takes place, else, the sphere remains undeformed.

For a sphere made of a hyperelastic material with the strain-energy function given by (4.49) and (4.56) takes the form

$$W(\lambda) = \frac{\mu_1}{2m^2}\left(\lambda^{-4m} + 2\lambda^{2m} - 3\right) + \frac{\mu_2}{2n^2}\left(\lambda^{-4n} + 2\lambda^{2n} - 3\right). \tag{4.62}$$

For the onset of cavitation, the critical dead-load traction, defined by (4.60), is equal to

$$P_0 = \frac{2\mu_1}{m} \int_1^\infty \frac{\lambda^{2m-1} - \lambda^{-4m-1}}{\lambda^3 - 1} d\lambda + \frac{2\mu_2}{n} \int_1^\infty \frac{\lambda^{2n-1} - \lambda^{-4n-1}}{\lambda^3 - 1} d\lambda, \tag{4.63}$$

or equivalently, by the change of variable $x = \lambda^3 - 1$,

$$
\begin{aligned}
P_0 = &\frac{2\mu_1}{3m} \int_0^\infty \frac{(x+1)^{(2m-3)/3} - (x+1)^{-(4m+3)/3}}{x} dx \\
&+ \frac{2\mu_2}{3n} \int_0^\infty \frac{(x+1)^{(2n-3)/3} - (x+1)^{-(4n+3)/3}}{x} dx.
\end{aligned} \tag{4.64}
$$

When μ_1 and μ_2 are positive, P_0 given by (4.64) is finite. In this case, a spherical cavity forms if and only if the following conditions are simultaneously satisfied: $2m - 3 < 0$, $-4m - 3 < 0$, $2n - 3 < 0$, $-4n - 3 < 0$, or equivalently [97, 265] (see also Example 5.1 of [39]), if and only if

$$-3/4 < m, n < 3/2. \tag{4.65}$$

In particular, cavitation is found in a neo-Hookean sphere (with $m = 1$ and $n = 0$), but not in a Mooney–Rivlin sphere (with $m = 1$ and $n = -1$). The special cases when $m \in \{-1/2, 1\}$ and $n = 0$ are given as examples in [39], and when $m \in \{1/2, 3/4, 1, 5/4\}$ and $n = 0$, the explicit critical loads are provided in [97]. When these bounds and the BE inequalities are satisfied, the critical pressure P_0 is finite and the problem is to find the behaviour of the cavity in a neighbourhood of this critical value. In each of those previously studied cases (see, e.g., Figure 2 of [97]), cavitation forms from zero radius and then presents itself as a supercritical bifurcation with stable cavitation (i.e., the new bifurcated solution exists locally for values of $P > P_0$, and the radius of the cavity monotonically increases with the applied load post-bifurcation).

Another theoretical possibility is that the bifurcation could be subcritical (i.e., the cavitated solution exists locally for values less than P_0 and is unstable). Explicitly, in [359], it was shown that, depending on the model parameters, the family of materials (4.56) can exhibit both behaviours. General conditions for a given material to exhibit either a subcritical or a supercritical bifurcation are provided by the following result.

Proposition 4.3.1 *Let $\mathcal{W}(\lambda)$ be twice differentiable at $\lambda = 1$, and*

$$P(c) = \left(1 + c^n\right)^{(n-1)/n} \int_{(1+c^n)^{1/n}}^{\infty} \frac{d\mathcal{W}}{d\lambda} \frac{d\lambda}{\lambda^n - 1},$$

where $n > 1$. Then

$$\lim_{c \to 0_+} \frac{dP}{dc} = 0, \tag{4.66}$$

and if

$$\lim_{c \to 0_+} P(c) - \lim_{\lambda \to 1} \frac{1}{n(n-1)} \frac{d^2\mathcal{W}}{d\lambda^2} > 0, \tag{4.67}$$

then $dP/dc > 0$ for sufficiently small $c > 0$, i.e., the bifurcation is supercritical, while if

$$\lim_{c \to 0_+} P(c) - \lim_{\lambda \to 1} \frac{1}{n(n-1)} \frac{d^2\mathcal{W}}{d\lambda^2} < 0, \tag{4.68}$$

then $dP/dc < 0$ for sufficiently small $c > 0$, i.e., the bifurcation is subcritical.

Proof Denoting $\theta = (1 + c^n)^{(n-1)/n}$ and $\widehat{P}(\theta) = P(c)$, we have

$$\widehat{P}(\theta) = \theta \int_{\theta^{1/(n-1)}}^{\infty} \frac{d\mathcal{W}}{d\lambda} \frac{d\lambda}{\lambda^n - 1}$$

and

$$\frac{\mathrm{d}P}{\mathrm{d}c} = \frac{\mathrm{d}\theta}{\mathrm{d}c}\frac{\mathrm{d}\widehat{P}}{\mathrm{d}\theta} = (1 + c^n)^{1/n} \frac{c^{n-1}}{n-1}\frac{\mathrm{d}\widehat{P}}{\mathrm{d}\theta},$$

where

$$\frac{\mathrm{d}\widehat{P}}{\mathrm{d}\theta} = \int_{\theta^{1/(n-1)}}^{\infty} \frac{\mathrm{d}\mathcal{W}}{\mathrm{d}\lambda}\frac{\mathrm{d}\lambda}{\lambda^n - 1} - \frac{\theta^{1/(n-1)}}{n-1}\left(\frac{\mathrm{d}\mathcal{W}}{\mathrm{d}\lambda}\frac{1}{\lambda^n - 1}\right)\Big|_{\lambda = \theta^{1/(n-1)}}.$$

Then

$$\lim_{\theta \to 1}\frac{\mathrm{d}\widehat{P}}{\mathrm{d}\theta} = \lim_{\theta \to 1}\int_{\theta^{1/(n-1)}}^{\infty} \frac{\mathrm{d}\mathcal{W}}{\mathrm{d}\lambda}\frac{\mathrm{d}\lambda}{\lambda^n - 1} - \lim_{\theta \to 1}\frac{\theta^{1/(n-1)}}{n-1}\frac{\mathrm{d}\mathcal{W}}{\mathrm{d}\lambda}\frac{1}{\lambda^n - 1}\Big|_{\lambda = \theta^{1/(n-1)}}$$

$$= \lim_{c \to 0_+} P(c) - \lim_{\theta \to 1}\frac{\theta^{1/(n-1)}}{n-1}\left(\frac{\mathrm{d}\mathcal{W}}{\mathrm{d}\lambda}\frac{1}{\lambda^n - 1}\right)\Big|_{\lambda = \theta^{1/(n-1)}}$$

$$= \lim_{c \to 0_+} P(c) - \lim_{\lambda \to 1}\frac{1}{n(n-1)}\frac{\mathrm{d}\mathcal{W}}{\mathrm{d}\lambda}\frac{1}{\lambda - 1}$$

$$= \lim_{c \to 0_+} P(c) - \lim_{\lambda \to 1}\frac{1}{n(n-1)}\frac{\mathrm{d}^2\mathcal{W}}{\mathrm{d}\lambda^2}.$$

Since $\mathcal{W}$ is twice differentiable at $\lambda = 1$, provided that $\lim_{c \to 0_+} P(c)$ is finite, the above limit is also finite. Hence, (4.66) follows, and, for sufficiently small $c > 0$, $\mathrm{d}P/\mathrm{d}c > 0$ (respectively, $\mathrm{d}P/\mathrm{d}c < 0$) if and only if (4.67) (respectively, (4.68)) holds. This concludes the proof.

We illustrate the different behaviours in the case when $m = 1$ and $n = -1/2$ in (4.49), such that

$$W(\alpha_1, \alpha_2, \alpha_3) = \frac{\mu_1}{2}\left(\alpha_1^2 + \alpha_2^2 + \alpha_3^2 - 3\right) + 2\mu_2\left(\alpha_1^{-1} + \alpha_2^{-1} + \alpha_3^{-1} - 3\right). \tag{4.69}$$

Then (4.56) takes the form

$$\mathcal{W}(\lambda) = \frac{\mu_1}{2}\left(\lambda^{-4} + 2\lambda^2 - 3\right) + 2\mu_2\left(\lambda^2 + 2\lambda^{-1} - 3\right). \tag{4.70}$$

In this case, under the deformation (4.50), the BE inequalities (2.53) are reduced to

$$\mu_1 + 2\mu_2\frac{\lambda^3}{1 + \lambda^3} > 0. \tag{4.71}$$

Inequality (4.71) implies that, when $\lambda \to 1$, the shear modulus must be positive, i.e., $\mu = \mu_1 + \mu_2 > 0$, while if $\lambda \to \infty$, then $\mu_1 + 2\mu_2 > 0$. Noting that the function of λ on the left-hand side is monotonically increasing when μ_2 is positive and decreasing if μ_2 is negative, and taking $\mu_1 > 0$, the two limits imply that the BE inequalities are satisfied for all $\lambda \in (1, \infty)$ when

$$0 < \frac{\mu_1}{\mu} < 2. \tag{4.72}$$

For sufficiently small c/B, the corresponding dead-load traction, defined by (4.59), is equal to

$$P = 2\mu_1 \left[\left(1 + \frac{c^3}{B^3} \right)^{1/3} + \frac{1}{4} \left(1 + \frac{c^3}{B^3} \right)^{-2/3} \right] + 4\mu_2 \left(1 + \frac{c^3}{B^3} \right)^{1/3}$$

$$= 4\mu \left(1 + \frac{c^3}{B^3} \right)^{1/3} - 2\mu_1 \left[\left(1 + \frac{c^3}{B^3} \right)^{1/3} - \frac{1}{4} \left(1 + \frac{c^3}{B^3} \right)^{-2/3} \right]. \tag{4.73}$$

Then, the critical dead load given by (4.60) takes the form

$$P_0 = 4\mu - \frac{3\mu_1}{2} \tag{4.74}$$

and is positive when $0 < \mu_1/\mu < 8/3$, which is guaranteed if (4.72) holds.

The problem then is to find the possible behaviour of the cavity opening c as a function of P in a neighbourhood of P_0. On differentiating (4.73) with respect to c/B, we obtain

$$\frac{dP}{d(c/B)} = 2\frac{c^2}{B^2} \left\{ 2\mu \left(1 + \frac{c^3}{B^3} \right)^{-2/3} - \mu_1 \left[\left(1 + \frac{c^3}{B^3} \right)^{-2/3} + \frac{1}{2} \left(1 + \frac{c^3}{B^3} \right)^{-5/3} \right] \right\}. \tag{4.75}$$

Hence, by Proposition 4.3.1 (with $n = 3$), when

$$0 < \frac{\mu_1}{\mu} < \frac{4}{3} = \inf_{0 < c/B < 1} \left[2 \left(1 + \frac{c^3}{B^3} \right) \left(\frac{3}{2} + \frac{c^3}{B^3} \right)^{-1} \right], \tag{4.76}$$

the bifurcation is supercritical and the radius of the cavity monotonically increases as the tensile dead load increases. However, if there exists $c_0 > 0$, such that

$$2 \left(1 + \frac{c_0^3}{B^3} \right) \left(\frac{3}{2} + \frac{c_0^3}{B^3} \right)^{-1} < \frac{\mu_1}{\mu} < 2, \tag{4.77}$$

then the bifurcation is subcritical and the required applied load starts to decrease at $c = c_0$, where there is a sudden jump in the opening of cavity. In particular, if (4.77) holds for $c_0 = 0$, i.e.,

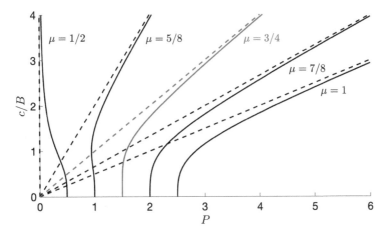

Fig. 4.11 The influence of parameter values on the behaviour of a unit sphere ($B = 1$) of material model (4.70) when $\mu_1 = 1$ and $\mu = \mu_1 + \mu_2$ changes. Note the critical case occurring when $\mu = 3/4$, i.e., for $\mu_2 = -1/4$. The dashed line indicates the asymptotic behaviour for large values of P and is given by $P = (4\mu - 2\mu_1)c/B$

$$\frac{4}{3} < \frac{\mu_1}{\mu} < 2, \tag{4.78}$$

then (4.72) is valid while the cavitation becomes unstable.

Thus, $dP/d(c/B) \to 0$ as $c \to 0_+$, and by Proposition 4.3.1, the bifurcation at the critical load, P_0, is supercritical (respectively, subcritical) if $dP/d(c/B) > 0$ (respectively, $dP/d(c/B) < 0$) for arbitrarily small c/B. Examples of both these behaviours are illustrated in Fig. 4.11. Note that, if $2 < \mu_1/\mu < 8/3$, then $P_0 > 0$ and $dP/d(c/B) < 0$, i.e., the cavitation is still unstable, but the BE inequalities no longer hold for all $\lambda \in (0, \infty)$. However, if $\mu_1/\mu > 8/3$, then $P_0 < 0$.

We now turn our attention to the stochastic model described by (4.49), with $m = 1$ and $n = -1/2$, and the other parameters drawn from probability distributions. In this case, for μ, we choose a Gamma distribution $g(u; \rho_1, \rho_2)$, such that $\underline{\mu} = \rho_1\rho_2 > \underline{\mu}_1/2$, where $\underline{\mu}$ and $\underline{\mu}_1$ are the mean values of μ and μ_1, respectively. Then, the probability of stable cavitation is

$$P\left(\mu > \frac{3\mu_1}{4}\right) = 1 - \int_0^{\frac{3\mu_1}{4}} g(u; \rho_1, \rho_2)du \tag{4.79}$$

and that of unstable cavitation is

$$P\left(\mu < \frac{3\mu_1}{4}\right) = \int_0^{\frac{3\mu_1}{4}} g(u; \rho_1, \rho_2)du. \tag{4.80}$$

Fig. 4.12 Probability distributions of whether cavitation is stable or not in a sphere of stochastic material described by (4.49) with $m = 1$ and $n = -1/2$, when the shear modulus, μ, follows a Gamma distribution with $\rho_1 = 405$ and $\rho_2 = 0.01$. Dark coloured lines represent analytically derived solutions, given by equations (4.79)–(4.80), whereas the lighter versions represent stochastically generated data. The vertical black line at the critical value $\mu_1 = 4\underline{\mu}/3 = 5.4$ separates the expected regions based only on the mean value of the shear modulus, $\underline{\mu} = \rho_1\rho_2 = 4.05$ [359]

For example, taking $\rho_1 = 405$ and $\rho_2 = 0.01$, the mean value of the shear modulus is $\underline{\mu} = \rho_1\rho_2 = 4.05$, and the probability distributions given by equations (4.79)–(4.80) are illustrated numerically in Fig. 4.12. In this case, if $\underline{\mu}_1 = 5 < 5.4 = 4\underline{\mu}/3$ say, then stable cavitation is expected, but there is also about 10% chance that unstable snap cavitation occurs. Similarly, when $4\underline{\mu}/3 = 5.4 < \underline{\mu}_1 = 5.8 < 8.1 = 2\underline{\mu}$, unstable cavitation is expected, but there is also about 10% chance that the cavitation is stable. Stable and unstable cavitation of a stochastic sphere are illustrated numerically in Fig. 4.13. Specifically,

(i) In Fig. 4.13a, $b = 0$ in (3.8), and the random variable $R_1 = \mu_1/\mu$ is drawn from a Beta distribution with $\xi_1 = 287$ and $\xi_2 = 36$. In this case, $\underline{\mu}_1 = 3.6 < 5.4 = 4\underline{\mu}/3$, and stable cavitation, with supercritical bifurcation after the spherical cavity opens, is expected.
(ii) In Fig. 4.13b, $b = -3$ in (3.8), and the random variable $R_1 = (\mu_1 + 3)/(\mu + 6)$ draws its values from a Beta distribution with $\xi_1 = 325$ and $\xi_2 = 10$. Thus, $4\underline{\mu}/3 = 5.4 < \underline{\mu}_1 = 6.75 < 8.1 = 2\underline{\mu}$, and unstable cavitation, with subcritical bifurcation after the spherical cavity forms, is expected.

For the numerical examples shown in Fig. 4.13 also, the critical dead load is $P_0 = 4\mu - 3\mu_1/2$, as given by (4.74), with μ and μ_1 following probability distributions. In each case, the expectation is that the onset of cavitation occurs at the mean value $\underline{P}_0 = 4\underline{\mu} - 3\underline{\mu}_1/2$, found at the intersection of the dashed black line with the horizontal axis. However, there is a chance that cavity can form under smaller or greater critical loads than the expected load value, as shown by the coloured interval about the mean value along the horizontal axis.

To summarise, for a stochastic elastic sphere under uniform tensile dead load, we obtain the probabilities of stable or unstable cavitation, given that the material parameters are generated from known probability density functions. In the deterministic elastic case, there is a single critical parameter value that strictly separates the cases where the initiation of either stable or unstable cavitation occurs. By contrast,

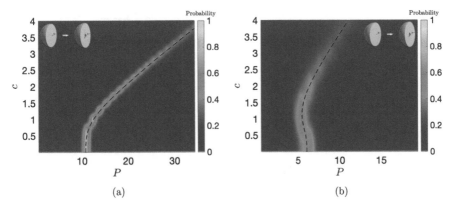

Fig. 4.13 Probability distribution of the applied dead-load traction P causing cavitation of radius c in a unit sphere (where $B = 1$) of stochastic material described by (4.49) with $m = 1$ and $n = -1/2$, when μ follows a Gamma distribution with $\rho_1 = 405$ and $\rho_2 = 0.01$, while (**a**) $R_1 = \mu_1/\mu$ follows a Beta distribution with $\xi_1 = 287$ and $\xi_2 = 36$ and (**b**) $R_1 = (\mu_1+3)/(\mu+6)$ follows a Beta distribution with $\xi_1 = 325$ and $\xi_2 = 10$. The black line corresponds to the expected bifurcation based only on mean parameter values [359]

in the stochastic case, there is a probabilistic interval, containing the deterministic critical value, where there is always a competition between the stable and unstable states in the sense that both have a quantifiable chance to be found. For the onset of cavitation, there is also a probabilistic interval where a cavity may form, with a given probability, under smaller or greater loads than the expected critical value.

4.4 Inflation and Perversion of Fibre-Reinforced Tubes

We consider a circular cylindrical tube, occupying the reference domain $(R, \Theta, Z) \in [A, B] \times [-\pi, \pi) \times [0, H]$, where A, B and H are positive constants, subject to the following combined deformation consisting of simple torsion superposed on axial stretch [558, pp. 184-186]:

$$r = \sqrt{a^2 + \frac{R^2 - A^2}{\zeta}}, \qquad \theta = \Theta + \tau\zeta Z, \qquad z = \zeta Z, \qquad (4.81)$$

where $(r, \theta, z) \in [a, b] \times [-\pi, \pi) \times [0, h]$ are the cylindrical polar coordinates in the deformed configuration, a, τ and ζ are given positive constants, $b = \sqrt{a^2 + (B^2 - A^2)/\zeta}$, and $h = \zeta H$ (see Fig. 4.14).

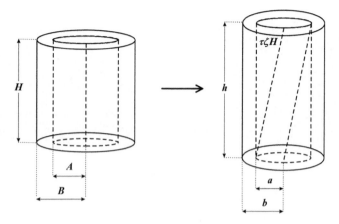

Fig. 4.14 Schematic of circular cylindrical tube (left) deformed by combined stretch and torsion (right) [373]

Through the deformation (4.81), the circular plane section at $Z = 0$ remains fixed and each circular plane section normal to the central axis remains plane and rotates by an angle $\tau \zeta Z$. Denoting

$$\lambda = \frac{r}{R} = \frac{1}{R} \sqrt{a^2 + \frac{R^2 - A^2}{\zeta}}, \qquad (4.82)$$

the deformation gradient in terms of the current cylindrical polar coordinates (r, θ, z) is equal to

$$\mathbf{A} = \begin{bmatrix} \partial r/\partial R & 0 & 0 \\ 0 & (r/R)\partial\theta/\partial\Theta & r\partial\theta/\partial Z \\ 0 & 0 & \partial z/\partial Z \end{bmatrix} = \begin{bmatrix} 1/(\lambda\zeta) & 0 & 0 \\ 0 & \lambda & \tau\zeta r \\ 0 & 0 & \zeta \end{bmatrix}, \qquad (4.83)$$

and the left and right Cauchy–Green tensors are, respectively,

$$\mathbf{B} = \mathbf{A}\mathbf{A}^T = \begin{bmatrix} 1/\left(\lambda^2\zeta^2\right) & 0 & 0 \\ 0 & \lambda^2 + \tau^2\zeta^2 r^2 & \tau\zeta^2 r \\ 0 & \tau\zeta^2 r & \zeta^2 \end{bmatrix}, \qquad (4.84)$$

$$\mathbf{C} = \mathbf{A}^T \mathbf{A} = \begin{bmatrix} 1/\left(\lambda^2\zeta^2\right) & 0 & 0 \\ 0 & \lambda^2 & \tau\zeta r\lambda \\ 0 & \tau\zeta r\lambda & \zeta^2\left(\tau^2 r^2 + 1\right) \end{bmatrix}. \qquad (4.85)$$

Denoting $\alpha_a = a/A$, by (4.82), we can write

$$\alpha_b = \frac{b}{B} = \frac{1}{B}\sqrt{a^2 + \frac{B^2 - A^2}{\zeta}} \tag{4.86}$$

and

$$r = A\lambda\sqrt{1 - \alpha_a^2\zeta}\left(1 - \lambda^2\zeta\right). \tag{4.87}$$

Next, assuming that the components of the Cauchy stress, $\mathbf{T}$, are independent of θ and z, the equilibrium equations are reduced to [558, p. 185]

$$\frac{\partial T_{rr}}{\partial r} + \frac{T_{rr} - T_{\theta\theta}}{r} = 0,$$

$$\frac{\partial T_{r\theta}}{\partial r} + 2\frac{T_{r\theta}}{r} = 0, \tag{4.88}$$

$$\frac{\partial T_{rz}}{\partial r} + \frac{T_{rz}}{r} = 0.$$

Then, by further assuming that $T_{r\theta} = 0$ and $T_{rz} = 0$, it follows that

$$\frac{dT_{rr}}{dr} = \frac{T_{\theta\theta} - T_{rr}}{r}, \tag{4.89}$$

hence,

$$T_{\theta\theta} = \frac{d\,(rT_{rr})}{dr}. \tag{4.90}$$

In this case, when a uniform internal pressure is applied, the outer surface of the cylinder may be rendered free of traction, i.e., $T_{rr} = 0$ on $r = B$. Thus, for an internally pressurised tube which is free on the outer surface,

$$T_{rr}|_{r=a} = -P, \qquad T_{rr}|_{r=b} = 0, \tag{4.91}$$

where $P > 0$ is constant. Integration of equation (4.89), with respect to r, followed by substitution in the first condition of (4.91), then gives

$$P = \int_a^b \frac{T_{\theta\theta} - T_{rr}}{r}dr. \tag{4.92}$$

For the deformation (4.81), by (4.90), the resultant normal force acting upon the plane $a \leq r \leq b$, z-constant, $|\theta| \leq \pi$ is calculated as follows [558, pp. 185]:

$$N = 2\pi \int_a^b T_{zz} r \, dr,$$

$$= 2\pi \int_a^b (T_{zz} - T_{rr}) \, r \, dr + 2\pi \int_a^b T_{rr} r \, dr,$$

$$= 2\pi \int_a^b (T_{zz} - T_{rr}) \, r \, dr + \pi \int_a^b T_{rr} r^2 \, dr + \pi \int_a^b (T_{rr} - T_{\theta\theta}) \, r \, dr, \qquad (4.93)$$

$$= \pi P a^2 + \pi \int_a^b (2T_{zz} - T_{rr} - T_{\theta\theta}) \, r \, dr.$$

Equivalently,

$$N = \pi P a^2 + F, \qquad (4.94)$$

where

$$F = \pi \int_a^b (2T_{zz} - T_{rr} - T_{\theta\theta}) \, r \, dr. \qquad (4.95)$$

Similarly, the resulting twisting moment is equal to [558, p. 190]

$$T = 2\pi \int_a^b T_{\theta z} r^2 \, dr. \qquad (4.96)$$

After applying the change of variable (4.87), the expressions of P, F, and T, given by (4.92), (4.95), and (4.96), respectively, take the following equivalent forms:

$$P = \int_{\alpha_a}^{\alpha_b} \frac{T_{\theta\theta} - T_{rr}}{\lambda \left(1 - \lambda^2 \zeta\right)} d\lambda, \qquad (4.97)$$

$$F = \pi A^2 \int_{\alpha_a}^{\alpha_b} \lambda \left(2T_{zz} - T_{rr} - T_{\theta\theta}\right) \frac{1 - \alpha_a^2 \zeta}{\left(1 - \lambda^2 \zeta\right)^2} d\lambda, \qquad (4.98)$$

$$T = 2\pi A^3 \int_{\alpha_a}^{\alpha_b} \lambda^2 T_{\theta z} \frac{\left(1 - \alpha_a^2 \zeta\right)^{3/2}}{\left(1 - \lambda^2 \zeta\right)^{5/2}} d\lambda. \qquad (4.99)$$

4.4.1 Stochastic Anisotropic Hyperelastic Tubes

There are many applications of cylindrical structures in the biomedical sciences and engineering. These applications range from soft tissues [57, 182, 307] and cardiovascular systems [200, 451] to soft actuators [110, 111] and dielectric elastomers [242]. For fibrous composites, crossed fibre arrays are rarely oriented at right angles to the longitudinal axis of the cylinder, and the fibres wind helically around the cylinder.

We analyse here a tube of stochastic homogeneous anisotropic incompressible hyperelastic material for which the model parameters are random variables characterised by probability density functions. Namely, we assume that the tube material has two preferred directions with respect to the reference configuration, induced by two families of aligned extensible fibres embedded in an isotropic matrix. The two preferred directions are given as follows [211], [205, pp. 328-336]:

$$
\mathbf{M}_1 = \begin{bmatrix} M_{1r} \\ M_{1\theta} \\ M_{1z} \end{bmatrix} = \begin{bmatrix} 0 \\ \cos\Phi \\ \sin\Phi \end{bmatrix}, \qquad \mathbf{M}_2 = \begin{bmatrix} M_{2r} \\ M_{2\theta} \\ M_{2z} \end{bmatrix} = \begin{bmatrix} 0 \\ -\cos\Psi \\ \sin\Psi \end{bmatrix}, \qquad (4.100)
$$

where $\Phi, \Psi \in [0, \pi/2]$. Under the deformation (4.81), the stretch ratios of the fibres, α_4 and α_6, respectively, are

$$
I_4 = \alpha_4^2 = (\mathbf{CM}_1) \cdot \mathbf{M}_1, \qquad I_6 = \alpha_6^2 = (\mathbf{CM}_2) \cdot \mathbf{M}_2, \qquad (4.101)
$$

where $\mathbf{C}$ is the right Cauchy–Green tensor defined by (4.85).

Specifically, we focus our attention on the stochastic incompressible anisotropic hyperelastic materials with two families of extensible fibres embedded in a neo-Hookean matrix. In this case, we define the following strain-energy function:

$$
\mathcal{W}(I_1, I_4, I_6) = \frac{\mu}{2}(I_1 - 3) + \frac{\mu_4}{4}(I_4 - 1)^2 + \frac{\mu_6}{4}(I_6 - 1)^2, \qquad (4.102)
$$

where μ, μ_4, and μ_6 are positive random variables, which we assume to be stochastically independent [522, 523]. This assumption enables us to extend directly the analytical calculations from the deterministic case treated in [211], where μ, μ_4, and μ_6 were single-valued constants, to the stochastic problem, where these parameters are characterised by probability distributions. From the continuum mechanics perspective, in general, anisotropic materials have distinct material properties in different directions. For our model example, the shear moduli are μ in two directions and a linear combination of μ, μ_4, and μ_6 in the third direction. Therefore, assuming that μ, μ_4, and μ_6 are independent random variables allows for the shear moduli in different directions. The shear moduli of other anisotropic hyperelastic materials can be treated analogously. However, the physical behaviour of stochastic anisotropic materials deserves further attention, and we hope that our theoretical analysis may serve as a motivation for future experimental work.

For a cylindrical tube of stochastic anisotropic hyperelastic material, with the strain-energy density function described by (4.102) and subject to the deformation (4.81), the Cauchy stress tensor takes the form

$$
\mathbf{T} = -p\mathbf{I} + \beta_1\mathbf{B} + \beta_4\mathbf{AM}_1 \otimes \mathbf{AM}_1 + \beta_6\mathbf{AM}_2 \otimes \mathbf{AM}_2, \qquad (4.103)
$$

where $\mathbf{B}$ is the left Cauchy–Green tensor described by (4.84), $\beta_i = 2\partial\mathcal{W}/\partial I_i$, $i = 1, 4, 6$, are the material response coefficients, and p is the Lagrange multiplier for the incompressibility constraint, $\det\mathbf{A} = 1$. The preferred directions given by (4.100) are deformed into the following directions, respectively,

$$\mathbf{m}_1 = \mathbf{AM}_1 = \begin{bmatrix} 0 \\ \lambda\cos\Phi + \tau\zeta r\sin\Phi \\ \zeta\sin\Phi \end{bmatrix}, \quad \mathbf{m}_2 = \mathbf{AM}_2 = \begin{bmatrix} 0 \\ -\lambda\cos\Psi + \tau\zeta r\sin\Psi \\ \zeta\sin\Psi \end{bmatrix}.$$
$$(4.104)$$

Thus, the non-zero components of the stress tensor given by (4.103) take the form

$$T_{rr} = -p + \frac{\beta_1}{\lambda^2\zeta^2},$$

$$T_{\theta\theta} = -p + \beta_1\left(\lambda^2 + \tau^2\zeta^2 r^2\right) + \beta_4\left(\lambda\cos\Phi + \tau\zeta r\sin\Phi\right)^2 + \beta_6\left(\lambda\cos\Psi - \tau\zeta r\sin\Psi\right)^2,$$

$$T_{\theta z} = \beta_1\tau\zeta^2 r + \beta_4\zeta\sin\Phi\left(\lambda\cos\Phi + \tau\zeta r\sin\Phi\right) - \beta_6\zeta\sin\Psi\left(\lambda\cos\Psi - \tau\zeta r\sin\Psi\right),$$

$$T_{zz} = -p + \beta_1\zeta^2 + \beta_4\zeta^2\sin^2\Phi + \beta_6\zeta^2\sin^2\Psi.$$
$$(4.105)$$

Assuming that the tube wall is thin (see Fig. 4.15), we set $A = 1$ and $B = 1 + \epsilon$ and represent P, F, and T, given by (4.97), (4.98), and (4.99), respectively, as the following series expansions [211]:

$$P = \qquad P^{(0)} + P^{(1)}\epsilon + P^{(2)}\epsilon^2 + \cdots, \qquad (4.106)$$

$$F = \qquad F^{(0)} + F^{(1)}\epsilon + F^{(2)}\epsilon^2 + \cdots, \qquad (4.107)$$

$$T = \qquad T^{(0)} + T^{(1)}\epsilon + T^{(2)}\epsilon^2 + \cdots, \qquad (4.108)$$

where we assume $P^{(0)} = F^{(0)} = T^{(0)} = 0$. We then truncate the series given by (4.106), (4.107), and (4.108), respectively, to first order in ϵ, as follows:

$$P = \frac{1}{\zeta}\left[\mu\left(\zeta^2\tau^2 + 1 - \frac{1}{\lambda^4\zeta^2}\right) + \mu_4 J_4\left(\cos\Phi + \tau\zeta\sin\Phi\right)^2 + \mu_6 J_6\left(\cos\Psi - \tau\zeta\sin\Psi\right)^2\right],$$
$$(4.109)$$

Fig. 4.15 Schematic of cylindrical shell of anisotropic material, showing the orientation of the preferred directions induced by two families of aligned fibres tangential to the cylindrical surface

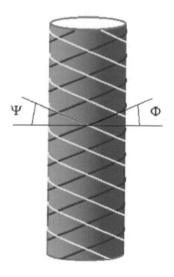

$$F = -\frac{\pi}{\zeta}\left[\mu\left(\lambda^2\zeta^2\tau^2 - 2\zeta^2 + \lambda^2 + \frac{1}{\lambda^2\zeta^2}\right)\right. \tag{4.110}$$

$$\left.+\mu_4 J_4\left(J_4 + 1 - 3\zeta\sin^2\Phi\right) + \mu_6 J_6\left(J_6 + 1 - 3\zeta\sin^2\Psi\right)\right], \tag{4.111}$$

$$T = 2\pi\lambda\left[\mu\lambda\zeta\tau + \mu_4 J_4\lambda\sin\Phi\left(\cos\Phi + \zeta\tau\sin\Phi\right) + \mu_6 J_6\lambda\sin\Psi\left(\cos\Psi - \zeta\tau\sin\Psi\right)\right], \tag{4.112}$$

where

$$J_4 = I_4 - 1 = \lambda^2\cos^2\Phi + 2\lambda^2\zeta\tau\cos\Phi\sin\Phi + \zeta^2\sin^2\Phi\left(\lambda^2\tau^2 + 1\right) - 1, \tag{4.113}$$

$$J_6 = I_6 - 1 = \lambda^2\cos^2\Psi - 2\lambda^2\zeta\tau\cos\Psi\sin\Psi + \zeta^2\sin^2\Psi\left(\lambda^2\tau^2 + 1\right) - 1. \tag{4.114}$$

Next, we define the following Jacobian matrix [211]:

$$\mathbf{J} = (J_{ij})_{i,j=1,2,3} = \begin{bmatrix} \partial P/\partial\lambda & \partial P/\partial\zeta & \partial P/\partial\tau \\ \partial F/\partial\lambda & \partial F/\partial\zeta & \partial F/\partial\tau \\ \partial T/\partial\lambda & \partial T/\partial\zeta & \partial T/\partial\tau \end{bmatrix} \tag{4.115}$$

and concentrate our attention on infinitesimal deformations near the reference configuration, with $(\lambda, \zeta, \tau) = (1, 1, 0)$, where $\mathbf{J}|_{(1,1,0)}$ has the following components:

$$J_{11}|_{(1,1,0)} = \frac{\partial P}{\partial\lambda}\Big|_{(1,1,0)} = 4\mu + 2\mu_4\cos^4\Phi + 2\mu_6\cos^4\Psi,$$

$$J_{12}|_{(1,1,0)} = \frac{\partial P}{\partial\zeta}\Big|_{(1,1,0)} = 2\mu + 2\mu_4\cos^2\Phi\sin^2\Phi + 2\mu_6\cos^2\Psi\sin^2\Psi,$$

$$J_{13}|_{(1,1,0)} = \frac{\partial P}{\partial\tau}\Big|_{(1,1,0)} = 2\mu_4\cos^3\Phi\sin\Phi - 2\mu_6\cos^3\Psi\sin\Psi,$$

$$J_{21}|_{(1,1,0)} = \frac{\partial F}{\partial\lambda}\Big|_{(1,1,0)} = 2\pi\mu_4\cos^2\Phi\left(3\sin^2\Phi - 1\right) + 2\pi\mu_6\cos^2\Psi\left(3\sin^2\Psi - 1\right),$$

$$J_{22}|_{(1,1,0)} = \frac{\partial F}{\partial\zeta}\Big|_{(1,1,0)} = 6\pi\mu + 2\pi\mu_4\sin^2\Phi\left(3\sin^2\Phi - 1\right) + 2\pi\mu_6\sin^2\Psi\left(3\sin^2\Psi - 1\right),$$

$$J_{23}|_{(1,1,0)} = \frac{\partial F}{\partial\tau}\Big|_{(1,1,0)} = 2\pi\mu_4\cos\Phi\sin\Phi\left(3\sin^2\Phi - 1\right) - 2\pi\mu_6\cos\Psi\sin\Psi\left(3\sin^2\Psi - 1\right),$$

$$J_{31}|_{(1,1,0)} = \frac{\partial T}{\partial\lambda}\Big|_{(1,1,0)} = 4\pi\mu_4\cos^3\Phi\sin\Phi - 4\pi\mu_6\cos^3\Psi\sin\Psi,$$

$$J_{32}|_{(1,1,0)} = \frac{\partial T}{\partial\zeta}\Big|_{(1,1,0)} = 4\pi\mu_4\cos\Phi\sin^3\Phi - 4\pi\mu_6\cos\Psi\sin^3\Psi,$$

$$J_{33}|_{(1,1,0)} = \frac{\partial T}{\partial\tau}\Big|_{(1,1,0)} = 2\pi\mu + 4\pi\mu_4\cos^2\Phi\sin^2\Phi + 4\pi\mu_6\cos^2\Psi\sin^2\Psi. \tag{4.116}$$

Assuming that $\det \mathbf{J}|_{(1,1,0)} \neq 0$, we can define the matrix inverse

$$\bar{\mathbf{A}} = (\bar{A}_{ij})_{i,j=1,2,3} = \mathbf{J}^{-1}|_{(1,1,0)}. \tag{4.117}$$

This will be useful when exploring the critical points where an inversion in the deformation occurs.

In order to obtain clear explicit results that will show the role played by the stochastic parameters, henceforth, we limit our investigation to the case where the anisotropic material has the same mechanical properties in the two preferred directions, i.e., $\mu_4 = \mu_6$. Recalling that μ follows a Gamma probability distribution $g(u; \rho_1, \rho_2)$ and μ_4 follows a Gamma distribution, $g_4\left(u; \rho_1^{(4)}, \rho_2^{(4)}\right)$, the probability distribution of each entry of the Jacobian matrix (4.115) can be computed using the summation formula for independent Gamma-distributed random variables [346, 387] (see Appendix B.5).

4.4.2   Inflation of Stochastic Anisotropic Tubes

First, we assume that $\Phi = \Psi$ [211] and consider the critical point for radial stretch λ, such that

$$\bar{A}_{11} = 0, \quad \text{where} \quad \bar{A}_{11} = \frac{\mu_4 \left[3\cos^2(2\phi) - 4\cos(2\phi) + 1\right] + 6\mu}{8\mu \left[3\mu_4 \cos^2(2\phi) + \mu_4 + 3\mu\right]}. \tag{4.118}$$

The radial stretch increases if $\bar{A}_{11} > 0$ and decreases if $\bar{A}_{11} < 0$, and equation (4.118) is equivalent to the following quadratic equation in $\cos(2\Phi)$:

$$\mu_4 \left[3\cos^2(2\Phi) - 4\cos(2\Phi) + 1\right] + 6\mu = 0, \tag{4.119}$$

which has real solutions when $0 < \mu \leq \mu_4/18$. As the denominator of $\bar{A}_{11}$ is positive, it follows that, as the internal pressure increases, the radial stretch, λ, increases if $\mu > \mu_4/18$ and decreases if $0 < \mu < \mu_4/18$.

In this case, we can express the probability distribution of stable inflation, such that the radial stretch monotonically increases when the internal pressure increases, as

$$P_1(\mu_4) = 1 - \int_0^{\mu_4/18} g(u; \rho_1, \rho_2) du, \tag{4.120}$$

and that of unstable inflation, whereby the radial stretch starts to decrease under increasing pressure, as

$$P_2(\mu_4) = 1 - P_1(\mu_4). \tag{4.121}$$

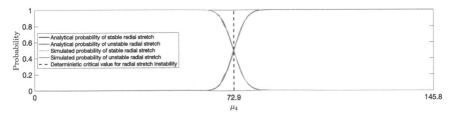

Fig. 4.16 Probability distributions of whether radial stretch instability can occur or not for an anisotropic cylindrical tube of stochastic hyperelastic material described by (4.102), when the shear modulus, μ, follows a Gamma distribution with $\rho_1 = 405$ and $\rho_2 = 0.01$, while μ_4 follows a Gamma distribution with $\rho_1^{(4)} = 405$ and $\rho_2^{(4)} = 0.2$. Dark coloured lines represent analytically derived solutions, given by Eqs. (4.120)–(4.121), whereas the lighter versions represent stochastically generated data. The vertical black line at the critical value, $\mu_4 = 72.9$, separates the expected regions based only on mean parameter values, $\underline{\mu} = \rho_1 \rho_2 = 4.05$ [373]

For example, taking $\rho_1 = 405$ and $\rho_2 = 0.01$, the mean value of the shear modulus is $\underline{\mu} = \rho_1 \rho_2 = 4.05$, and the probability distributions given by equations (4.120)–(4.121) are illustrated in Fig. 4.16 (blue lines for P_1 and red lines for P_2). To plot those results, the interval $\left(0, 27\underline{\mu}\right)$ was discretised into 100 representative points, and then for each value of μ_4, 100 random values of μ were numerically generated from the specified Gamma distribution and compared with the inequalities defining the two intervals for values of μ_4. For the deterministic elastic case, which is based on the mean value of the shear modulus, $\underline{\mu} = \rho_1 \rho_2 = 4.05$, the critical value of $\mu_4 = 18\underline{\mu} = 72.9$ strictly separates the cases where radial stretch instability can, and cannot, occur. For the stochastic problem, for the same critical value, there is, by definition, exactly 50% chance that radial stretch is stable (blue lines) and 50% chance that instability occurs (red lines). To increase the probability of stable radial stretch ($P_1 \approx 1$), one must consider values of μ_4 that are sufficiently smaller than the expected critical value, whereas an instability is almost certain to occur ($P_2 \approx 1$) for values of μ_4 which are sufficiently greater than the expected critical value. However, the inherent variability in the probabilistic system means that there exist events where there is competition between the two cases.

In Fig. 4.17, we represent the probability distribution of $\bar{A}_{11}$ as a function of the angle Φ when μ follows a Gamma distribution with hyperparameters $\rho_1 = 405$ and $\rho_2 = 0.01$, while μ_4 follows a Gamma distribution with $\rho_1^{(4)} = 405$ and $\rho_2^{(4)} = 0.2$. In this case, $\underline{\mu}_4 = 20\underline{\mu} < 81$, and unstable radial stretch is expected. Nevertheless, the probability distribution suggests that there is also around 2% chance that the radial stretch is stable.

For the axial stretch ζ, the critical value satisfies the equation

$$\bar{A}_{21} = 0 \qquad \text{where} \qquad \bar{A}_{21} = \frac{\mu_4 \left[3 \cos^2(2\phi) + 2\cos(2\phi) - 1\right]}{8\mu \left[3\mu_4 \cos^2(2\phi) + \mu_4 + 3\mu\right]}, \qquad (4.122)$$

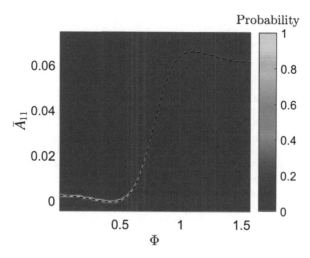

Fig. 4.17 Probability distribution of stochastic $\bar{A}_{11}$, for the inflation of an anisotropic cylindrical tube of stochastic hyperelastic material, given by (4.102), where the shear modulus, μ, is drawn from a Gamma distribution with $\rho_1 = 405$ and $\rho_2 = 0.01$, while μ_4 is drawn from a Gamma distribution with $\rho_1^{(4)} = 405$ and $\rho_2^{(4)} = 0.2$. The black line corresponds to the expected value of $\bar{A}_{11}$ based only on mean parameter values, $\underline{\mu} = \rho_1 \rho_2 = 4.05$ and $\underline{\mu}_4 = \rho_1^{(4)} \rho_2^{(4)} = 81$ [373]

and the axial stretch increases if $\bar{A}_{21} > 0$ and decreases if $\bar{A}_{21} < 0$. Solving Eq. (4.122), we obtain

$$\Phi = \Phi_m = \frac{1}{2} \arccos \frac{1}{3} \approx 35.3°, \qquad (4.123)$$

where Φ_m denotes the so-called "magic angle" [205, p. 337] (see also [260, 262]). In this case, as the denominator of $\bar{A}_{21}$ is always positive and (4.123) is independent of the random material parameters μ and μ_4, there is no uncertainty to be resolved regarding this instability. For the deterministic cases where $\mu_4 = 1$ and $\mu \in \{0.01, 0.02, \cdots, 0.1, 0.2, \cdots, 1\}$, the values of $\bar{A}_{11}$ and $\bar{A}_{21}$ are presented graphically in Fig. 4.18.

4.4.3 Perversion of Stochastic Anisotropic Tubes

Next, we assume that the angle of a preferred direction is kept fixed, while the other angle can vary [211]. When the radius and axial length of the tube do not change, the condition that the torsion parameter τ may start to decrease when the internal pressure increases is

$$\bar{A}_{31} = 0. \qquad (4.124)$$

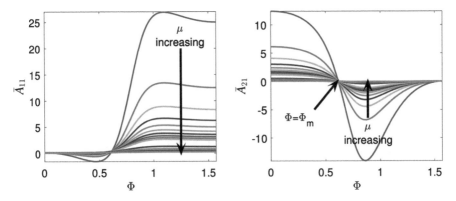

Fig. 4.18 Deterministic values of $\bar{A}_{11}$ (left) and $\bar{A}_{21}$ (right), for $\mu_4 = 1$ and $\mu \in \{0.01, 0.02, \cdots, 0.1, 0.2, \cdots, 1\}$ (the direction of increasing values of μ is indicated by arrow) [373]

This is equivalent to

$$6\mu\,[2\sin(2\Phi) + \sin(4\Phi) - 2\sin(2\Psi) - \sin(4\Psi)]$$
$$= \mu_4\,[3\sin(2\Psi) + \sin(4\Psi) - 3\sin(2\Phi) - \sin(4\Phi) \qquad (4.125)$$
$$+2\sin(2\Phi - 2\Psi) - 3\sin(2\Phi + 4\Psi) + 3\sin(4\Phi + 2\Psi)],$$

and it is clearly satisfied when $\Phi = \Psi$ or if $\Phi \in \{0, \pi/2\}$ and $\Psi \in \{0, \pi/2\}$.
    When $\Phi \neq \Psi$ and either $\Phi \notin \{0, \pi/2\}$ or $\Psi \notin \{0, \pi/2\}$, we define

$$\xi = \frac{3\sin(2\Psi) + \sin(4\Psi) - 3\sin(2\Phi) - \sin(4\Phi)}{6\,[2\sin(2\Phi) + \sin(4\Phi) - 2\sin(2\Psi) - \sin(4\Psi)]}$$
$$+ \frac{2\sin(2\Phi - 2\Psi) - 3\sin(2\Phi + 4\Psi) + 3\sin(4\Phi + 2\Psi)}{6\,[2\sin(2\Phi) + \sin(4\Phi) - 2\sin(2\Psi) - \sin(4\Psi)]} \qquad (4.126)$$

and distinguish the following cases:

(*i*) When $\Psi = \Psi_0$ is fixed and $\Phi \in (0, \pi/2) \setminus \{\Psi_0\}$:

    (i_1) If $\Psi_0 \in \{0, \pi/2\}$, then $\bar{A}_{31} < 0$, i.e., there is left-handed (clockwise) twist.
    (i_2) If $\Psi_0 = \Psi_m$, where Ψ_m is the magic angle given by (4.123), then $\bar{A}_{31} > 0$, i.e., there is right-handed (anti-clockwise) twist.
    (i_3) If $0 < \Psi_0 < \Psi_m$, then $\bar{A}_{31} > 0$ (right-handed twist) when $\Phi < \Psi_0$, and there exists $\Phi^* > \Psi_0$, such that $\bar{A}_{31} < 0$ (left-handed twist) when $\Psi_0 < \Phi < \Phi^*$ and $\bar{A}_{31} > 0$ (right-handed twist) when $\Phi^* < \Psi < \pi/2$. Hence, a first inversion from right-handed to left-handed twist occurs at $\Phi = \Psi_0$, and a second one from left-handed to right-handed twist takes place at $\Phi = \Phi^*$. Then, if $\Psi_0 < \Phi < \pi/2$, the torsion parameter τ increases (i.e., $\bar{A}_{31} > 0$) when $\mu > \mu_4/\xi$ and decreases when $0 < \mu < \mu_4/\xi$,

and if $\Phi < \Psi$, the torsion parameter increases when $0 < \mu < \mu_4/\xi$ and decreases (i.e., $\bar{A}_{31} < 0$) when $\mu > \mu_4/\xi$. Thus, the probability distribution of right-handed torsion, such that τ monotonically increases as the internal pressure increases, is

$$P_3(\mu_4) = 1 - \int_0^{\mu_4/\xi} g(u; \rho_1, \rho_2)\,du \qquad (4.127)$$

and that of left-handed torsion, such that τ decreases as the pressure increases, is

$$P_4(\mu_4) = 1 - P_3(\mu_4). \qquad (4.128)$$

(i_4) If $\Psi_m < \Psi_0 < \pi/2$, then $\bar{A}_{31} > 0$ (right-handed twist) when $\Phi > \Psi_0$, and there exists $\Phi^* < \Psi_0$, such that $\bar{A}_{31} > 0$ (right-handed twist) when $0 < \Phi < \Phi^*$ and $\bar{A}_{31} < 0$ (left-handed twist) when $\Phi^* < \Psi < \Psi_0$. In this case, the first inversion from right-handed to left-handed twist occurs at $\Phi = \Phi^*$, and the second one from left-handed to right-handed twist at $\Phi = \Psi_0$. Hence, for $0 < \Phi < \Psi_0$, the probability of right-handed torsion is $P_4(\mu_4)$, given by (4.128), and that of left-handed torsion is $P_3(\mu_4)$, given by (4.127).

(ii) When $\Phi = \Phi_0$ is fixed and $\Psi \in (0, \pi/2) \setminus \{\Phi_0\}$:

  (ii_1) If $\Phi_0 \in \{0, \pi/2\}$, then $\bar{A}_{31} > 0$, i.e., there is right-handed twist.
  (ii_2) If $\Phi_0 = \Phi_m$, where Φ_m is the magic angle given by (4.123), then $\bar{A}_{31} < 0$, i.e., there is left-handed twist.
  (ii_3) If $0 < \Phi_0 < \Phi_m$, then $\bar{A}_{31} < 0$ (left-handed twist) when $0 < \Psi < \Phi_0$, and there exists $\Psi^* > \Phi_0$, such that $\bar{A}_{31} > 0$ (right-handed twist) when $\Phi_0 < \Psi < \Psi^*$ and $\bar{A}_{31} < 0$ (left-handed twist) when $\Psi^* < \Psi < \pi/2$. Hence, a first inversion from left-handed to right-handed twist takes place at $\Psi = \Phi_0$ and a second one from right-handed to left-handed twist at $\Psi = \Psi^*$. Then, for $\Psi_0 < \Phi < \pi/2$, the probability of right-handed torsion is $P_4(\mu_4)$, given by (4.128), and that of left-handed torsion is $P_3(\mu_4)$, given by (4.127).
  (ii_4) If $\Phi_m < \Phi_0 < \pi/2$, then $\bar{A}_{31} < 0$ (left-handed twist) when $\Phi_0 < \Psi < \pi/2$, and there exists $\Psi^* < \Phi_0$, such that $\bar{A}_{31} < 0$ (left-handed twist) when $0 < \Psi < \Psi^*$ and $\bar{A}_{31} > 0$ (right-handed twist) when $\Psi^* < \Psi < \Phi_0$. In this case, the first inversion from left-handed to right-handed twist occurs at $\Psi = \Psi^*$ and the second one from left-handed to right-handed twist at $\Psi = \Phi_0$. Hence, for $0 < \Phi < \Psi_0$, the probability of right-handed torsion is $P_3(\mu_4)$, given by (4.127), and that of left-handed torsion is $P_4(\mu_4)$, given by (4.128).

For example, when $\rho_1 = 405$ and $\rho_2 = 0.01$, the probability distributions given by Eqs. (4.127)–(4.128) are illustrated numerically in Fig. 4.19, where the different

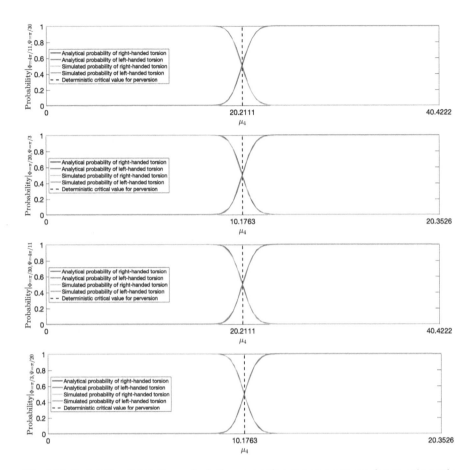

Fig. 4.19 Probability distributions of whether perversion can occur or not for an anisotropic cylindrical tube of stochastic hyperelastic material described by (4.102), when the shear modulus, μ, follows a Gamma distribution with $\rho_1 = 405$ and $\rho_2 = 0.01$, and μ_4 follows a Gamma distribution with $\rho_1^{(4)} = 405$ and $\rho_2^{(4)} = 0.2$, for the cases where (from top to bottom): $\Phi = 4\pi/11$ and $\Psi = \pi/30$; $\Phi = \pi/20$ and $\Psi = \pi/3$; $\Phi = \pi/30$ and $\Psi = 4\pi/11$; and $\Phi = \pi/3$ and $\Psi = \pi/20$. Dark coloured lines represent analytically derived solutions, given by Eqs. (4.127)–(4.128), whereas the lighter versions represent stochastically generated data. The vertical black line separates the expected regions based only on mean parameter value, $\underline{\mu} = \rho_1 \rho_2 = 4.05$ [373]

plots, from top to bottom, correspond, respectively, to the cases: (i_3) with $\Phi = 4\pi/11$ and $\Psi = \pi/30$, (i_4) with $\Phi = \pi/20$ and $\Psi = \pi/3$, (ii_3) with $\Phi = \pi/30$ and $\Psi = 4\pi/11$, and (ii_4) with $\Phi = \pi/3$ and $\Psi = \pi/20$ (blue lines for P_3 and red lines for P_4). In each case, the interval $(0, 2\xi\mu)$ was discretised into 100 representative points, and then for each value of μ_4, 100 random values of μ were numerically generated from the specified Gamma distribution and compared with the inequalities defining the two intervals for values of μ_4. For example, if μ_4 follows a Gamma distribution with $\rho_1^{(4)} = 405$ and $\rho_2^{(4)} = 0.2$, then the change

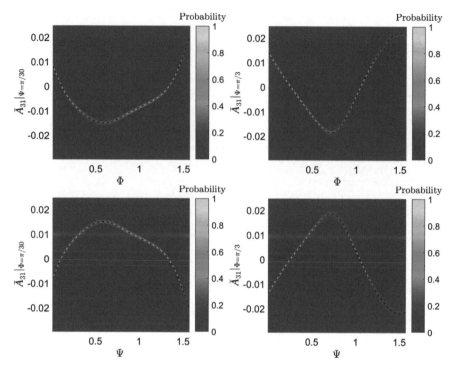

Fig. 4.20 Probability distribution of stochastic $\bar{A}_{31}$ for the torsion of an anisotropic cylindrical tube of stochastic hyperelastic material, given by (4.102), when μ is drawn from a Gamma distribution with $\rho_1 = 405$ and $\rho_2 = 0.01$, and μ_4 is drawn from a Gamma distribution with $\rho_1^{(4)} = 405$ and $\rho_2^{(4)} = 0.2$, for the cases where $\Psi_0 = \pi/30$ (top-left), $\Psi_0 = \pi/3$ (top-right), $\Phi_0 = \pi/30$ (bottom left), and $\Phi_0 = \pi/3$ (bottom right). The black line corresponds to the expected value of $\bar{A}_{31}$ based only on mean parameter values, $\underline{\mu} = \rho_1\rho_2 = 4.05$ and $\underline{\mu}_4 = \rho_1^{(4)}\rho_2^{(4)} = 81$ [373]

of chirality (perversion) will take place with certainty at $\Phi = \Psi$ and is expected to occur also, with a given probability, at a different angle, as seen from Fig. 4.20.

For the deterministic case with $\mu_4 = 1$ and $\mu \in \{0.1, 0.2, \cdots, 1\}$, the values of $\bar{A}_{31}$ are indicated in Fig. 4.21, where $\Psi_0 \in \{0, \pi/30, \Psi_m, \pi/3, \pi/2\}$ (left) and $\Phi_0 \in \{0, \pi/30, \Phi_m, \pi/3, \pi/2\}$ (right). These plots suggest that a pressurised tube with a right-handed family of fibres deforms by a left-handed torsion if the other fibres are kept either horizontal or vertical (top and bottom left), and a tube with a left-handed family of fibres deforms by a right-handed torsion if the other fibres are horizontal or vertical (top and bottom right). This is also the expected behaviour when there is only one family of fibres [211] (see Fig. 4.22).

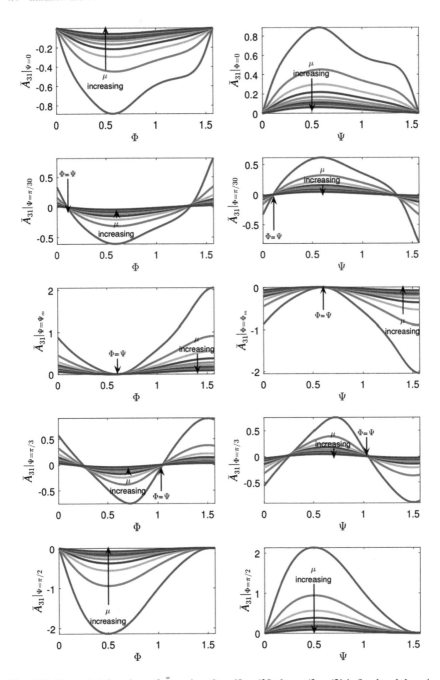

Fig. 4.21 Deterministic values of $\bar{A}_{31}$ when $\Psi \in \{0, \pi/30, \Psi_m, \pi/3, \pi/2\}$ is fixed and Φ varies (left) and when $\Phi \in \{0, \pi/30, \Phi_m, \pi/3, \pi/2\}$ is fixed and Ψ varies (right), for $\mu_4 = 1$ and $\mu \in \{0.1, 0.2, \cdots, 1\}$ (the direction of increasing values of μ is indicated by arrow) [373]

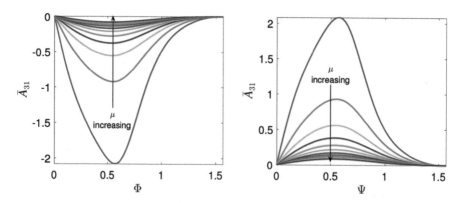

Fig. 4.22 The values of $\bar{A}_{31}$ in the case of a single family of right-handed fibres (left) or left-handed fibres (right), for $\mu_4 = 1$ and $\mu \in \{0.1, 0.2, \cdots, 1\}$ (the arrow shows the direction of increasing values of μ) [373]

Chapter 5
Oscillatory Motions

When appropriate initial conditions are given, we can predict with certainty the future, or 'retrodict' the past. Once instability is included, this is no longer the case, and the meaning of the laws of nature changes radically, for they now express possibilities or probabilities.

–I. Prigogine [448, p. 4]

Oscillatory motions of spherical or cylindrical shells of elastic material [306, 329, 330, 458] have generated extensive experimental, theoretical, and computational studies [11, 13, 14, 70, 144]. In contrast, time-dependent finite oscillations of cylindrical tubes and spherical shells of nonlinear hyperelastic material, relevant to the modelling of physical responses in many biological and synthetic systems [8, 27, 45, 135, 229, 230, 239, 293, 310], have been less investigated. In finite nonlinear elasticity, such work has focused mostly on the static stability of pressurised shells [2, 62, 74, 75, 85, 184, 203, 206, 217, 340, 389, 466, 495, 584, 620] or on wave-type solutions in infinite media [274, 424].

Internally pressurised hollow cylinders and spheres are relevant in many biological and engineering structures [205, 574]. For rubber spherical and tubular balloons, the first experimental observations of inflation instabilities under internal pressure were reported in [337]. Cylindrical tubes of homogeneous isotropic incompressible hyperelastic material subject to finite symmetric inflation and stretching were theoretically analysed for the first time in [466]. The finite radially symmetric inflation of elastic spherical shells was initially addressed in [217] and then in [2, 495]. For both elastic tubular and spherical shells, it was shown in [85] that, depending on the material model, the internal pressure may increase monotonically, or increase and then decrease, or increase, decrease, and then increase again. For different material constitutive laws, these deformations were examined in [19, 206, 389, 620]. Localised bulging in long inflated isotropic hyperelastic tubes of arbitrary thickness was analysed in [153, 184, 185, 247, 613].

Large amplitude oscillations of cylindrical tubes and spherical shells of homogeneous isotropic incompressible nonlinear hyperelastic material were formulated as special cases of *quasi-equilibrated motions* in [555], and reviewed in [558]. These

© Springer Nature Switzerland AG 2022

L. A. Mihai, *Stochastic Elasticity*, Interdisciplinary Applied Mathematics 55,

https://doi.org/10.1007/978-3-031-06692-4_5

are the class of motions for which the deformation is circulation preserving, and at every time instant, the deformed configuration is a possible static configuration under the given forces.

Free and forced axially symmetric radial oscillations of infinitely long, isotropic incompressible circular cylindrical tubes, with arbitrary wall thickness, were described for the first time in [301, 302]. In [249, 303, 579], free and forced oscillations of spherical shells were derived analogously. For the combined radial–axial large amplitude oscillations of hyperelastic cylindrical tubes, in [488], the surface tractions necessary to maintain the periodic motions were discussed, and the results were applied to a tube sealed at both ends and filled with an incompressible fluid. The dynamic deformation of cylindrical tubes of Mooney–Rivlin material in finite amplitude radial oscillation was obtained in [488–490].

For a wide class of hyperelastic materials, the static and dynamic cavitation of homogeneous spheres were analysed in [39]. Oscillatory motion caused by the dynamic cavitation of a neo-Hookean sphere was considered in [98]. A numerical solution to the nonlinear problem of large amplitude oscillations of a hyperelastic sphere of Mooney–Rivlin material containing a cavity was proposed in [37].

Theoretical and experimental studies of cylindrical and spherical shells of rubber-like material under external pressure were presented in [580]. The finite amplitude radial oscillations of homogeneous isotropic incompressible hyperelastic spherical and cylindrical shells under a constant pressure difference between the inner and outer surfaces were studied theoretically in [82].

The finite longitudinal, or "telescopic", oscillations of infinitely long cylindrical tubes were investigated in [405]. In [404], the oscillatory motions of cylindrical and prismatic bodies of incompressible hyperelastic material under dynamic finite shear deformation were analysed. Other dynamic shear deformations were treated in [578], where it was emphasised that such shear motions were not quasi-equilibrated.

In [271], the dynamic problem of axially symmetric oscillations of cylindrical tubes of transversely isotropic incompressible material, with radial transverse isotropy, was considered. The dynamic deformation of a longitudinally anisotropic thin-walled cylindrical tube under radial oscillations was obtained in [487]. Radial oscillations of non-homogeneous thick-walled cylindrical and spherical shells of neo-Hookean material, with a material parameter varying continuously along the radial direction, were examined in [158].

In [10], for pressurised homogeneous isotropic compressible hyperelastic tubes of arbitrary wall thickness under uniform radial dead-load traction, the stability of the finitely deformed state and small radial vibrations about this state were treated using the theory of small deformations superposed on large elastic deformations, while the governing equations were solved numerically. In [570], the dynamic inflation of hyperelastic spherical membranes of Mooney–Rivlin material subjected to a uniform step pressure was considered, and the absence of damping in these models was discussed. It was concluded there that, as the amplitude and period of oscillations are strongly influenced by the rate of internal pressure, if the pressure was suddenly imposed and the inflation process was short, then sustained oscillations due to the dominant elastic effects could be observed. For many systems under slowly increasing pressure, strong damping would typically prevent oscillatory motion [127].

The dynamic response of incompressible hyperelastic cylindrical and spherical shells subjected to periodic loading was examined in [459, 460]. Radial oscillations of cylindrical tubes and spherical shells of different hyperelastic materials were analysed in [51, 52]. There, it was found that, in general, both the amplitude and period of oscillations decrease when the stiffness of the material increases. The influence of material constitutive law on the dynamic behaviour of cylindrical and spherical shells was investigated also in [26, 28, 473, 619], where the results for Yeoh [616] and Mooney–Rivlin material models were compared. In [504], the nonlinear static and dynamic behaviour of a spherical membrane of neo-Hookean or Mooney–Rivlin material, subject to a uniformly distributed radial pressure on the inner surface, was analysed, and the influence of the material constants was discussed.

Axisymmetric vibrations of soft electro-active tubes were studied in [147, 629]. In [71], the static and dynamic behaviours of circular cylindrical shells of homogeneous isotropic incompressible hyperelastic material modelling arterial walls were considered. General presentations on the theory of nonlinear oscillations applicable to a diverse range of engineering and physical systems can be found in [159, 397, 482].

In this chapter, we first analyse explicitly the cavitation and finite amplitude oscillations under radially symmetric finite deformation of a sphere of stochastic hyperelastic material, subject to either a uniform tensile surface dead load (which is constant in the reference configuration) or an impulse traction (which is maintained constant in the current configuration), when each of them is applied uniformly in the radial direction. As in the previous chapter, cavitation represents the opening of a spherical cavity at the centre of the sphere as a bifurcated solution from the trivial solution, which becomes unstable at a critical value of the applied load. Moreover, the two different types of loads lead to different static and dynamic solutions [374]. Besides the stochastic homogeneous material models, for which the elastic parameters are spatially independent random variables, a class of stochastic inhomogeneous models is introduced, where the parameters are spatially dependent non-Gaussian random fields. Universally, deformations in inhomogeneous isotropic nonlinear elastic solids are analysed in [612]. For radially inhomogeneous spheres, we find that the critical load at which a non-trivial cavitated solution appears is the same as for the homogeneous sphere composed entirely of the material found at its centre. However, there are important differences in the post-cavitation behaviours under the different constitutive formulations and loading conditions:

- For a neo-Hookean sphere under dead-load traction, cavitation is stable and the motion is oscillatory (i.e., the radius of the cavity increases up to a point, then decreases, then increases again, and so on), while under impulse traction, cavitation is non-oscillatory (i.e., the cavity radius increases monotonically) and unstable.

- In composite spheres with two concentric neo-Hookean phases subject to either dead-load or impulse traction, both stable and unstable cavitations are presented, depending on the material parameters of the two phases. However, oscillatory motion only occurs when cavitation is stable under dead-load traction.
- For radially inhomogeneous spheres under dead-load traction, stable or unstable cavitation is obtained, and oscillatory motion is found when cavitation is stable under dead-load traction. Under impulse traction, cavitation is unstable and the motion is non-oscillatory.

We further extend our stochastic approach to the dynamic inflation and finite amplitude oscillatory motion of cylindrical tubes of stochastic hyperelastic material. These tubes are deformed by radially symmetric uniform inflation and are subject to an impulse traction applied uniformly in the radial direction. We consider homogeneous tubes with an arbitrary wall thickness of stochastic Mooney–Rivlin material, composite tubes with two concentric stochastic homogeneous neo-Hookean phases, and inhomogeneous tubes of stochastic neo-Hookean material with constitutive parameters varying continuously in the radial direction [354, 358]. Our stochastic analysis can also be extended to other inflation instabilities, such as localised bulging in inflated circular cylindrical tubes, which is likely to occur for all isotropic material models when the axial stretch is fixed and below a certain threshold value that is dependent on the material model. We conclude this chapter with the analysis of finite shear oscillations of a stochastic homogeneous cuboid, which are not quasi-equilibrated.

For the finite dynamic deformations considered here, particular attention is given to the periodic (oscillatory) motion and the time-dependent stresses where the stochastic model parameters are taken into account. It is generally concluded that the amplitude and period of the oscillation are characterised by probability distributions, and there is a parameter interval where both the oscillatory and non-oscillatory motions can be observed with a given probability.

The particular values of the stochastic parameters in our numerical calculations are for illustrative purposes only, while the propagation of stochastic variation from input material parameters to output mechanical behaviour is mathematically traceable, offering valuable insights into how probabilistic approaches can be incorporated into the nonlinear elasticity theory. Other stochastic hyperelastic models, such as those defined in [371], can also be used. However, for compressible materials, the theorem on quasi-equilibrated dynamics is not applicable [558, p. 209]. As our analytical approach relies on the notion of quasi-equilibrated motion for incompressible cylindrical tubes and spherical shells, the same approach cannot be used for the compressible case. Nevertheless, the standard elastodynamic problems can be formulated and then treated numerically.

5.1   Cavitation and Radial Quasi-equilibrated Motion of Homogeneous Spheres

We consider first the static and dynamic radially symmetric deformations of a stochastic homogeneous hyperelastic sphere, where the shear modulus μ of the stochastic model given by (4.49) is a space-invariant random variable. Hence, the model is of stochastic neo-Hookean type [519], with the shear modulus μ characterised by the Gamma distribution. The radial motion of spheres of neo-Hookean material was treated deterministically in [98], while inhomogeneous cylindrical and spherical shells of neo-Hookean-like material with a radially varying material parameter were analysed in [158]. For homogeneous spheres of stochastic isotropic incompressible hyperelastic material, the static cavitation under uniform tensile dead load was investigated in [359], while radial oscillatory motions of homogeneous cylindrical and spherical shells of stochastic Mooney–Rivlin and neo-Hookean material, respectively, were treated in [358].

A sphere of stochastic hyperelastic material defined by (4.49) is subject to the following radially symmetric dynamic deformation [358]:

$$r^3 = R^3 + c^3, \qquad \theta = \Theta, \qquad \phi = \Phi, \tag{5.1}$$

where (R, Θ, Φ) and $(r, \theta, \phi) = (r(t), \theta(t), \phi(t))$, with t the time variable, are the spherical polar coordinates in the reference and current configurations, respectively, such that $0 \leq R \leq B$, B is the radius of the undeformed sphere, $c = c(t) \geq 0$ is the cavity radius to be calculated, and $b = b(t) = \sqrt[3]{B^3 + c(t)^3}$ is the radius of the deformed sphere at time t (see Fig. 4.10).

By the governing equations (5.1), condition (2.168) is valid for $\mathbf{x} = (r, \theta, \phi)^T$ (see Appendix B.2). Therefore, (5.1) describes a quasi-equilibrated motion, such that

$$-\frac{\partial \xi}{\partial r} = \ddot{r} = \frac{2c\dot{c}^2 + c^2\ddot{c}}{r^2} - \frac{2c^4\dot{c}^2}{r^5}, \tag{5.2}$$

where ξ is the acceleration potential satisfying (2.167). Integrating (5.2) gives

$$-\xi = -\frac{2c\dot{c}^2 + a^2\ddot{c}}{r} + \frac{c^4\dot{c}^2}{2r^4} = -r\ddot{r} - \frac{3}{2}\dot{r}^2. \tag{5.3}$$

For the deformation (5.1), the gradient tensor with respect to the polar coordinates (R, Θ, Φ) is equal to

$$\mathbf{A} = \text{diag}\left(\frac{R^2}{r^2}, \frac{r}{R}, \frac{r}{R}\right), \tag{5.4}$$

and the corresponding Cauchy–Green tensor is

$$\mathbf{B} = \mathbf{A}^2 = \text{diag}\left(\frac{R^4}{r^4}, \frac{r^2}{R^2}, \frac{r^2}{R^2}\right), \tag{5.5}$$

with the principal invariants

$$I_1 = \text{tr}\,(\mathbf{B}) = \frac{R^4}{r^4} + 2\frac{r^2}{R^2},$$

$$I_2 = \frac{1}{2}\left[(\text{tr}\,\mathbf{B})^2 - \text{tr}\,\left(\mathbf{B}^2\right)\right] = \frac{r^4}{R^4} + 2\frac{R^2}{r^2}, \tag{5.6}$$

$$I_3 = \det \mathbf{B} = 1.$$

The associated Cauchy stress tensor then takes the form [216, pp. 87-91]

$$\mathbf{T}^{(0)} = -p^{(0)}\mathbf{I} + \beta_1\mathbf{B} + \beta_{-1}\mathbf{B}^{-1}, \tag{5.7}$$

where the scalar $p^{(0)}$, which is commonly referred to as the arbitrary hydrostatic pressure [205, pp. 286–287] and [415, pp. 198–201], is the Lagrange multiplier for the internal constraint $I_3 = 1$ of incompressibility (i.e., all deformations are isochoric for incompressible materials) [558, pp. 71–72], and the coefficients

$$\beta_1 = 2\frac{\partial W}{\partial I_1}, \qquad \beta_{-1} = -2\frac{\partial W}{\partial I_2}, \tag{5.8}$$

are nonlinear material parameters, with $W = W(I_1, I_2, I_3) = \mathcal{W}(\alpha_1, \alpha_2, \alpha_2)$, and I_1, I_2, I_3 defined by (5.6). For this stress tensor, the principal components at time t are as follows:

$$T_{rr}^{(0)} = -p^{(0)} + \beta_1\frac{R^4}{r^4} + \beta_{-1}\frac{r^4}{R^4},$$

$$T_{\theta\theta}^{(0)} = T_{rr}^{(0)} + \left(\beta_1 - \beta_{-1}\frac{r^2}{R^2}\right)\left(\frac{r^2}{R^2} - \frac{R^4}{r^4}\right), \tag{5.9}$$

$$T_{\phi\phi}^{(0)} = T_{\theta\theta}^{(0)}.$$

As the stress components depend only on the radius r, the system of equilibrium equations reduces to

$$\frac{\partial T_{rr}^{(0)}}{\partial r} = 2\frac{T_{\theta\theta}^{(0)} - T_{rr}^{(0)}}{r}. \tag{5.10}$$

Hence, by (5.9) and (5.10), the radial Cauchy stress for the equilibrium state at t is equal to

$$T_{rr}^{(0)}(r, t) = 2 \int \left(\beta_1 - \beta_{-1} \frac{r^2}{R^2} \right) \left(\frac{r^2}{R^2} - \frac{R^4}{r^4} \right) \frac{dr}{r} + \psi(t), \tag{5.11}$$

where $\psi = \psi(t)$ is an arbitrary function of time. Substitution of (5.3) and (5.11) into (2.169) then gives the following principal Cauchy stresses at time t:

$$T_{rr}(r, t) = -\rho \left(\frac{c^2 \ddot{c} + 2c\dot{c}^2}{r} - \frac{c^4 \dot{c}^2}{2r^4} \right) + 2 \int \left(\beta_1 - \beta_{-1} \frac{r^2}{R^2} \right) \left(\frac{r^2}{R^2} - \frac{R^4}{r^4} \right) \frac{dr}{r} + \psi(t),$$

$$T_{\theta\theta}(r, t) = T_{rr}(r, t) + \left(\beta_1 - \beta_{-1} \frac{r^2}{R^2} \right) \left(\frac{r^2}{R^2} - \frac{R^4}{r^4} \right),$$

$$T_{\phi\phi}(r, t) = T_{\theta\theta}(r, t).$$
$$\tag{5.12}$$

When $R^2/r^2 \to 1$, all the stress components defined in (5.12) are equal.

5.1.1 Oscillatory Motion of Stochastic Neo-Hookean Spheres Under Dead-Load Traction

For a sphere of stochastic neo-Hookean material subject to the quasi-equilibrated motion (5.1), we denote the inner and outer radial pressures acting on the curvilinear surfaces, $r = c(t)$ and $r = b(t)$ at time t, as $T_1(t)$ and $T_2(t)$, respectively [558, pp. 217–219].

Evaluating $T_1(t) = -T_{rr}(c, t)$ and $T_2(t) = -T_{rr}(b, t)$, using (5.12), with $r = c$ and $r = b$, respectively, and subtracting the results, gives

$$T_1(t) - T_2(t) = \rho \left[\left(c^2 \ddot{c} + 2c\dot{c}^2 \right) \left(\frac{1}{c} - \frac{1}{b} \right) - \frac{c^4 \dot{c}^2}{2} \left(\frac{1}{c^4} - \frac{1}{b^4} \right) \right]$$

$$+ 2 \int_c^b \mu \left(\frac{r^2}{R^2} - \frac{R^4}{r^4} \right) \frac{dr}{r}$$

$$= \rho \left[\left(c\ddot{c} + 2\dot{c}^2 \right) \left(1 - \frac{c}{b} \right) - \frac{\dot{c}^2}{2} \left(1 - \frac{c^4}{b^4} \right) \right]$$
$$\tag{5.13}$$

$$+ 2 \int_c^b \mu \left(\frac{r^2}{R^2} - \frac{R^4}{r^4} \right) \frac{dr}{r}$$

$$= \rho B^2 \left[\left(\frac{c}{B} \frac{\ddot{c}}{B} + 2 \frac{\dot{c}^2}{B^2} \right) \left(1 - \frac{c}{b} \right) - \frac{\dot{c}^2}{2B^2} \left(1 - \frac{c^4}{b^4} \right) \right]$$

$$+ 2 \int_c^b \mu \left(\frac{r^2}{R^2} - \frac{R^4}{r^4} \right) \frac{dr}{r}.$$

By setting the notation

$$u = \frac{r^3}{R^3} = \frac{r^3}{r^3 - c^3}, \qquad x = \frac{c}{B}, \tag{5.14}$$

we can rewrite

$$\left(\frac{c}{B}\frac{\ddot{c}}{B} + 2\frac{\dot{c}^2}{B^2}\right)\left(1 - \frac{c}{b}\right) - \frac{\dot{c}^2}{2B^2}\left(1 - \frac{c^4}{b^4}\right)$$

$$= \left(\ddot{x}x + 2\dot{x}^2\right)\left[1 - \left(1 + \frac{1}{x^3}\right)^{-1/3}\right] - \frac{\dot{x}^2}{2}\left[1 - \left(1 + \frac{1}{x^3}\right)^{-4/3}\right]$$

$$= \left(\ddot{x}x + \frac{3}{2}\dot{x}^2\right)\left[1 - \left(1 + \frac{1}{x^3}\right)^{-1/3}\right] - \frac{\dot{x}^2}{2}\frac{1}{x^3}\left(1 + \frac{1}{x^3}\right)^{-4/3}$$

$$= \frac{1}{2x^2}\frac{d}{dx}\left\{\dot{x}^2 x^3\left[1 - \left(1 + \frac{1}{x^3}\right)^{-1/3}\right]\right\}$$

and

$$\int_c^b \mu\left(\frac{r^2}{R^2} - \frac{R^4}{r^4}\right)\frac{dr}{r} = \int_c^b \mu\left[\left(\frac{r^3}{r^3 - c^3}\right)^{2/3} - \left(\frac{r^3 - c^3}{r^3}\right)^{4/3}\right]\frac{dr}{r}$$

$$= \frac{1}{3}\int_{x^3+1}^\infty \mu\frac{1+u}{u^{7/3}}du.$$

Thus (5.13) is equivalent to

$$2x^2\frac{T_1(t) - T_2(t)}{\rho B^2} = \frac{d}{dx}\left\{\dot{x}^2 x^3\left[1 - \left(1 + \frac{1}{x^3}\right)^{-1/3}\right]\right\} + \frac{4x^2}{3\rho B^2}\int_{x^3+1}^\infty \mu\frac{1+u}{u^{7/3}}du.$$
$$\tag{5.15}$$

Next, assuming that the cavity surface is traction-free, $T_1(t) = 0$ and (5.15) is equivalent to

$$2x^2\frac{T_{rr}(b, t)}{\rho B^2} = \frac{d}{dx}\left\{\dot{x}^2 x^3\left[1 - \left(1 + \frac{1}{x^3}\right)^{-1/3}\right]\right\} + \frac{4x^2}{3\rho B^2}\int_{x^3+1}^\infty \mu\frac{1+u}{u^{7/3}}du.$$
$$\tag{5.16}$$

We denote

$$H(x) = \frac{4}{3\rho B^2}\int_0^x \left(\zeta^2 \int_{\zeta^3+1}^\infty \mu\frac{1+u}{u^{7/3}}du\right)d\zeta \tag{5.17}$$

and set the uniform dead-load traction (see also eq. (2.7) of [98])

$$P_{rr}(B) = \left(x^3 + 1\right)^{2/3} \quad T_{rr}(b, t) = \begin{cases} 0 & \text{if } t \leq 0, \\ p_0 & \text{if } t > 0, \end{cases} \tag{5.18}$$

where p_0 is constant, and $x = x(t)$ is the dimensionless cavity radius, as denoted in (5.14).

Integrating (5.16) once gives

$$\dot{x}^2 x^3 \left[1 - \left(1 + \frac{1}{x^3}\right)^{-1/3} \right] + H(x) = \frac{2p_0}{\rho B^2} \left(x^3 + 1\right)^{1/3} + C_0, \tag{5.19}$$

with $H(x)$ defined by (5.17), and

$$C_0 = \dot{x}_0^2 x_0^3 \left[1 - \left(1 + \frac{1}{x_0^3}\right)^{-1/3} \right] + H(x_0) - \frac{2p_0}{\rho B^2} \left(x_0^3 + 1\right)^{1/3}. \tag{5.20}$$

Then, after setting the initial conditions $x_0 = x(0) = 0$ and $\dot{x}_0 = \dot{x}(0) = 0$, Eq. (5.19) takes the form

$$\dot{x}^2 x^3 \left[1 - \left(1 + \frac{1}{x^3}\right)^{-1/3} \right] + H(x) = \frac{2p_0}{\rho B^2} \left[\left(x^3 + 1\right)^{1/3} - 1 \right]. \tag{5.21}$$

From (5.21), we obtain the velocity

$$\dot{x} = \pm \sqrt{\frac{\frac{2p_0}{\rho B^2} \left[\left(x^3 + 1\right)^{1/3} - 1 \right] - H(x)}{x^3 \left[1 - \left(1 + \frac{1}{x^3}\right)^{-1/3} \right]}}. \tag{5.22}$$

It is useful to note that this nonlinear elastic system is analogous to the motion of a point mass with energy

$$E = \frac{1}{2} m(x)\dot{x}^2 + V(x), \tag{5.23}$$

where the energy is $E = C_0$, the potential is given by $V(x) = H(x) - \frac{2p_0}{\rho B^2} \left(x^3 + 1\right)^{1/3}$, and the position-dependent mass is $m(x) = 2x^3 \left[1 - \left(1 + \frac{1}{x^3}\right)^{-1/3} \right]$. Due to the constraints on the function H, the system has simple dynamics, and the only solutions of interest are either static or periodic solutions.

For the given sphere to undergo an oscillatory motion, the following equation

$$H(x) = \frac{2p_0}{\rho B^2} \left[\left(x^3 + 1\right)^{1/3} - 1 \right] \tag{5.24}$$

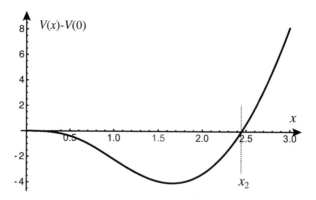

Fig. 5.1 Example of potential $V(x)$ with $p_0 = 15$ and $\mu = 4.05$, $\rho = B = 1$. The periodic orbits lie between $x_1 = 0$ and $x_2 \approx 2.45$ [374]

must have exactly two finite distinct positive roots x_1, x_2 such that $0 \leq x_1 < x_2 < \infty$, as shown in Fig. 5.1. Then, the minimum and maximum radii of the cavity in the oscillation are $x_1 B$ and $x_2 B$, respectively, and the period of oscillation is equal to

$$T = 2 \left| \int_{x_1}^{x_2} \frac{dx}{\dot{x}} \right| = 2 \left| \int_{x_1}^{x_2} \sqrt{\frac{x^3 \left[1 - \left(1 + \frac{1}{x^3}\right)^{-1/3} \right]}{\frac{2p_0}{\rho B^2} \left[(x^3 + 1)^{1/3} - 1 \right] - H(x)}} \, dx \right|. \tag{5.25}$$

For the stochastic sphere, the amplitude and period of the oscillation are random variables characterised by probability distributions.

Evaluating the integral in (5.17) gives

$$H(x) = \frac{\mu}{\rho B^2} \left[2 \left(x^3 + 1\right)^{2/3} - \frac{1}{(x^3 + 1)^{1/3}} - 1 \right], \tag{5.26}$$

and assuming that the shear modulus, μ, which is a random variable, is bounded from below, i.e.,

$$\mu > \mu_0, \tag{5.27}$$

for some constant $\mu_0 > 0$, it follows that $H(0) = 0$ and $\lim_{x \to \infty} H(x) = \infty$.

When $p_0 \neq 0$, substitution of (5.26) in (5.24) gives

$$p_0 \left[\left(x^3 + 1\right)^{1/3} - 1 \right] = \frac{\mu}{2} \left[2 \left(x^3 + 1\right)^{2/3} - \frac{1}{(x^3 + 1)^{1/3}} - 1 \right]. \tag{5.28}$$

Equation (5.28) has one solution at $x_1 = 0$, while the second solution, x_2, is a root of

$$p_0 = \frac{\mu}{2} \left[2 \left(x^3 + 1 \right)^{1/3} + \frac{1}{\left(x^3 + 1 \right)^{1/3}} + 2 \right]. \tag{5.29}$$

As the right-hand side of Eq. (5.29) is an increasing function of x, this equation has a solution, $x_2 > 0$, if and only if (see also eq. (2.7) of [98])

$$p_0 > \frac{5\mu}{2} = \lim_{x \to 0_+} \frac{\mu}{2} \left[2 \left(x^3 + 1 \right)^{1/3} + \frac{1}{\left(x^3 + 1 \right)^{1/3}} + 2 \right]. \tag{5.30}$$

Then,

$$\frac{2p_0}{\rho B^2} \left[\left(x^3 + 1 \right)^{1/3} - 1 \right] - H(x) \begin{cases} \geq 0 \text{ if } x_1 \leq x \leq x_2, \\ < 0 \text{ if } x > x_2. \end{cases} \tag{5.31}$$

In this case, at the centre of the given sphere, a spherical cavity forms and expands until its radius reaches the value $c = x_2 B$, where x_2 is the root of (5.29), and then contracts again to zero radius and repeats the cycle.

The critical dead load for the onset of cavitation is

$$\lim_{x \to 0_+} p_0 = \frac{5\mu}{2}. \tag{5.32}$$

Then, by (5.27) and (5.30), for the motion to be oscillatory, the following condition must hold

$$\mu_0 < \mu < \frac{2p_0}{5}. \tag{5.33}$$

As μ follows a Gamma distribution, the probability distribution of oscillatory motion occurring is

$$P_1(p_0) = \int_0^{\frac{2p_0}{5}} g(u; \rho_1, \rho_2) du, \tag{5.34}$$

and that of non-oscillatory motion is

$$P_2(p_0) = 1 - P_1(p_0) = 1 - \int_0^{\frac{2p_0}{5}} g(u; \rho_1, \rho_2) du. \tag{5.35}$$

For example, when μ satisfies the Gamma distribution with $\rho_1 = 405$ and $\rho_2 = 0.01$, the probability distributions given by (5.34) and (5.35) are shown in Fig. 5.2

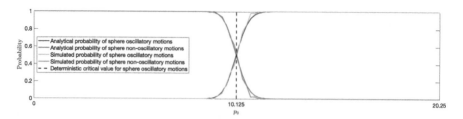

Fig. 5.2 Probability distributions of whether oscillatory motion can occur or not for a sphere of stochastic neo-Hookean material under dead-load traction, when the shear modulus, μ, follows a Gamma distribution with $\rho_1 = 405$, $\rho_2 = 0.01$. Dark coloured lines represent analytically derived solutions, given by Eqs. (5.34) and (5.35), whereas the lighter versions represent stochastically generated data. The vertical black line at the critical value $p_0 = 10.125$ separates the expected regions based only on mean value, $\underline{\mu} = \rho_1 \rho_2 = 4.05$ [374]

(with blue lines for P_1 and red lines for P_2). Specifically, the interval $(0, 5\underline{\mu})$, where $\underline{\mu} = \rho_1 \rho_2 = 4.05$ is the mean value of μ, was divided into 100 steps; then for each value of p_0, 100 random values of μ were numerically generated from the specified Gamma distribution and compared with the inequalities defining the two intervals for values of p_0. For the deterministic elastic sphere, the critical value $p_0 = 5\underline{\mu}/2 = 10.125$ strictly divides the cases of oscillations occurring or not. For the stochastic problem, for the same critical value, there is, by definition, exactly 50% chance that the motion is oscillatory, and 50% chance that is not. To increase the probability of oscillatory motion ($P_1 \approx 1$), one must apply a sufficiently small traction, p_0, below the expected critical point, whereas a non-oscillatory motion is certain to occur ($P_2 \approx 1$) if p_0 is sufficiently large. However, the inherent variability in the probabilistic system means that there will also exist events where there is competition between the two cases.

In Fig. 5.3, we illustrate the stochastic function $H(x)$, defined by (5.26), intersecting the curve $\frac{2p_0}{\rho B^2}\left[\left(x^3 + 1\right)^{1/3} - 1\right]$, with $p_0 = 15$, to find the two distinct solutions of equation (5.24), and the associated velocity, given by (5.22), assuming that $\rho = 1$, $B = 1$, and μ is drawn from the Gamma distribution with $\rho_1 = 405$ and $\rho_2 = 0.01$. Each figure displays a probability histogram at each value of x. The histogram comprises 1000 stochastic simulations, and the colour bar defines the probability of finding a given value of $H(x)$, or of the associated velocity, respectively, at a given value of x. The dashed black line corresponds to the expected values based only on the mean value $\underline{\mu} = \rho_1 \rho_2 = 4.05$, of μ.

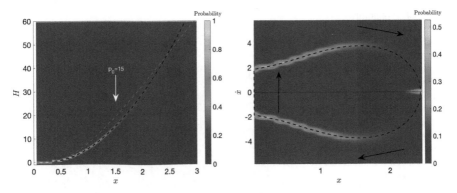

Fig. 5.3 The function $H(x)$, defined by (5.26), intersecting the (dashed red) curve $\frac{2p_0}{\rho B^2}\left[(x^3+1)^{1/3}-1\right]$, with $p_0 = 15$ (left), and the associated velocity, given by (5.22) (right), for a dynamic sphere of stochastic neo-Hookean material under dead-load traction, when $\rho = 1$, $B = 1$, and μ is drawn from the Gamma distribution with $\rho_1 = 405$ and $\rho_2 = 0.01$. The black lines correspond to the expected values based only on mean value, $\underline{\mu} = \rho_1\rho_2 = 4.05$ [374]

5.1.2   Static Deformation of Stochastic Neo-Hookean Spheres Under Dead-Load Traction

We review here the cavitation of a static sphere of stochastic neo-Hookean material, with the shear modulus μ following a Gamma distribution. Incompressible spheres of different stochastic homogeneous hyperelastic materials were treated in detail in [359] (see also [97] for the deterministic spheres). In this case, if the surface of the cavity is traction-free, then $T_1 = 0$ and (5.16) reduces to

$$T_{rr}(b) = \frac{2}{3}\int_{x^3+1}^{\infty}\mu\frac{1+u}{u^{7/3}}du. \qquad (5.36)$$

After evaluating the integral in (5.36), the required uniform dead-load traction at the outer surface, $R = B$, in the reference configuration takes the form

$$P = \left(x^3+1\right)^{2/3}T_{rr}(b) = 2\mu\left[\left(x^3+1\right)^{1/3}+\frac{1}{4\left(x^3+1\right)^{2/3}}\right] \qquad (5.37)$$

and increases as x increases. The critical dead load for the onset of cavitation is then

$$P_0 = \lim_{x\to 0_+} P = \frac{5\mu}{2} \qquad (5.38)$$

and is equal to that given by (5.32) for the dynamic sphere [39, 98].

To analyse the stability of this cavitation, we study the behaviour of the cavity opening, with radius c as a function of P, in a neighbourhood of P_0. After

Fig. 5.4 Probability distribution of the applied dead-load traction P causing cavitation of radius c in a static unit sphere (with $B = 1$) of stochastic neo-Hookean material, when the shear modulus, μ, follows a Gamma distribution with $\rho_1 = 405$ and $\rho_2 = 0.01$. The black line corresponds to the expected bifurcation based only on mean value, $\underline{\mu} = \rho_1\rho_2 = 4.05$ [374]

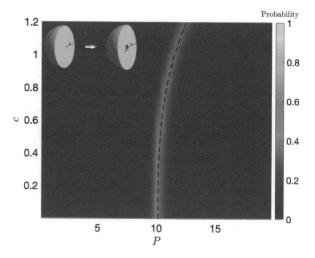

differentiating the function given by (5.37), with respect to the dimensionless cavity radius $x = c/B$, we have

$$\frac{\mathrm{d}P}{\mathrm{d}x} = 2\mu x^2 \left[\frac{1}{\left(x^3 + 1\right)^{2/3}} - \frac{1}{2\left(x^3 + 1\right)^{5/3}} \right] > 0, \qquad (5.39)$$

i.e., the cavitation is stable (with 100% certainty), regardless of the material parameter $\mu > 0$ [359]. For example, the post-cavitation stochastic behaviour of the static unit sphere when the shear modulus, μ, follows a Gamma distribution with $\rho_1 = 405$ and $\rho_2 = 0.01$ is shown in Fig. 5.4.

5.1.3 Non-oscillatory Motion of Stochastic Neo-Hookean Spheres Under Impulse Traction

For a sphere subject to the radially symmetric dynamic deformation (5.1), an impulse (suddenly applied) traction, expressed in terms of the Cauchy stresses, is prescribed as follows [358]:

$$2\frac{T_{rr}(b, t)}{\rho B^2} = \begin{cases} 0 & \text{if } t \leq 0, \\ p_0 & \text{if } t > 0, \end{cases} \qquad (5.40)$$

where p_0 is constant in time. Introducing the dimensionless cavity radius $x(t) = c(t)/B$ and setting the initial conditions $x_0 = x(0) = 0$ and $\dot{x}_0 = \dot{x}(0) = 0$, we obtain the following differential equation:

$$\dot{x}^2 x^3 \left[1 - \left(1 + \frac{1}{x^3} \right)^{-1/3} \right] + H(x) = \frac{p_0}{3} x^3, \qquad (5.41)$$

where $H(x)$ is given by (5.26). From (5.41), we obtain the velocity

$$\dot{x} = \pm \sqrt{\frac{\frac{p_0}{3} x^3 - H(x)}{x^3 \left[1 - \left(1 + \frac{1}{x^3} \right)^{-1/3} \right]}}, \qquad (5.42)$$

assuming that $\frac{p_0}{3} x^3 - H(x) \geq 0$. For an oscillatory motion to be obtained, the equation

$$H(x) = \frac{p_0}{3} x^3 \qquad (5.43)$$

must have exactly two distinct solutions, $x = x_1$ and $x = x_2$, such that $0 \leq x_1 < x_2 < \infty$.

When $p_0 \neq 0$, substitution of (5.26) in (5.43) gives

$$\frac{p_0}{3} x^3 = \frac{\mu}{\rho B^2} \left[2 \left(x^3 + 1 \right)^{2/3} - \frac{1}{\left(x^3 + 1 \right)^{1/3}} - 1 \right], \qquad (5.44)$$

which has a solution at $x_1 = 0$. The second solution, $x_2 > 0$, is a root of

$$p_0 = \frac{3\mu}{\rho B^2} \left[\frac{1}{\left(x^3 + 1 \right)^{1/3}} + \frac{\left(x^3 + 1 \right)^{1/3} + 1}{\left(x^3 + 1 \right)^{2/3} + \left(x^3 + 1 \right)^{1/3} + 1} \right]. \qquad (5.45)$$

As the right-hand side of Eq. (5.45) is a decreasing function of x, this equation has a solution, $x_2 > 0$, if and only if

$$p_0 > 0 = \lim_{x \to \infty} \frac{3\mu}{\rho B^2} \left[\frac{1}{\left(x^3 + 1 \right)^{1/3}} + \frac{\left(x^3 + 1 \right)^{1/3} + 1}{\left(x^3 + 1 \right)^{2/3} + \left(x^3 + 1 \right)^{1/3} + 1} \right] \qquad (5.46)$$

and

$$p_0 < \frac{5\mu}{\rho B^2} = \lim_{x \to 0_+} \frac{3\mu}{\rho B^2} \left[\frac{1}{\left(x^3 + 1 \right)^{1/3}} + \frac{\left(x^3 + 1 \right)^{1/3} + 1}{\left(x^3 + 1 \right)^{2/3} + \left(x^3 + 1 \right)^{1/3} + 1} \right]. \qquad (5.47)$$

However, since this implies

$$\frac{p_0}{3} x^3 - H(x) \begin{cases} < 0 \text{ if } x_1 < x < x_2, \\ > 0 \text{ if } x > x_2, \end{cases} \qquad (5.48)$$

Fig. 5.5 The function $H(x)$, defined by (5.26), intersecting the (dashed red) curve $p_0 x^3/3$, with $p_0 = 9$, for a dynamic sphere of stochastic neo-Hookean material under impulse traction, when $\rho = 1$, $B = 1$, and μ is drawn from the Gamma distribution with $\rho_1 = 405$ and $\rho_2 = 0.01$. The black line corresponds to the expected values based only on mean value, $\underline{\mu} = \rho_1 \rho_2 = 4.05$ [374]

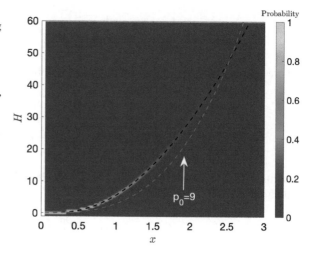

the sphere cannot oscillate when p_0 is constant in time, as assumed in (5.40) (see Fig. 5.5).

By (5.40) and (5.45), the tensile traction, $T = T_{rr}(b, t)$, takes the form

$$T = \frac{3\mu}{2} \left[\frac{1}{(x^3 + 1)^{1/3}} + \frac{(x^3 + 1)^{1/3} + 1}{(x^3 + 1)^{2/3} + (x^3 + 1)^{1/3} + 1} \right] \qquad (5.49)$$

and decreases as x increases. Thus, the critical tension for the onset of cavitation is

$$T_0 = \lim_{x \to 0_+} T = \frac{5\mu}{2}, \qquad (5.50)$$

as found also under dead loading.

5.1.4   Static Deformation of Stochastic Neo-Hookean Spheres Under Impulse Traction

For the static sphere subject to a uniform constant surface load in the current configuration, given in terms of the Cauchy stresses, we have

$$T = 2\mu \left[\frac{1}{(x^3 + 1)^{1/3}} + \frac{1}{4(x^3 + 1)^{4/3}} \right], \qquad (5.51)$$

which decreases as x increases. Then, the critical tension for cavitation initiation is also given by (5.50).

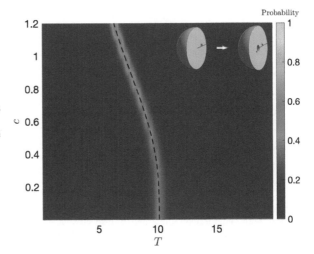

Fig. 5.6 Probability distribution of the applied impulse traction T causing cavitation of radius c in a static unit sphere (with $B = 1$) of stochastic neo-Hookean material, when the shear modulus, μ, follows a Gamma distribution with $\rho_1 = 405$ and $\rho_2 = 0.01$. The black line corresponds to the expected bifurcation based only on the mean value, $\underline{\mu} = \rho_1\rho_2 = 4.05$ [374]

The post-cavitation behaviour can be inferred from the sign of the derivative of T in a neighbourhood of T_0. By differentiating (5.51) with respect to x, we obtain

$$\frac{\mathrm{d}T}{\mathrm{d}x} = -2\mu x^2 \left[\frac{1}{\left(x^3 + 1\right)^{4/3}} + \frac{1}{\left(x^3 + 1\right)^{7/3}} \right] < 0, \tag{5.52}$$

i.e., the cavitation is unstable, regardless of the material parameter μ [39]. The post-cavitation stochastic behaviour of the static homogeneous sphere is shown in Fig. 5.6, for the unit sphere (with $B = 1$) of stochastic neo-Hookean material, where the shear modulus, μ, is drawn from a Gamma distribution with shape and scale parameters $\rho_1 = 405$ and $\rho_2 = 0.01$, respectively (see also Figure 2 of [39]).

5.2   Cavitation and Radial Quasi-equilibrated Motion of Concentric Homogeneous Spheres

Next, we investigate the behaviour under the quasi-equilibrated radial motion (5.1) of a composite formed from two concentric homogeneous spheres (see Fig. 5.7). We restrict our attention to composite spheres with two stochastic neo-Hookean phases, similar to those containing two concentric spheres of different neo-Hookean materials treated deterministically in [264] and [498]. In this case, we define the following strain-energy function:

$$W(\alpha_1, \alpha_2, \alpha_3) = \begin{cases} \frac{\mu^{(1)}}{2} \left(\alpha_1^2 + \alpha_2^2 + \alpha_3^2 - 3\right), & 0 < R < A, \\ \frac{\mu^{(2)}}{2} \left(\alpha_1^2 + \alpha_2^2 + \alpha_3^2 - 3\right), & A < R < B, \end{cases} \tag{5.53}$$

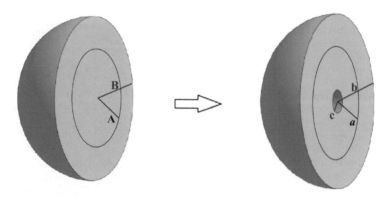

Fig. 5.7 Schematic of cross-section of a composite sphere made of two concentric homogeneous spheres, with undeformed radii A and B, respectively, showing the reference state (left), and the deformed state, with cavity radius c and radii a and b of the concentric spheres, respectively (right)

where $0 < R < A$ and $A < R < B$ denote the radii of the inner and outer spheres in the reference configuration, and the corresponding shear moduli $\mu^{(1)}$ and $\mu^{(2)}$ are spatially independent random variables characterised by the Gamma distributions $g(u; \rho_1^{(1)}, \rho_2^{(1)})$ and $g(u; \rho_1^{(2)}, \rho_2^{(2)})$, respectively.

5.2.1 Oscillatory Motion of Spheres with Two Stochastic Neo-Hookean Phases Under Dead-Load Traction

For the deformed composite sphere, we denote the radial pressures acting on the curvilinear surfaces $r = c(t)$ and $r = b(t)$ at time t as $T_1(t)$ and $T_2(t)$, respectively, and impose the continuity condition for the stress components across their interface, $r = a(t)$. By analogy to (5.13), we obtain

$$T_1(t) - T_2(t) = \rho B^2 \left[\left(\frac{c}{B} \frac{\ddot{c}}{B} + 2 \frac{\dot{c}^2}{B^2} \right) \left(1 - \frac{c}{b} \right) - \frac{\dot{c}^2}{2B^2} \left(1 - \frac{c^4}{b^4} \right) \right]$$

$$+ 2 \int_c^a \mu^{(1)} \left(\frac{r^2}{R^2} - \frac{R^4}{r^4} \right) \frac{dr}{r} + 2 \int_a^b \mu^{(2)} \left(\frac{r^2}{R^2} - \frac{R^4}{r^4} \right) \frac{dr}{r}.$$

$$(5.54)$$

Using the notation (5.14), we can write (5.54) equivalently as follows:

$$2x^2 \frac{T_1(t) - T_2(t)}{\rho B^2} = \frac{d}{dx} \left\{ \dot{x}^2 x^3 \left[1 - \left(1 + \frac{1}{x^3} \right)^{-1/3} \right] \right\}$$

$$+ \frac{4x^2}{3\rho B^2} \int_{x^3+1}^{x^3 B^3/A^3+1} \mu^{(2)} \frac{1+u}{u^{7/3}} du + \frac{4x^2}{3\rho B^2} \int_{x^3 B^3/A^3+1}^{\infty} \mu^{(1)} \frac{1+u}{u^{7/3}} du.$$

$$(5.55)$$

In this case, we denote

$$
H(x) = \frac{4}{3\rho B^2} \int_0^x \left(\zeta^2 \int_{\zeta^3+1}^{\zeta^3 B^3/A^3+1} \mu^{(2)} \frac{1+u}{u^{7/3}} du \right) d\zeta
$$
$$
+ \frac{4}{3\rho B^2} \int_0^x \left(\zeta^2 \int_{\zeta^3 B^3/A^3+1}^{\infty} \mu^{(1)} \frac{1+u}{u^{7/3}} du \right) d\zeta,
$$

(5.56)

and, after evaluating the integrals, we obtain the equivalent form

$$
H(x) = \frac{\mu^{(2)}}{\rho B^2} \left[2 \left(x^3 + 1 \right)^{2/3} - \frac{1}{\left(x^3 + 1 \right)^{1/3}} - 1 \right]
$$
$$
+ \frac{\mu^{(1)} - \mu^{(2)}}{\rho B^2} \frac{A^3}{B^3} \left[2 \left(x^3 \frac{B^3}{A^3} + 1 \right)^{2/3} - \frac{1}{\left(x^3 \frac{B^3}{A^3} + 1 \right)^{1/3}} - 1 \right].
$$

(5.57)

Then, substitution of (5.57) in (5.24) gives

$$
p_0 \left[\left(x^3 + 1 \right)^{1/3} - 1 \right] = \frac{\mu^{(2)}}{2} \left[2 \left(x^3 + 1 \right)^{2/3} - \frac{1}{\left(x^3 + 1 \right)^{1/3}} - 1 \right]
$$
$$
+ \frac{\mu^{(1)} - \mu^{(2)}}{2} \frac{A^3}{B^3} \left[2 \left(x^3 \frac{B^3}{A^3} + 1 \right)^{2/3} - \frac{1}{\left(x^3 \frac{B^3}{A^3} + 1 \right)^{1/3}} - 1 \right].
$$

(5.58)

This equation has one solution at $x_1 = 0$, while the second solution, $x_2 > 0$, is a root of

$$
p_0 = \frac{\mu^{(2)}}{2} \left[2 \left(x^3 + 1 \right)^{1/3} + \frac{1}{\left(x^3 + 1 \right)^{1/3}} + 2 \right]
$$
$$
- \frac{\mu^{(2)}}{2} \frac{\left(x^3 + 1 \right)^{2/3} + \left(x^3 + 1 \right)^{1/3} + 1}{\left(x^3 \frac{B^3}{A^3} + 1 \right)^{2/3} + \left(x^3 \frac{B^3}{A^3} + 1 \right)^{1/3} + 1} \left[2 \left(x^3 \frac{B^3}{A^3} + 1 \right)^{1/3} + \frac{1}{\left(x^3 \frac{B^3}{A^3} + 1 \right)^{1/3}} + 2 \right]
$$
$$
+ \frac{\mu^{(1)}}{2} \frac{\left(x^3 + 1 \right)^{2/3} + \left(x^3 + 1 \right)^{1/3} + 1}{\left(x^3 \frac{B^3}{A^3} + 1 \right)^{2/3} + \left(x^3 \frac{B^3}{A^3} + 1 \right)^{1/3} + 1} \left[2 \left(x^3 \frac{B^3}{A^3} + 1 \right)^{1/3} + \frac{1}{\left(x^3 \frac{B^3}{A^3} + 1 \right)^{1/3}} + 2 \right].
$$

(5.59)

Next, we expand p_0, given by (5.59), to the first order in x^3 to obtain

$$
p_0 \approx \frac{5\mu^{(1)}}{2} + \frac{x^3}{3} \left[2 \left(\frac{B^3}{A^3} - 1 \right) \left(\mu^{(2)} - \mu^{(1)} \right) + \frac{\mu^{(1)}}{2} \right].
$$

(5.60)

The critical tension for cavity initiation is then

$$\lim_{x \to 0_+} p_0 = \frac{5\mu^{(1)}}{2},$$ (5.61)

and a comparison with (5.32) shows that it is the same as for the homogeneous sphere made of the same material as the inner sphere.

Considering the random shear parameters of the two concentric spheres, we distinguish the following two cases:

(i) If

$$\frac{\mu^{(2)}}{\mu^{(1)}} > 1 - \frac{1}{4}\left(\frac{B^3}{A^3} - 1\right)^{-1},$$ (5.62)

then the right-hand side of (5.60) is an increasing function of x, and

$$p_0 > \lim_{x \to 0_+} \frac{5\mu^{(1)}}{2} + \frac{x^3}{3}\left[2\left(\frac{B^3}{A^3} - 1\right)\left(\mu^{(2)} - \mu^{(1)}\right) + \frac{\mu^{(1)}}{2}\right] = \frac{5\mu^{(1)}}{2}.$$ (5.63)

By (5.63), for the motion of the composite sphere to be oscillatory, the shear modulus of the inner sphere must satisfy

$$\mu_0 < \mu^{(1)} < \frac{2p_0}{5},$$ (5.64)

where the upper bound follows from (5.63), while the lower bound is assumed to be strictly positive, i.e., $\mu_0 > 0$. Thus, the probability distribution of oscillatory motion occurring is given by (5.34) and that of non-oscillatory motion by (5.35), with $g(u; \rho_1^{(1)}, \rho_2^{(1)})$ instead of $g(u; \rho_1, \rho_2)$ (see Fig. 5.2).

As the material parameters, $\mu^{(1)}$ and $\mu^{(2)}$, are positive, (5.62) is possible for any values $0 < A < B$. In particular, if

$$\frac{B}{A} < \left(\frac{5}{4}\right)^{1/3},$$ (5.65)

then (5.62) always holds, regardless of the material constants.

(ii) If

$$\frac{\mu^{(2)}}{\mu^{(1)}} < 1 - \frac{1}{4}\left(\frac{B^3}{A^3} - 1\right)^{-1},$$ (5.66)

then the right-hand side in (5.60) decreases as x increases; hence,

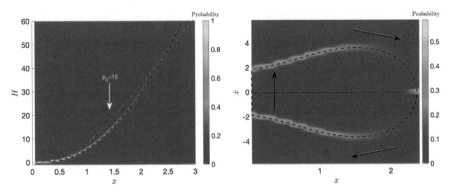

Fig. 5.8 The function $H(x)$, defined by (5.57), intersecting the (dashed red) curve $\frac{2p_0}{\rho B^2}\left[\left(x^3+1\right)^{1/3}-1\right]$, with $p_0 = 15$ (left), and the associated velocity, given by (5.22) (right), for a dynamic composite sphere with two concentric stochastic neo-Hookean phases, with radii $A = 1/2$ and $B = 1$, respectively, under dead-load traction, when the shear modulus of the inner phase, $\mu^{(1)}$, follows a Gamma distribution with $\rho_1^{(1)} = 405$ and $\rho_2^{(1)} = 4.05/\rho_1^{(1)} = 0.01$, while the shear modulus of the outer phase, $\mu^{(2)}$, is drawn from a Gamma distribution with $\rho_1^{(2)} = 405$ and $\rho_2^{(2)} = 4.2/\rho_1^{(2)}$. The black lines correspond to the expected values based only on mean values, $\underline{\mu}^{(1)} = 4.05$ and $\underline{\mu}^{(2)} = 4.2$ [374]

$$0 < p_0 < \frac{5\mu^{(1)}}{2}. \tag{5.67}$$

As the material constants, $\mu^{(1)}$ and $\mu^{(2)}$, are positive, (5.66) is possible if and only if

$$\frac{B}{A} > \left(\frac{5}{4}\right)^{1/3}. \tag{5.68}$$

However, in this case,

$$\frac{2p_0}{\rho B^2}\left[\left(x^3+1\right)^{1/3}-1\right] - H(x) \begin{cases} < 0 \text{ if } x_1 < x < x_2, \\ > 0 \text{ if } x > x_2, \end{cases} \tag{5.69}$$

and the sphere cannot oscillate.

In Fig. 5.8, we represent the stochastic function $H(x)$, defined by (5.57), intersecting the curve $\frac{2p_0}{\rho B^2}\left[\left(x^3+1\right)^{1/3}-1\right]$, with $p_0 = 15$, to find the two distinct solutions to (5.24), and the associated velocity, given by (5.22), for a composite sphere with two concentric stochastic neo-Hookean phases, with radii $A = 1/2$ and $B = 1$, respectively, when the shear modulus of the inner phase, $\mu^{(1)}$, follows a Gamma distribution with $\rho_1^{(1)} = 405$, $\rho_2^{(1)} = 4.05/\rho_1^{(1)} =$

0.01, while the shear modulus of the outer phase, $\mu^{(2)}$, is drawn from a Gamma distribution with $\rho_1^{(2)} = 405$, $\rho_2^{(2)} = 4.2/\rho_1^{(2)}$.

5.2.2   Static Deformation of Spheres with Two Stochastic Neo-Hookean Phases Under Dead-Load Traction

For the static composite sphere, if the surface of the cavity is traction-free, then $T_1 = 0$ and (5.55) reduces to

$$T_{rr}(b) = \frac{2}{3} \int_{x^3+1}^{x^3 B^3/A^3 + 1} \mu^{(2)} \frac{1+u}{u^{7/3}} du + \frac{2}{3} \int_{x^3 B^3/A^3 + 1}^{\infty} \mu^{(1)} \frac{1+u}{u^{7/3}} du. \qquad (5.70)$$

After evaluating the integrals in (5.70), the required dead-load traction at the outer surface, $R = B$, in the reference configuration is equal to

$$P = \left(x^3 + 1\right)^{2/3} T_{rr}(b)$$

$$= 2\mu^{(2)} \left[\left(x^3 + 1\right)^{1/3} + \frac{1}{4\left(x^3 + 1\right)^{2/3}} \right]$$

$$+ 2\left(\mu^{(1)} - \mu^{(2)}\right)\left(x^3 + 1\right)^{2/3} \left[\frac{1}{\left(x^3 \frac{B^3}{A^3} + 1\right)^{1/3}} + \frac{1}{4\left(x^3 \frac{B^3}{A^3} + 1\right)^{4/3}} \right]. \qquad (5.71)$$

The critical dead load for the cavity formation is then

$$P_0 = \lim_{x \to 0_+} P = \frac{5\mu^{(1)}}{2} \qquad (5.72)$$

and is the same as for the homogeneous sphere made entirely of the material of the inner sphere [264], as found also in the dynamic case.

To study the stability of this cavitation, we examine the behaviour of the cavity opening, with radius c as a function of P, in a neighbourhood of P_0. Expanding the expression of P, given by (5.71), to the first order in x^3, we obtain [264]

$$P \approx \frac{5\mu^{(1)}}{2} + \frac{x^3}{3} \left[4\left(\frac{B^3}{A^3} - 1\right)\left(\mu^{(2)} - \mu^{(1)}\right) + \mu^{(1)} \right]. \qquad (5.73)$$

After differentiating the above function with respect to the dimensionless cavity radius $x = c/B$, we have

$$\frac{\mathrm{d}P}{\mathrm{d}x} \approx x^2 \left[4 \left(\frac{B^3}{A^3} - 1 \right) \left(\mu^{(2)} - \mu^{(1)} \right) + \mu^{(1)} \right]. \tag{5.74}$$

Then, $\mathrm{d}P/\mathrm{d}x \to 0$ as $x \to 0_+$, and the bifurcation at the critical dead load, P_0, is supercritical (respectively, subcritical) if $\mathrm{d}P/\mathrm{d}x > 0$ (respectively, $\mathrm{d}P/\mathrm{d}x < 0$) for arbitrarily small x. From (5.74), the following two cases are distinguished (see also the discussion in [264]):

(i) If (5.62) is valid, then the solution presents itself as a supercritical bifurcation, and the cavitation is stable, in the sense that a new bifurcated solution exists locally for values of $P > P_0$, and the cavity radius monotonically increases with the applied load post-bifurcation. When $\mu^{(1)} = \mu^{(2)}$, the problem reduces to the case of a homogeneous sphere made of neo-Hookean material, for which stable cavitation is known to occur [359].

(ii) If (5.66) holds, then the bifurcation is subcritical, with the cavitation being unstable, in the sense that the required dead load starts to decrease post-bifurcation, causing a snap cavitation, i.e., a sudden jump in the cavity opening.

Thus, on the one hand, when (5.65) is satisfied, (5.62) is valid regardless of the values of $\mu^{(1)}$ and $\mu^{(2)}$, and the cavitation is guaranteed to be stable (with 100% certainty). On the other hand, if (5.68) holds, then, as either of the two inequalities (5.62) and (5.66) is possible, the probability distribution of stable cavitation in the composite sphere is equal to

$$P_1(\mu^{(2)}) = \int_0^{\mu^{(2)} / \left[1 - \frac{1}{4} \left(\frac{B^3}{A^3} - 1 \right)^{-1} \right]} g(u; \rho_1^{(1)}, \rho_2^{(1)}) \mathrm{d}u, \tag{5.75}$$

while that of unstable (snap) cavitation is

$$P_2(\mu^{(2)}) = 1 - P_1(\mu^{(2)}) = 1 - \int_0^{\mu^{(2)} / \left[1 - \frac{1}{4} \left(\frac{B^3}{A^3} - 1 \right)^{-1} \right]} g(u; \rho_1^{(1)}, \rho_2^{(1)}) \mathrm{d}u. \tag{5.76}$$

For example, setting $A = 1/2$ and $B = 1$, if $\rho_1^{(1)} = 405$, $\rho_2^{(1)} = 4.05/\rho_1^{(1)} = 0.01$ for the inner sphere, and $\rho_1^{(2)} = 405$, $\rho_2^{(2)} = 4.2/\rho_1^{(2)}$ for the outer sphere, then the probability distributions given by Eqs. (5.75) and (5.76) are illustrated numerically in Fig. 5.9 (with blue lines for P_1 and red lines for P_2). For the numerical realisations in these plots, the interval $(0, 8.1) = \left(0, 2\mu^{(1)} \right)$ was discretised into 100 representative points; then for each value of $\mu^{(2)}$, 100 random values of $\mu^{(1)}$ were numerically generated from the specified Gamma distribution and compared with the inequalities defining the two intervals for values of $\mu^{(2)}$. For the deterministic elastic case, which is based on the mean value of the shear modulus, $\underline{\mu}^{(1)} = \rho_1^{(1)} \rho_2^{(1)} = 4.05$, and the critical value of $27\underline{\mu}^{(1)}/28 = 3.9054$

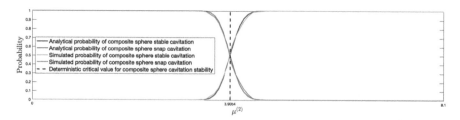

Fig. 5.9 Probability distributions of stable or unstable cavitation in a static sphere made of two concentric stochastic neo-Hookean phases, with radii $A = 1/2$ and $B = 1$, respectively, under dead-load traction, when the shear modulus of the inner sphere, $\mu^{(1)}$, follows a Gamma distribution with $\rho_1^{(1)} = 405$ and $\rho_2^{(1)} = 0.01$. Dark coloured lines represent analytically derived solutions, given by Eqs. (5.75) and (5.76), whereas the lighter versions represent stochastically generated data. The vertical black line at the critical value $\mu^{(2)} = 27\mu^{(1)}/28 = 3.9054$ separates the expected regions based only on mean value, $\underline{\mu} = \rho_1^{(1)}\rho_2^{(1)} = 4.05$ [374]

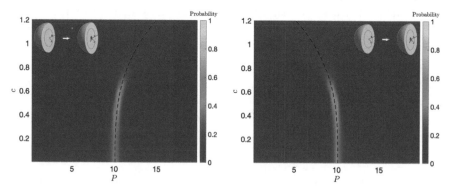

Fig. 5.10 Probability distribution of the applied dead-load traction P causing cavitation of radius c in a static sphere of two concentric stochastic neo-Hookean phases, with radii $A = 1/2$ and $B = 1$, respectively, when the shear modulus of the inner phase, $\mu^{(1)}$, follows a Gamma distribution with $\rho_1^{(1)} = 405$ and $\rho_2^{(1)} = 4.05/\rho_1^{(1)} = 0.01$, while the shear modulus of the outer phase, $\mu^{(2)}$, is drawn from Gamma distribution with $\rho_1^{(2)} = 405$ and $\rho_2^{(2)} = 4.2/\rho_1^{(2)}$ (left) or $\rho_1^{(2)} = 405$ and $\rho_2^{(2)} = 3.6/\rho_1^{(2)}$ (right). The black line corresponds to the expected bifurcation based only on mean parameter values [374]

strictly separates the cases where cavitation instability can, and cannot, occur. However, in the stochastic case, the two states compete. For example, at the same critical value, there is, by definition, exactly 50% chance that the cavitation is stable, and 50% chance that is not.

The post-cavitation stochastic behaviours shown in Fig. 5.10 correspond to two different static composite spheres where the shear modulus of the inner phase, $\mu^{(1)}$, follows a Gamma distribution with $\rho_1^{(1)} = 405$, $\rho_2^{(1)} = 4.05/\rho_1^{(1)} = 0.01$, while the shear modulus of the outer phase, $\mu^{(2)}$, is drawn from a Gamma distribution with $\rho_1^{(2)} = 405$, $\rho_2^{(2)} = 4.2/\rho_1^{(2)}$ (in Fig. 5.10-left) or $\rho_1^{(2)} = 405$, $\rho_2^{(2)} = 3.6/\rho_1^{(2)}$ (in

Fig. 5.10-right). In each case, unstable or stable cavitation is expected, respectively, but there is also about 5% chance that the opposite behaviour is presented.

5.2.3 Non-oscillatory Motion of Spheres with Two Stochastic Neo-Hookean Phases Under Impulse Traction

Setting the initial conditions $x_0 = x(0) = 0$ and $\dot{x}_0 = \dot{x}(0) = 0$, we obtain Eq. (5.41), where $H(x)$ is given by (5.57). Then, substitution of (5.57) in (5.43) gives

$$\frac{p_0}{3}x^3 = \frac{\mu^{(2)}}{\rho B^2}\left[2\left(x^3 + 1\right)^{2/3} - \frac{1}{\left(x^3 + 1\right)^{1/3}} - 1\right]$$

$$+ \frac{\mu^{(1)} - \mu^{(2)}}{\rho B^2}\frac{A^3}{B^3}\left[2\left(x^3\frac{B^3}{A^3} + 1\right)^{2/3} - \frac{1}{\left(x^3\frac{B^3}{A^3} + 1\right)^{1/3}} - 1\right]. \tag{5.77}$$

Equivalently, by (5.40), in terms of the tensile traction $T = T_{rr}(b, t)$,

$$\frac{2T}{3}x^3 = \mu^{(2)}\left[2\left(x^3 + 1\right)^{2/3} - \frac{1}{\left(x^3 + 1\right)^{1/3}} - 1\right]$$

$$+ \left(\mu^{(1)} - \mu^{(2)}\right)\frac{A^3}{B^3}\left[2\left(x^3\frac{B^3}{A^3} + 1\right)^{2/3} - \frac{1}{\left(x^3\frac{B^3}{A^3} + 1\right)^{1/3}} - 1\right]. \tag{5.78}$$

This equation has one solution at $x_1 = 0$, while the second solution, $x_2 > 0$, is a root of

$$T = \frac{3\mu^{(2)}}{2}\left[\frac{\left(x^3 + 1\right)^{1/3} + 1}{\left(x^3 + 1\right)^{2/3} + \left(x^3 + 1\right)^{1/3} + 1} + \frac{1}{\left(x^3 + 1\right)^{1/3}}\right]$$

$$+ \frac{3\left(\mu^{(1)} - \mu^{(2)}\right)}{2}\left[\frac{\left(x^3\frac{B^3}{A^3} + 1\right)^{1/3} + 1}{\left(x^3\frac{B^3}{A^3} + 1\right)^{2/3} + \left(x^3\frac{B^3}{A^3} + 1\right)^{1/3} + 1} + \frac{1}{\left(x^3\frac{B^3}{A^3} + 1\right)^{1/3}}\right]. \tag{5.79}$$

After expanding T, given by (5.79), to the first order in x^3, we have

$$T \approx \frac{5\mu^{(1)}}{2} + x^3\left[\frac{B^3}{A^3}\left(\mu^{(2)} - \mu^{(1)}\right) - \mu^{(2)}\right]. \tag{5.80}$$

The associated critical tension for cavity initiation is

$$T_0 = \lim_{x \to 0_+} T = \frac{5\mu^{(1)}}{2}, \tag{5.81}$$

and a comparison with that given by (5.50) shows that it is the same as for the homogeneous sphere made of the same material as the inner sphere.

We now distinguish the following two cases:

(i) If

$$\frac{\mu^{(2)}}{\mu^{(1)}} > \left(1 - \frac{A^3}{B^3}\right)^{-1}, \tag{5.82}$$

and then the right-hand side of (5.80) is an increasing function of x; hence,

$$T > \lim_{x \to 0_+} \frac{5\mu^{(1)}}{2} + x^3 \left[\frac{B^3}{A^3}\left(\mu^{(2)} - \mu^{(1)}\right) - \mu^{(2)}\right] = \frac{5\mu^{(1)}}{2}. \tag{5.83}$$

By (5.83), for the motion of the composite sphere to be oscillatory, the shear modulus of the inner sphere must satisfy

$$\mu_0 < \mu^{(1)} < \frac{2T}{5}, \tag{5.84}$$

where the lower bound is assumed to be strictly positive, i.e., $\mu_0 > 0$. Equivalently, by (5.40), in terms of the constant p_0,

$$\mu_0 < \mu^{(1)} < \frac{p_0}{5}\rho B^2. \tag{5.85}$$

However, in this case,

$$\frac{p_0}{3}x^3 - H(x) > 0, \tag{5.86}$$

for all $x > 0$, and the sphere does not oscillate (see Fig. 5.11-left).

(ii) If

$$\frac{\mu^{(2)}}{\mu^{(1)}} < \left(1 - \frac{A^3}{B^3}\right)^{-1}, \tag{5.87}$$

then the right-hand side of (5.80) decreases as x increases; hence,

$$0 < T < \frac{5\mu^{(1)}}{2}, \tag{5.88}$$

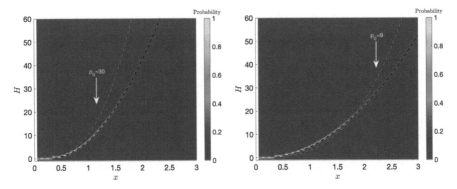

Fig. 5.11 The function $H(x)$, defined by (5.57), and the (dashed red) curve $\frac{p_0 x^3}{3}$, with different values of p_0, for a dynamic sphere made of two concentric stochastic neo-Hookean phases, with radii $A = 1/2$ and $B = 1$, respectively, under impulse traction, when $\rho = 1$ and the shear modulus of the inner sphere, $\mu^{(1)}$, follows a Gamma distribution with $\rho_1^{(1)} = 405$ and $\rho_2^{(1)} = 0.01$, while for the outer sphere, the shear modulus, $\mu^{(2)}$, follows a Gamma distribution with $\rho_1^{(2)} = 405$ and $\rho_2^{(1)} = 0.02$ (left) or $\rho_1^{(2)} = 405$ and $\rho_2^{(1)} = 0.005$ (right). The black line corresponds to the expected values based only on mean value parameters [374]

or equivalently, by (5.40),

$$0 < p_0 < \frac{5\mu^{(1)}}{\rho B^2}.$$ (5.89)

However, in this case,

$$\frac{p_0}{3}x^3 - H(x) \begin{cases} < 0 \text{ if } x_1 < x < x_2, \\ > 0 \text{ if } x > x_2, \end{cases}$$ (5.90)

and the sphere cannot oscillate (see Fig. 5.11-right).

Note that, since the material parameters, $\mu^{(1)}$ and $\mu^{(2)}$, are positive, both (5.82) and (5.87) are possible for any values $0 < A < B$.

5.2.4   Static Deformation of Spheres with Two Stochastic Neo-Hookean Phases Under Impulse Traction

For the static composite sphere subject to a uniform constant surface load in the current configuration, given in terms of the Cauchy stresses, the tensile traction takes the form

$$T = 2\mu^{(2)} \left[\frac{1}{\left(x^3 + 1\right)^{1/3}} + \frac{1}{4\left(x^3 + 1\right)^{4/3}} \right]$$

$$+ 2\left(\mu^{(1)} - \mu^{(2)}\right) \left[\frac{1}{\left(x^3 \frac{B^3}{A^3} + 1\right)^{1/3}} + \frac{1}{4\left(x^3 \frac{B^3}{A^3} + 1\right)^{4/3}} \right].$$

$$(5.91)$$

The critical traction for the onset of cavitation is thus also given by (5.81).

To investigate the stability of this cavitation, we examine the behaviour of the cavity opening, with scaled radius x as a function of T, in a neighbourhood of T_0. Expanding the expression of T, given by (5.91), to the first order in x^3, we have

$$T \approx \frac{5\mu^{(1)}}{2} + \frac{4x^3}{3} \left[\frac{B^3}{A^3} \left(\mu^{(2)} - \mu^{(1)}\right) - \mu^{(2)} \right] \tag{5.92}$$

and

$$\frac{dT}{dx} \approx 4x^2 \left[\frac{B^3}{A^3} \left(\mu^{(2)} - \mu^{(1)}\right) - \mu^{(2)} \right]. \tag{5.93}$$

Then, $dT/dx \to 0$ as $x \to 0_+$, and the bifurcation at the critical impulse load, T_0, is supercritical (respectively, subcritical) if $dT/dx > 0$ (respectively, $dT/dx < 0$) for arbitrarily small x. From (5.93), the following two cases arise:

(i) If (5.82) holds, then the solution after bifurcation is supercritical, and the cavitation is stable, in the sense that a new bifurcated solution exists locally for values of $T > T_0$, and the cavity radius monotonically increases with the applied load post-bifurcation.

(ii) If (5.87) holds, then the bifurcated solution is subcritical, with the cavitation being unstable, in the sense that the required dead load starts to decrease post-bifurcation, causing a snap cavitation, i.e., a sudden jump in the cavity opening. When $\mu^{(1)} = \mu^{(2)}$, the problem reduces to the case of a homogeneous sphere of neo-Hookean material, for which unstable cavitation always occurs (see Sect. 5.1).

As either of the two inequalities (5.82) and (5.87) is possible, the probability distribution of stable cavitation in the composite sphere is equal to (see Fig. 5.12)

$$P_1(\mu^{(2)}) = \int_0^{\mu^{(2)}\left(1 - \frac{A^3}{B^3}\right)} g(u; \rho_1^{(1)}, \rho_2^{(1)}) \, du, \tag{5.94}$$

while that of unstable (snap) cavitation is

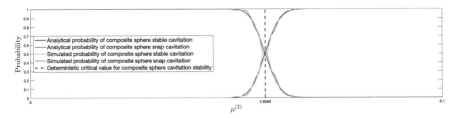

Fig. 5.12 Probability distributions of stable or unstable cavitation in a static sphere made of two concentric stochastic neo-Hookean phases, with radii $A = 1/2$ and $B = 1$, respectively, under impulse traction, when the shear modulus of the inner sphere, $\mu^{(1)}$, follows a Gamma distribution with $\rho_1^{(1)} = 405$ and $\rho_2^{(1)} = 0.01$. Dark coloured lines represent analytically derived solutions, given by Eqs. (5.94) and (5.95), whereas the lighter versions represent stochastically generated data. The vertical black line at the critical value $\mu^{(2)} = 8\mu^{(1)}/7 = 4.6286$ separates the expected regions based only on mean value, $\underline{\mu} = \rho_1^{(1)}\rho_2^{(1)} = 4.05$ [374]

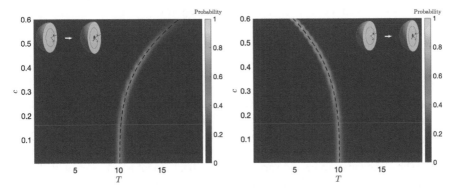

Fig. 5.13 Probability distribution of the applied impulse traction T causing cavitation of radius c in a static sphere of two concentric stochastic neo-Hookean phases, with radii $A = 1/2$ and $B = 1$, respectively, when the shear modulus of the inner phase, $\mu^{(1)}$, follows a Gamma distribution with $\rho_1^{(1)} = 405$, $\rho_2^{(1)} = 4.05/\rho_1^{(1)} = 0.01$, while the shear modulus of the outer phase, $\mu^{(2)}$, is drawn Gamma distribution with $\rho_1^{(2)} = 405$ and $\rho_2^{(2)} = 0.02$ (left) or $\rho_1^{(2)} = 405$, $\rho_2^{(2)} = 0.005$ (right). The black line corresponds to the expected bifurcation based only on mean parameter values [374]

$$P_2(\mu^{(2)}) = 1 - P_1(\mu^{(2)}) = 1 - \int_0^{\mu^{(2)}\left(1-\frac{A^3}{B^3}\right)} g(u; \rho_1^{(1)}, \rho_2^{(1)})\, du. \qquad (5.95)$$

The post-cavitation stochastic behaviours shown in Fig. 5.13 correspond to two different static composite sphere where the shear modulus of the inner phase, $\mu^{(1)}$, follows a Gamma distribution with $\rho_1^{(1)} = 405$, $\rho_2^{(1)} = 4.05/\rho_1^{(1)} = 0.01$, while the shear modulus of the outer phase, $\mu^{(2)}$, is drawn from a Gamma distribution with $\rho_1^{(2)} = 405$, $\rho_2^{(2)} = 0.02$ (in Fig. 5.13-left) or $\rho_1^{(2)} = 405$, $\rho_2^{(2)} = 0.005$ (in Fig. 5.13-right). Note that, in each case, unstable or stable cavitation is expected, respectively, but there is other values of the parameters that lead to the opposite behaviour.

5.3   Cavitation and Radial Quasi-equilibrated Motion of Inhomogeneous Spheres

We also examine the cavitation of radially inhomogeneous, incompressible spheres of stochastic hyperelastic material. The radially inhomogeneous sphere can be regarded as an extension of the composite with two concentric spheres to the case with infinitely many concentric layers and continuous inhomogeneity. Then one can build directly on the analytical results available for homogeneous spheres to derive analytical results. Cavitation of radially inhomogeneous hyperelastic spheres in static equilibrium was investigated in [295, 498]. Here, we adopt a neo-Hookean-like model, where the constitutive parameter varies continuously along the radial direction. Our inhomogeneous model is similar to those proposed in [158], where the dynamic inflation of spherical shells was treated explicitly.

Specifically, we define the class of stochastic inhomogeneous hyperelastic models (4.49), with the shear modulus taking the form

$$\mu(R) = C_1 + C_2 \frac{R^3}{B^3},\tag{5.96}$$

where $\mu(R) > 0$, for all $0 \le R \le B$. For simplicity, we further assume that, for any fixed R, the mean value $\underline{\mu}$ of $\mu = \mu(R)$ is independent of R (namely, the expected value of C_2 is 0), whereas its variance $\mathrm{Var}[\mu]$ changes with R. Then, $C_1 = \mu(0) > 0$ is a single-valued (deterministic) constant and C_2 is a random value, defined by a given probability distribution (see Fig. 5.14). Alternative modelling formulations where both the mean value and variance of the shear modulus may vary with R will be discussed at the end of this section.

Fig. 5.14 Examples of Gamma distribution, with hyperparameters $\rho_1 = 405 \cdot B^6/R^6$ and $\rho_2 = 0.01 \cdot R^6/B^6$, for the shear modulus $\mu(R)$ given by (5.96). In this case, $C_1 = \underline{\mu} = \rho_1 \rho_2 = 4.05$ and $C_2 = \overline{\mu}(B) - C_1$ [374]

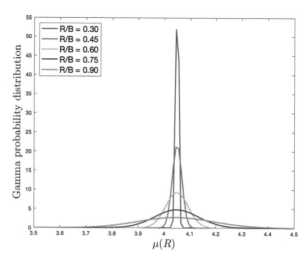

When the mean value of the shear modulus $\mu(R)$, described by (5.96), does not depend on R, it follows that, for any fixed R,

$$\underline{\mu} = C_1, \qquad \text{Var}[\mu] = \text{Var}[C_2]\frac{R^6}{B^6}, \tag{5.97}$$

where $\text{Var}[C_2]$ is the variance of C_2. Note that, at $R = 0$, the value $\mu(0) = C_1$ represents the material property at the centre of the sphere, which is a unique point, whereas at any fixed $R > 0$, the standard deviation $\|\mu\| = \sqrt{\text{Var}[\mu]}$ is proportional to the volume $4\pi R^3/3$ of the sphere with radius R.

The hyperparameters of the corresponding Gamma distribution take the form

$$\rho_1 = \frac{C_1}{\rho_2}, \qquad \rho_2 = \frac{\text{Var}[\mu]}{C_1} = \frac{\text{Var}[C_2]}{C_1}\frac{R^6}{B^6}. \tag{5.98}$$

For example, one can choose two constant values, $C_1 > 0$ and $C_1' > 0$, and set the hyperparameters for the Gamma distribution at any given R as follows:

$$\rho_1 = \frac{C_1}{C_1'}\frac{B^6}{R^6}, \qquad \rho_2 = C_1'\frac{R^6}{B^6}. \tag{5.99}$$

By (5.96), $C_2 = (\mu(R) - C_1) B^3/R^3$ is the shifted Gamma-distributed random variable with mean value $\underline{C_2} = 0$ and variance $\text{Var}[C_2] = \rho_1\rho_2^2 B^6/R^6 = C_1 C_1'$. When R decreases towards zero, by (5.99), ρ_1 increases, while ρ_2 decreases, and the Gamma distribution converges to a normal distribution [167, 357]. In Fig. 5.14, we show these distributions when $\rho_1 = 405 \cdot B^6/R^6$ and $\rho_2 = 0.01 \cdot R^6/B^6$. For this case, by (5.96), $C_1 = \underline{\mu} = \rho_1\rho_2 = 4.05$ and $C_2 = \mu(B) - C_1$.

5.3.1 Oscillatory Motion of Stochastic Radially Inhomogeneous Spheres Under Dead-Load Traction

The shear modulus of the form given by (5.96) can be expressed equivalently as follows:

$$\mu(u) = C_1 + C_2\frac{x^3}{u - 1}, \tag{5.100}$$

where $u = r^3/R^3$ and $x = c/B$, as denoted in (5.14).

Writing the invariants given by (5.6) in the equivalent forms

$$I_1 = u^{-4/3} + 2u^{2/3}, \qquad I_2 = u^{4/3} + 2u^{-2/3}, \qquad I_3 = 1, \tag{5.101}$$

and substituting these in (5.8), implies

$$\beta_1 = 2\frac{\partial W}{\partial I_1} = \mu + \frac{d\mu}{du}\frac{du}{dI_1}(I_1 - 3)$$

$$= C_1 + \frac{3C_2}{4}\frac{x^3}{u-1}\left[\frac{4}{3} - \frac{2u^3 - 3u^{7/3} + u}{(u-1)^2(u+1)}\right]$$

$$(5.102)$$

and

$$\beta_{-1} = -2\frac{\partial W}{\partial I_2} = -\frac{d\mu}{du}\frac{du}{dI_2}(I_1 - 3)$$

$$= \frac{3C_2}{4}\frac{x^3}{u-1}\frac{2u^{7/3} - 3u^{5/3} + u^{1/3}}{(u-1)^2(u+1)}.$$

$$(5.103)$$

Hence,

$$\beta_1 - \beta_{-1}\frac{r^2}{R^2} = C_1 + \frac{3C_2}{2}\frac{x^3}{u-1}\left[\frac{2}{3} - \frac{2u^3 - 3u^{7/3} + u}{(u-1)^2(u+1)}\right].$$

$$(5.104)$$

After calculations similar to those performed in Sect. 5.1, but with $\beta_1 - \beta_{-1}\frac{r^2}{R^2}$ instead of μ in (5.13), (5.15), and (5.16), we denote

$$H(x) = \frac{4}{3\rho B^2}\int_0^x\left[\zeta^2\int_{\zeta^3+1}^{\infty}\left(\beta_1 - \beta_{-1}\frac{r^2}{R^2}\right)\frac{1+u}{u^{7/3}}du\right]d\zeta.$$

$$(5.105)$$

Equivalently, by (5.104),

$$H(x) = \frac{4C_1}{3\rho B^2}\int_0^x\left(\zeta^2\int_{\zeta^3+1}^{\infty}\frac{1+u}{u^{7/3}}du\right)d\zeta$$

$$+ \frac{2C_2}{\rho B^2}\int_0^x\left\{\zeta^5\int_{\zeta^3+1}^{\infty}\frac{1+u}{u^{7/3}(u-1)}\left[\frac{2}{3} - \frac{2u^3 - 3u^{7/3} + u}{(u-1)^2(u+1)}\right]du\right\}d\zeta$$

$$= \frac{C_1}{\rho B^2}\left[2\left(x^3+1\right)^{2/3} - \frac{1}{\left(x^3+1\right)^{1/3}} - 1\right]$$

$$- \frac{C_2}{\rho B^2}\left[\sqrt{3}\arctan\frac{2\left(x^3+1\right)^{1/3}+1}{\sqrt{3}} - \frac{\pi}{\sqrt{3}}\right]$$

$$+ \frac{3C_2}{2\rho B^2}\left\{\ln\left[\left(x^3+1\right)^{2/3} + \left(x^3+1\right)^{1/3} + 1\right] - \ln 3\right\}$$

$$- \frac{C_2}{\rho B^2}\left[\left(x^3+1\right)^{2/3} + \frac{1}{\left(x^3+1\right)^{1/3}} - 2\right].$$

$$(5.106)$$

Then, by setting the dead load (5.18) and substituting (5.106) in (5.24), we obtain

$$
p_0 \left[\left(x^3 + 1 \right)^{1/3} - 1 \right] = \frac{C_1}{2} \left[2 \left(x^3 + 1 \right)^{2/3} - \frac{1}{\left(x^3 + 1 \right)^{1/3}} - 1 \right]
$$
$$
- \frac{C_2}{2} \left[\sqrt{3} \arctan \frac{2 \left(x^3 + 1 \right)^{1/3} + 1}{\sqrt{3}} - \frac{\pi}{\sqrt{3}} \right]
$$
$$
+ \frac{3C_2}{4} \left\{ \ln \left[\left(x^3 + 1 \right)^{2/3} + \left(x^3 + 1 \right)^{1/3} + 1 \right] - \ln 3 \right\}
$$
$$
- \frac{C_2}{2} \left[\left(x^3 + 1 \right)^{2/3} + \frac{1}{\left(x^3 + 1 \right)^{1/3}} - 2 \right].
$$

(5.107)

Equation (5.107) has a solution at $x_1 = 0$, while the second solution, $x_2 > 0$, is a root of

$$
p_0 = \frac{C_1}{2} \left[2 \left(x^3 + 1 \right)^{1/3} + \frac{1}{\left(x^3 + 1 \right)^{1/3}} + 2 \right]
$$
$$
- \frac{C_2}{2} \left[\left(x^3 + 1 \right)^{1/3} - \frac{1}{\left(x^3 + 1 \right)^{1/3}} \right].
$$

(5.108)

To obtain the right-hand side of the above equation, we have assumed that, in (5.107), x is sufficiently small, such that, after expanding the respective functions to the second order in $(x^3 + 1)^{1/3}$,

$$
\left[\sqrt{3} \arctan \frac{2 \left(x^3 + 1 \right)^{1/3} + 1}{\sqrt{3}} - \frac{\pi}{\sqrt{3}} \right] \left[\left(x^3 + 1 \right)^{1/3} - 1 \right]^{-1} \approx \frac{3 - \left(x^3 + 1 \right)^{1/3}}{4}
$$

(5.109)

and

$$
\frac{\ln \left[\left(x^3 + 1 \right)^{2/3} + \left(x^3 + 1 \right)^{1/3} + 1 \right] - \ln 3}{\left(x^3 + 1 \right)^{1/3} - 1} \approx \frac{7 - \left(x^3 + 1 \right)^{1/3}}{6}.
$$

(5.110)

Next, expanding the right-hand side of the Eq. (5.108) to the first order in x^3 gives

$$
p_0 \approx \frac{5C_1}{2} + \frac{x^3}{6} \left(C_1 - 2C_2 \right).
$$

(5.111)

The critical dead load for the onset of cavitation is then

$$
\lim_{x \to 0_+} p_0 = \frac{5C_1}{2}
$$

(5.112)

and is the same as for the homogeneous sphere made entirely from the material found at the centre of the inhomogeneous sphere.

Considering the parameters C_1 and C_2, we have the following two cases:

(i) When $C_1 > 2C_2$, the right-hand side in (5.111) is an increasing function of x, and therefore,

$$p_0 > \frac{5C_1}{2}. \tag{5.113}$$

(ii) When $C_1 < 2C_2$, the right-hand side in (5.111) decreases as x increases; hence,

$$0 < p_0 < \frac{5C_1}{2}. \tag{5.114}$$

However, since

$$\frac{2p_0}{\rho B^2}\left[\left(x^3 + 1\right)^{1/3} - 1\right] - H(x) \begin{cases} < 0 \text{ if } x_1 < x < x_2, \\ > 0 \text{ if } x > x_2, \end{cases} \tag{5.115}$$

the sphere cannot oscillate.

In Fig. 5.15, we illustrate the stochastic function $H(x)$, defined by (5.106), intersecting the (dashed red) curve $\frac{2p_0}{\rho B^2}\left[\left(x^3 + 1\right)^{1/3} - 1\right]$, with $p_0 = 15$, to find the two distinct solutions to (5.24), and the associated velocity, given by (5.22), for a unit sphere (with $B = 1$) of stochastic radially inhomogeneous neo-Hookean-like material when the random field shear modulus $\mu(R)$ is given by (5.96), with

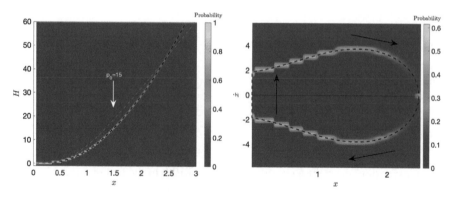

Fig. 5.15 The function $H(x)$, defined by (5.106), intersecting the (dashed red) curve $\frac{2p_0}{\rho B^2}\left[\left(x^3 + 1\right)^{1/3} - 1\right]$, with $p_0 = 15$ (left), and the associated velocity, given by (5.22) (right), for a dynamic sphere of stochastic radially inhomogeneous neo-Hookean-like material under dead-load traction, when $\rho = 1$, $B = 1$, and, for any fixed R, $\mu(R)$, given by (5.96), follows a Gamma distribution with $\rho_1 = 405/R^6$ and $\rho_2 = 0.01 \cdot R^6$. The black lines correspond to the expected values based only on mean parameter values [374]

$\rho_1 = 405/R^6$ and $\rho_2 = 0.01 \cdot R^6$ (see Fig. 5.14). In particular, $\rho_1 = 405$ and $\rho_2 = 0.01$ at $R = B$, hence, by (5.96), $C_1 = \mu = \rho_1\rho_2 = 4.05$ and $C_2 = \mu(B) - C_1$. For these functions, although the mean values, represented by black dashed lines, are the same as for those depicted in Fig. 5.3, their stochastic behaviours are quite different compared to the case of a stochastic neo-Hookean sphere, specifically, as we observe from.

5.3.2 Static Deformation of Stochastic Radially Inhomogeneous Spheres Under Dead-Load Traction

For the static inhomogeneous sphere, when the cavity surface is traction-free, at the outer surface, by analogy to (5.36), and using (5.104), we have

$$
T_{rr}(b) = \frac{2}{3} \int_{x^3+1}^{\infty} \left(\beta_1 - \beta_{-1} \frac{r^2}{R^2} \right) \frac{1+u}{u^{7/3}} du
$$

$$
= \frac{2C_1}{3} \int_{x^3+1}^{\infty} \frac{1+u}{u^{7/3}} du + C_2 x^3 \int_{x^3+1}^{\infty} \frac{1+u}{u^{7/3}(u-1)} \left[\frac{2}{3} - \frac{2u^3 - 3u^{7/3} + u}{(u-1)^2(u+1)} \right] du.
$$
(5.116)

Equivalently, after evaluating the integrals,

$$
T_{rr}(b) = 2C_1 \left[\frac{1}{(x^3+1)^{1/3}} + \frac{1}{4(x^3+1)^{4/3}} \right]
$$

$$
- \frac{C_2}{2} x^3 \frac{2(x^3+1)^{4/3} + 4(x^3+1) + 3(x^3+1)^{2/3} + 2(x^3+1)^{1/3} + 1}{(x^3+1)^{8/3} + 2(x^3+1)^{7/3} + 3(x^3+1)^2 + 2(x^3+1)^{5/3} + (x^3+1)^{4/3}}.
$$
(5.117)

After multiplying by $(x^3+1)^{2/3}$ the above Cauchy stress, and denoting $\alpha_b = b/B = (x^3+1)^{1/3}$, the required tensile dead load at the outer surface, $R = B$, in the reference configuration, takes the form

$$
P = 2C_1 \left(\alpha_b + \frac{1}{4\alpha_b^2} \right) - \frac{C_2}{2} \left(\alpha_b^3 - 1 \right) \frac{2\alpha_b^4 + 4\alpha_b^3 + 3\alpha_b^2 + 2\alpha_b + 1}{\alpha_b^6 + 2\alpha_b^5 + 3\alpha_b^4 + 2\alpha_b^3 + \alpha_b^2}.
$$
(5.118)

The critical dead-load traction for the initiation of cavitation is

$$
P_0 = \lim_{\alpha_b \to 1} P = \frac{5C_1}{2} = \frac{5\mu(0)}{2}
$$
(5.119)

and is the same as for the homogeneous sphere made entirely from the material found at its centre [498], as found also for the dynamic sphere.

To examine the post-cavitation behaviour, we take the first derivative of P, given by (5.118), with respect to α_b,

$$
\frac{\mathrm{d}P}{\mathrm{d}\alpha_b} = 2C_1 \left(1 - \frac{1}{2\alpha_b^3} \right)
$$

$$
- \frac{3C_2}{2} \frac{2\alpha_b^4 + 4\alpha_b^3 + 3\alpha_b^2 + 2\alpha_b + 1}{\alpha_b^4 + 2\alpha_b^3 + 3\alpha_b^2 + 2\alpha_b + 1}
$$

$$
- C_2 \left(\alpha_b^3 - 1 \right) \frac{4\alpha_b^3 + 6\alpha_b^2 + 3\alpha_b + 1}{\alpha_b^6 + 2\alpha_b^5 + 3\alpha_b^4 + 2\alpha_b^3 + \alpha_b^2}
$$

$$
- C_2 \left(\alpha_b^3 - 1 \right) \frac{\left(2\alpha_b^4 + 4\alpha_b^3 + 3\alpha_b^2 + 2\alpha_b + 1\right)\left(3\alpha_b^5 + 5\alpha_b^4 + 6\alpha_b^3 + 3\alpha_b^2 + \alpha_b\right)}{\left(\alpha_b^6 + 2\alpha_b^5 + 3\alpha_b^4 + 2\alpha_b^3 + \alpha_b^2\right)^2}.
$$

$$(5.120)$$

Letting $\alpha_b \to 1$ in (5.120), we obtain

$$
\lim_{\alpha_b \to 1} \frac{\mathrm{d}P}{\mathrm{d}\alpha_b} = C_1 - 2C_2. \tag{5.121}
$$

In Fig. 5.16, we represent the scaled dead load, P/C_2, with P given by (5.118) as a function of α_b, and its derivative with respect to α_b, for different values of the ratio C_1/C_2.

The following two types of cavitated solution are now possible:

(i) If $C_1 > 2C_2$, or equivalently, if $\mu(B) = C_1 + C_2 < 3C_1/2$, then cavitation is stable.

(ii) If $C_1 < 2C_2$, or equivalently, if $\mu(B) = C_1 + C_2 > 3C_1/2$, then snap cavitation is obtained.

Thus, the probability of stable cavitation (and also of oscillatory motion for the dynamic sphere) is equal to (see the example in Fig. 5.17)

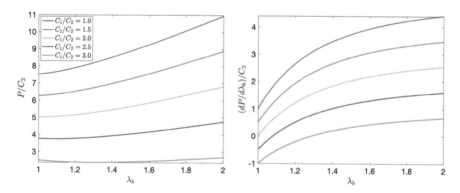

Fig. 5.16 Examples of the scaled dead load, P/C_2 (left), with P given by (5.118), and its derivative with respect to α_b (right), for a radially inhomogeneous sphere with different values of the ratio C_1/C_2 [374]

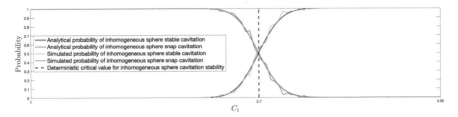

Fig. 5.17 Probability distributions of stable or unstable cavitation for a static stochastic sphere of radially inhomogeneous neo-Hookean-like material, under dead-load traction, when $\mu(B) = C_1 + C_2$, given by (5.96), follows a Gamma distribution with $\rho_1^{(1)} = 405$ and $\rho_2^{(1)} = 0.01$. Dark coloured lines represent analytically derived solutions, given by Eqs. (5.122) and (5.123), whereas the lighter versions represent stochastically generated data. The vertical black line at the critical value $C_1 = 2\underline{\mu}/3 = 2.7$ separates the expected regions based only on mean value, $\underline{\mu} = \rho_1 \rho_2 = 4.05$ [374]

Fig. 5.18 Probability distribution of the applied dead-load traction, P, causing cavitation of radius c in a radially inhomogeneous sphere, when $\rho = 1$, $B = 1$, and, for any fixed R, $\mu(R)$, given by (5.96), follows a Gamma distribution with $\rho_1 = 405/R^6$ and $\rho_2 = 0.01 \cdot R^6$. The black line corresponds to the expected bifurcation based only on mean value, $\underline{\mu} = C_1 = 4.05$ [374]

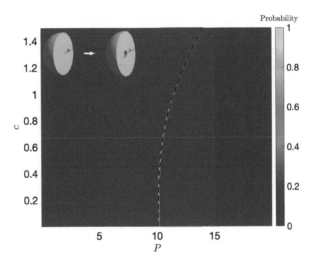

$$P_1(C_1) = \int_0^{3C_1/2} g(u; \rho_1, \rho_2)du, \tag{5.122}$$

and that of snap cavitation (and also of non-oscillatory motion for the dynamic sphere) is

$$P_2(C_1) = 1 - P_1(C_1) = 1 - \int_0^{3C_1/2} g(u; \rho_1, \rho_2)du. \tag{5.123}$$

An example of post-cavitation stochastic behaviour of the static inhomogeneous unit sphere (with $B = 1$) is shown in Fig. 5.18, for the case when the shear modulus $\mu(R)$ is given by (5.96), with $\rho_1 = 405/R^6$ and $\rho_2 = 0.01 \cdot R^6$ (see Fig. 5.14).

Thus, $\rho_1 = 405$ and $\rho_2 = 0.01$ at $R = B$, and by (5.96), $C_1 = \underline{\mu} = \rho_1 \rho_2 = 4.05$ and $C_2 = \mu(B) - C_1$.

5.3.3 Alternative Modelling of Radially Inhomogeneous Spheres

In the foregoing analysis, by following the approach of [508, 522], we have assumed that, for the inhomogeneous shear modulus $\mu(R)$, the mean value $\underline{\mu}(R) = \underline{\mu}$ is independent of R. Then, the shear modulus can be described by (5.96), where $\underline{C}_1$ is a deterministic constant and $\underline{C}_2$ is a (shifted) Gamma-distributed random variable with zero mean value. In this case, at the centre of the sphere, the shear modulus is a deterministic constant, while throughout the rest of the sphere, this modulus has constant mean value and non-constant variance [463].

An alternative model, where the mean value of the shear modulus varies throughout the sphere (see also [508]), can be constructed by taking C_2 as a Gamma-distributed random variable (with non-zero mean value, i.e., $\underline{C}_2 \neq 0$), while C_1 remains a deterministic constant. Then, by (5.96), for any fixed R, $\mu(R)$ is a shifted Gamma-distributed random variable, with the mean value and variance, respectively, equal to

$$\underline{\mu}(R) = C_1 + \underline{C}_2 \frac{R^3}{B^3}, \qquad \mathrm{Var}[\mu(R)] = \mathrm{Var}[C_2] \frac{R^6}{B^6}. \qquad (5.124)$$

A more general modelling approach, where the shear modulus is characterised by a probability distribution at the centre of the sphere as well, is to assume that C_1 and C_2 are described by independent Gamma distributions, with hyperparameters $\rho_1^{(1)}$, $\rho_2^{(1)}$, and $\rho_1^{(2)}$, $\rho_2^{(2)}$, respectively. Then, for any fixed R, the distribution of $\mu(R)$ is a linear combination of two Gamma distributions [346, 387] (see Appendix B.5), while at the centre of the sphere, the shear modulus $\mu(0) = C_1$ is Gamma-distributed.

In all of these cases, the analysis follows the exposition of Sects. 5.3.1 and 5.3.2, but the numerical results will differ in each case.

For the two alternative formulations, the probability of stable cavitation is equal to (see Fig. 5.19)

$$P_1(C_1) = \int_0^{C_1/2} g(u; \rho_1^{(2)}, \rho_2^{(2)})\mathrm{d}u, \qquad (5.125)$$

and that of snap cavitation is

$$P_2(C_1) = 1 - P_1(C_1) = 1 - \int_0^{C_1/2} g(u; \rho_1^{(2)}, \rho_2^{(2)})\mathrm{d}u. \qquad (5.126)$$

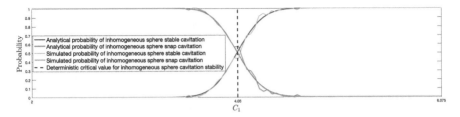

Fig. 5.19 Probability distributions of stable or unstable cavitation for a static stochastic sphere of radially inhomogeneous neo-Hookean-like material under dead-load traction, with $\mu(R) = C_1 + C_2 R^3/B^3$ given by (5.96), where C_2 follows a Gamma distribution with $\rho_1^{(2)} = 405$ and $\rho_2^{(2)} = 0.05$, while C_1 may be either deterministic or Gamma-distributed. Dark coloured lines represent analytically derived solutions, given by Eqs. (5.125) and (5.126), whereas the lighter versions represent stochastically generated data. The vertical black line at the critical value $C_1 = 2\underline{C}_2 = 4.05$ separates the expected regions based only on mean value, $\underline{C}_2 = \rho_1 \rho_2 = 2.025$ [374]

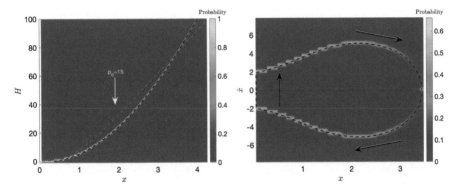

Fig. 5.20 The function $H(x)$, defined by (5.106), intersecting the (dashed red) curve $\frac{2p_0}{\rho B^2}\left[(x^3+1)^{1/3} - 1\right]$, with $p_0 = 15$ (left), and the associated velocity, given by (5.22) (right), for a dynamic sphere of stochastic radially inhomogeneous neo-Hookean-like material under dead-load traction, when $\rho = 1$, $B = 1$, and $\mu(R)$ is given by (5.96), where $C_1 = 4.05$ and C_2 follows a Gamma distribution with $\rho_1^{(2)} = 405$ and $\rho_2^{(2)} = 0.005$. The black lines correspond to the expected values based only on mean parameter values [374]

If $C_1 > 2C_2$, then the probability distribution of oscillatory motion occurring is given by (5.34) and that of non-oscillatory motion by (5.35), with C_1 instead of μ and $g(u; \rho_1^{(1)}, \rho_2^{(1)})$ instead of $g(u; \rho_1, \rho_2)$ (see Fig. 5.2).

For illustration, we choose a unit sphere with shear modulus $\mu(R)$ given by (5.96), where C_2 follows a Gamma distribution with shape and scale parameters $\rho_1^{(2)} = 405$ and $\rho_2^{(2)} = 0.05$, respectively, while C_1 is either the deterministic constant $C_1 = 4.05$, or a Gamma-distributed random variable, with hyperparameters $\rho_1^{(1)} = 405$ and $\rho_2^{(1)} = 0.01$. For the dynamic sphere, the associated function $H(x)$,

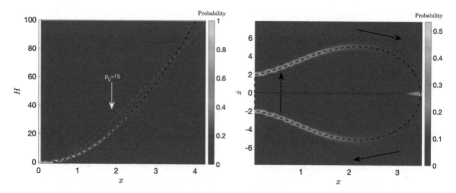

Fig. 5.21 The function $H(x)$, defined by (5.106), intersecting the (dashed red) curve $\frac{2p_0}{\rho B^2}\left[\left(x^3+1\right)^{1/3}-1\right]$, with $p_0 = 15$ (left), and the associated velocity, given by (5.22) (right), for a dynamic sphere of stochastic radially inhomogeneous neo-Hookean-like material under dead-load traction, when $\rho = 1$, $B = 1$, and $\mu(R)$ is given by (5.96), where C_1 follows a Gamma distribution with $\rho_1^{(1)} = 405$ and $\rho_2^{(1)} = 0.01$, while C_2 follows a Gamma distributions with $\rho_1^{(2)} = 405$ and $\rho_2^{(2)} = 0.005$. The black lines correspond to the expected values based only on mean parameter values [374]

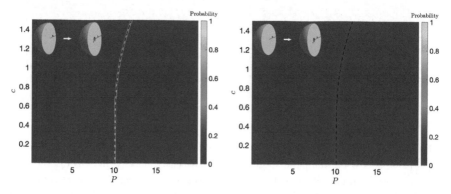

Fig. 5.22 Probability distribution of the applied dead-load traction, P, causing cavitation of radius c in a radially inhomogeneous sphere, when $\rho = 1$, $B = 1$, and $\mu(R)$ is given by (5.96), where C_2 follows a Gamma distribution with $\rho_1^{(2)} = 405$ and $\rho_2^{(2)} = 0.05$, while $C_1 = 4.05$ (left), or C_1 follows a Gamma distribution with $\rho_1^{(1)} = 405$ and $\rho_2^{(1)} = 0.01$ (right). The black line corresponds to the expected bifurcation based only on mean value parameters [374]

defined by (5.106), and the velocity, given by (5.22), are represented in Figs. 5.20 and 5.21, respectively. The corresponding post-cavitation behaviour of the static sphere is shown in Fig. 5.22.

5.3.4 Non-oscillatory Motion of Stochastic Radially Inhomogeneous Spheres Under Impulse Traction

We now consider the cavitation of radially inhomogeneous, incompressible spheres of stochastic hyperelastic material with the shear modulus of the form (5.96), such that, for any fixed R, both the mean value, μ, and variance, $\mathrm{Var}[\mu]$, of $\mu = \mu(R)$ may depend on R. Setting the initial conditions $x_0 = x(0) = 0$ and $\dot{x}_0 = \dot{x}(0) = 0$, we obtain Eq. (5.41), where $H(x)$ is given by (5.106). Then, substitution of (5.106) in (5.43), together with (5.40), implies

$$
\begin{aligned}
\frac{2T}{3}x^3 = {} & C_1 \left[2\left(x^3 + 1\right)^{2/3} - \frac{1}{\left(x^3 + 1\right)^{1/3}} - 1 \right] \\
& - C_2 \left[\sqrt{3}\arctan \frac{2\left(x^3 + 1\right)^{1/3} + 1}{\sqrt{3}} - \frac{\pi}{\sqrt{3}} \right] \\
& + \frac{3C_2}{2} \left\{ \ln \left[\left(x^3 + 1\right)^{2/3} + \left(x^3 + 1\right)^{1/3} + 1 \right] - \ln 3 \right\} \\
& - C_2 \left[\left(x^3 + 1\right)^{2/3} + \frac{1}{\left(x^3 + 1\right)^{1/3}} - 2 \right].
\end{aligned}
\tag{5.127}
$$

Equation (5.127) has a solution at $x_1 = 0$, while the second solution, $x_2 > 0$, is a root of

$$
\begin{aligned}
T = {} & \frac{3C_1}{2} \left[\frac{\left(x^3 + 1\right)^{1/3} + 1}{\left(x^3 + 1\right)^{2/3} + \left(x^3 + 1\right)^{1/3} + 1} + \frac{1}{\left(x^3 + 1\right)^{1/3}} \right] \\
& - \frac{C_2}{6}\left(x^3 - 3\right) - \frac{3C_2}{2} \left[2\frac{\left(x^3 + 1\right)^{1/3} + 1}{\left(x^3 + 1\right)^{2/3} + \left(x^3 + 1\right)^{1/3} + 1} - \frac{1}{\left(x^3 + 1\right)^{1/3}} \right].
\end{aligned}
\tag{5.128}
$$

To obtain the right-hand side of the above equation, we have assumed that, in (5.127), x is sufficiently small such that after expanding the respective functions to the second order in x^3,

$$
\left[\sqrt{3}\arctan \frac{2\left(x^3 + 1\right)^{1/3} + 1}{\sqrt{3}} - \frac{\pi}{\sqrt{3}} \right] \frac{1}{x^3} \approx \frac{2 - x^3}{12}
\tag{5.129}
$$

and

$$
\frac{\ln \left[\left(x^3 + 1\right)^{2/3} + \left(x^3 + 1\right)^{1/3} + 1 \right] - \ln 3}{x^3} \approx \frac{18 - 7x^3}{54}.
\tag{5.130}
$$

Next, expanding the right-hand side of the Eq. (5.128) to first order in x^3 gives

$$T \approx \frac{5C_1}{2} - \frac{x^3}{3}(2C_1 + C_2).$$

(5.131)

The critical dead load for the onset of cavitation is then

$$T_0 = \lim_{x \to 0_+} T = \frac{5C_1}{2}$$

(5.132)

and is the same as for the homogeneous sphere made entirely from the material found at the centre of the inhomogeneous sphere.

Noting that $2C_1 + C_2 = \mu(0) + \mu(B) > 0$, the right-hand side in (5.131) decreases as x increases; hence,

$$0 < T < \frac{5C_1}{2},$$

(5.133)

or equivalently, by (5.40),

$$0 < p_0 < \frac{5C_1}{\rho B^2}.$$

(5.134)

However,

$$\frac{2p_0}{\rho B^2}x^3 - H(x) \begin{cases} < 0 \text{ if } x_1 < x < x_2, \\ > 0 \text{ if } x > x_2, \end{cases}$$

(5.135)

and the sphere cannot oscillate.

5.3.5 Static Deformation of Stochastic Radially Inhomogeneous Spheres Under Impulse Traction

For the static inhomogeneous sphere, when the cavity surface is traction-free, at the outer surface, the tensile traction takes the form

$$T = 2C_1 \left[\frac{1}{\left(x^3 + 1\right)^{1/3}} + \frac{1}{4\left(x^3 + 1\right)^{4/3}} \right]$$
$$- \frac{C_2}{2}x^3 \frac{2\left(x^3 + 1\right)^{4/3} + 4\left(x^3 + 1\right) + 3\left(x^3 + 1\right)^{2/3} + 2\left(x^3 + 1\right)^{1/3} + 1}{\left(x^3 + 1\right)^{8/3} + 2\left(x^3 + 1\right)^{7/3} + 3\left(x^3 + 1\right)^2 + 2\left(x^3 + 1\right)^{5/3} + \left(x^3 + 1\right)^{4/3}}.$$

(5.136)

Denoting $\alpha_b = b/B = \left(x^3 + 1\right)^{1/3}$, the required tensile load at the outer surface, $R = B$, can be written equivalently as follows:

$$T = 2C_1 \left(\frac{1}{\alpha_b} + \frac{1}{4\alpha_b^4}\right) - \frac{C_2}{2}\left(\alpha_b^3 - 1\right)\frac{2\alpha_b^4 + 4\alpha_b^3 + 3\alpha_b^2 + 2\alpha_b + 1}{\alpha_b^8 + 2\alpha_b^7 + 3\alpha_b^6 + 2\alpha_b^5 + \alpha_b^4}. \tag{5.137}$$

Thus the critical load for the initiation of cavitation is

$$T_0 = \lim_{\alpha_b \to 1} T = \frac{5C_1}{2} = \frac{5\mu(0)}{2} \tag{5.138}$$

and is the same as for the homogeneous sphere made entirely from the material found at its centre, as found also for the dynamic sphere.

To examine the post-cavitation behaviour, we take the first derivative of T, given by (5.137), with respect to α_b,

$$\frac{dT}{d\alpha_b} = -2C_1\left(\frac{1}{\alpha_b^2} + \frac{1}{\alpha_b^5}\right)$$

$$- \frac{3C_2}{2}\frac{2\alpha_b^4 + 4\alpha_b^3 + 3\alpha_b^2 + 2\alpha_b + 1}{\alpha_b^6 + 2\alpha_b^5 + 3\alpha_b^4 + 2\alpha_b^3 + \alpha_b^2}$$

$$- C_2\left(\alpha_b^3 - 1\right)\frac{4\alpha_b^3 + 6\alpha_b^2 + 3\alpha_b + 1}{\alpha_b^8 + 2\alpha_b^7 + 3\alpha_b^6 + 2\alpha_b^5 + \alpha_b^4}$$

$$- C_2\left(\alpha_b^3 - 1\right)\frac{\left(2\alpha_b^4 + 4\alpha_b^3 + 3\alpha_b^2 + 2\alpha_b + 1\right)\left(4\alpha_b^7 + 7\alpha_b^6 + 9\alpha_b^5 + 5\alpha_b^4 + 2\alpha_b^3\right)}{\left(\alpha_b^8 + 2\alpha_b^7 + 3\alpha_b^6 + 2\alpha_b^5 + \alpha_b^4\right)^2}. \tag{5.139}$$

Letting $\alpha_b \to 1$ in (5.139), we obtain

$$\lim_{\alpha_b \to 1}\frac{dT}{d\alpha_b} = -4C_1 - 2C_2 = -2C_1 - 2\left(C_1 + C_2\right) = -2\mu(0) - 2\mu(B) < 0. \tag{5.140}$$

Hence, only snap cavitation can occur.

In Figs. 5.23 and 5.24, we show examples of post-cavitation stochastic behaviour of the static inhomogeneous unit sphere (with $B = 1$).

5.4   Oscillatory Motion of Circular Cylindrical Tubes

In this section, we analyse the stability and finite amplitude oscillations of a stochastic hyperelastic cylindrical tube of stochastic neo-Hookean or Mooney–Rivlin-type material, subject to the combined radial and axial quasi-equilibrated dynamic deformation [358].

Fig. 5.23 Probability distribution of the applied impulse traction, T, causing cavitation of radius c in a radially inhomogeneous sphere, when $\rho = 1$, $B = 1$, and, for any fixed R, $\mu(R)$, given by (5.96), follows a Gamma distribution with $\rho_1 = 405/R^6$ and $\rho_2 = 0.01 \cdot R^6$. The black line corresponds to the expected bifurcation based only on mean value, $\underline{\mu} = C_1 = 4.05$ [374]

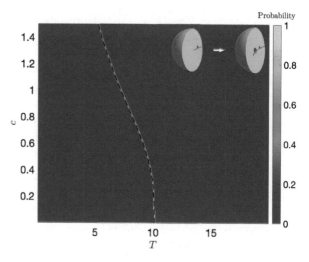

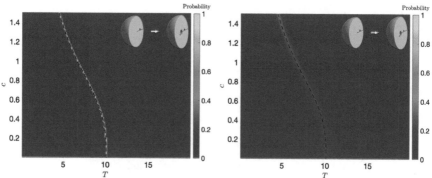

Fig. 5.24 Probability distribution of the applied impulse traction, T, causing cavitation of radius c in a radially inhomogeneous sphere, when $\rho = 1$, $B = 1$, and $\mu(R)$ is given by (5.96), where C_2 follows a Gamma distribution with $\rho_1^{(2)} = 405$ and $\rho_2^{(2)} = 0.05$, while $C_1 = 4.05$ (left), or C_1 follows a Gamma distribution with $\rho_1^{(1)} = 405$ and $\rho_2^{(1)} = 0.01$ (right). The black line corresponds to the expected bifurcation based only on mean value parameters [374]

For a circular cylindrical tube, the combined radial and axial motion is described by (see Fig. 5.25)

$$r^2 = a^2 + \frac{R^2 - A^2}{\alpha}, \qquad \theta = \Theta, \qquad z = \alpha Z, \qquad (5.141)$$

where (R, Θ, Z) and $(r, \theta, z) = (r(t), \theta(t), z(t))$, with t the time variable, are the cylindrical polar coordinates in the reference and current configurations, respectively, such that $A \leq R \leq B$, A and B are the inner and outer radii in the undeformed state, respectively, $a = a(t)$ and $b = b(t) = \sqrt{a^2 + \left(B^2 - A^2\right)/\alpha}$ are

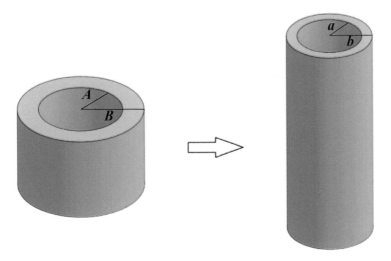

Fig. 5.25 Schematic of inflation of a cylindrical tube, showing the reference state, with inner radius A and outer radius B (left), and the deformed state, with inner radius a and outer radius b (right), respectively

the inner and outer radii at time t, respectively, and $\alpha > 0$ is a given constant (when $\alpha < 0$, the tube is everted, so that the inner surface becomes the outer surface). When $\alpha = 1$, the time-dependent deformation (5.141) simplifies to that studied also in [51, 301, 302]. The case when α is time-dependent was considered in [488].

The radial–axial motion (5.141) of the cylindrical tube is fully determined by the inner radius a at time t, which in turn is obtained from the initial conditions. Thus, the acceleration $\ddot{r}$ can be computed in terms of the acceleration $\ddot{a}$ on the inner surface. By the governing equations (5.141), the condition (2.168) is valid for $\mathbf{x} = (r, \theta, z)^T$ (see Appendix B.2). Hence, (5.141) describes a quasi-equilibrated motion, such that

$$-\frac{\partial \xi}{\partial r} = \ddot{r} = \frac{\dot{a}^2}{r} + \frac{a\ddot{a}}{r} - \frac{a^2 \dot{a}^2}{r^3}, \tag{5.142}$$

where ξ is the acceleration potential satisfying (2.167). Integrating (5.142) implies [558, p. 215]

$$-\xi = \dot{a}^2 \log r + a\ddot{a} \log r + \frac{a^2 \dot{a}^2}{2r^2} = \dot{r}^2 \log r + r\ddot{r} \log r + \frac{1}{2}\dot{r}^2. \tag{5.143}$$

For the deformation (5.141), the gradient tensor with respect to the polar coordinates (R, Θ, Z) is equal to

$$\mathbf{A} = \text{diag}\left(\frac{R}{\alpha r}, \frac{r}{R}, \alpha\right), \tag{5.144}$$

the Cauchy–Green deformation tensor is

$$\mathbf{B} = \mathbf{A}^2 = \mathrm{diag}\left(\frac{R^2}{\alpha^2 r^2}, \frac{r^2}{R^2}, \alpha^2\right), \tag{5.145}$$

and the principal invariants take the form

$$
\begin{aligned}
I_1 &= \mathrm{tr}\,(\mathbf{B}) = \frac{R^2}{\alpha^2 r^2} + \frac{r^2}{R^2} + \alpha^2, \\
I_2 &= \frac{1}{2}\left[(\mathrm{tr}\,\mathbf{B})^2 - \mathrm{tr}\left(\mathbf{B}^2\right)\right] = \frac{\alpha^2 r^2}{R^2} + \frac{R^2}{r^2} + \frac{1}{\alpha^2}, \\
I_3 &= \det \mathbf{B} = 1.
\end{aligned}
\tag{5.146}
$$

Thus, the principal components of the equilibrium Cauchy stress tensor at time t are

$$
\begin{aligned}
T_{rr}^{(0)} &= -p^{(0)} + \beta_1 \frac{R^2}{\alpha^2 r^2} + \beta_{-1}\frac{\alpha^2 r^2}{R^2}, \\
T_{\theta\theta}^{(0)} &= T_{rr}^{(0)} + \left(\beta_1 - \beta_{-1}\alpha^2\right)\left(\frac{r^2}{R^2} - \frac{R^2}{\alpha^2 r^2}\right), \\
T_{zz}^{(0)} &= T_{rr}^{(0)} + \left(\beta_1 - \beta_{-1}\frac{r^2}{R^2}\right)\left(\alpha^2 - \frac{R^2}{\alpha^2 r^2}\right),
\end{aligned}
\tag{5.147}
$$

where $p^{(0)}$ is the Lagrange multiplier for the incompressibility constraint ($I_3 = 1$), and

$$\beta_1 = 2\frac{\partial W}{\partial I_1}, \qquad \beta_{-1} = -2\frac{\partial W}{\partial I_2} \tag{5.148}$$

are the nonlinear material parameters, with I_1 and I_2 given by (5.146).

As the stress components depend only on the radius r, the system of equilibrium equations reduces to

$$\frac{\partial T_{rr}^{(0)}}{\partial r} = \frac{T_{\theta\theta}^{(0)} - T_{rr}^{(0)}}{r}. \tag{5.149}$$

Hence, by (5.147) and (5.149), the radial Cauchy stress for the equilibrium state at time t is equal to

$$T_{rr}^{(0)}(r, t) = \int \left(\beta_1 - \beta_{-1}\alpha^2\right)\left(\frac{r^2}{R^2} - \frac{R^2}{\alpha^2 r^2}\right)\frac{dr}{r} + \psi(t), \tag{5.150}$$

where $\psi = \psi(t)$ is an arbitrary function of time. Substitution of (5.143) and (5.150) into (2.169) then gives the principal Cauchy stress components at time t as follows:

$$T_{rr}(r,t) = \rho \left(a\ddot{a}\log r + \dot{a}^2 \log r + \frac{a^2\dot{a}^2}{2r^2} \right) + \int \left(\beta_1 - \beta_{-1}\alpha^2 \right) \left(\frac{r^2}{R^2} - \frac{R^2}{\alpha^2 r^2} \right) \frac{dr}{r} + \psi(t),$$

$$T_{\theta\theta}(r,t) = T_{rr}(r,t) + \left(\beta_1 - \beta_{-1}\alpha^2 \right) \left(\frac{r^2}{R^2} - \frac{R^2}{\alpha^2 r^2} \right),$$

$$T_{zz}(r,t) = T_{rr}(r,t) + \left(\beta_1 - \beta_{-1}\frac{r^2}{R^2} \right) \left(\alpha^2 - \frac{R^2}{\alpha^2 r^2} \right).$$

$$(5.151)$$

In (5.151), the function $\beta_1 - \beta_{-1}\alpha^2$ can be interpreted as the following nonlinear shear modulus, corresponding to the combined deformation of simple shear superposed on axial stretch, described by (2.98), with shear parameter $k = \sqrt{\alpha^2 R^2/r^2 + \alpha^4 r^2/R^2 - \alpha^6 - 1}$ and stretch parameter α [362],

$$\widetilde{\mu} = \beta_1 - \beta_{-1}\alpha^2. \tag{5.152}$$

As shown in [362], this modulus is positive if the BE inequalities (2.52) hold. In this case, the integrand is negative for $0 < r^2/R^2 < 1/\alpha$ and positive for $r^2/R^2 > 1/\alpha$. Using the first equation in (5.141), it is straightforward to show that $0 < r^2/R^2 < 1/\alpha$ (respectively, $r^2/R^2 > 1/\alpha$) is equivalent to $0 < a^2/A^2 < 1/\alpha$ (respectively, $a^2/A^2 > 1/\alpha$). When $\alpha = 1$, the modulus given by (5.152) coincides with the generalised shear modulus defined in [558, p. 174] and also in [51].

In the limiting case when $\alpha \to 1$ and $k \to 0$, the nonlinear shear modulus defined by (5.152) converges to the classical shear modulus from the infinitesimal theory [362],

$$\mu = \lim_{\alpha \to 1} \lim_{k \to 0} \widetilde{\mu}. \tag{5.153}$$

In this case, as $R^2/r^2 \to 1$, the three stress components defined by (5.151) are equal.

Next, for the cylindrical tube deformed by (5.141), we set the inner and outer radial pressures acting on the curvilinear surfaces $r = a(t)$ and $r = b(t)$ at time t (measured per unit area in the present configuration), as $T_1(t)$ and $T_2(t)$, respectively [558, pp. 214-217]. Evaluating $T_1(t) = -T_{rr}(a,t)$ and $T_2(t) = -T_{rr}(b,t)$, using (5.151), with $r = a$ and $r = b$, respectively, and then subtracting the results, gives

$$T_1(t) - T_2(t) = \frac{\rho}{2}\left[\left(a\ddot{a} + \dot{a}^2\right)\log\frac{b^2}{a^2} + \dot{a}^2\left(\frac{a^2}{b^2} - 1\right)\right] + \int_a^b \widetilde{\mu}\left(\frac{r^2}{R^2} - \frac{R^2}{\alpha^2 r^2}\right)\frac{dr}{r}$$

$$= \frac{\rho A^2}{2}\left[\left(\frac{a}{A}\frac{\ddot{a}}{A} + \frac{\dot{a}^2}{A^2}\right)\log\frac{b^2}{a^2} + \frac{\dot{a}^2}{A^2}\left(\frac{a^2}{b^2} - 1\right)\right] + \int_a^b \widetilde{\mu}\left(\frac{r^2}{R^2} - \frac{R^2}{\alpha^2 r^2}\right)\frac{dr}{r}.$$

$$(5.154)$$

Setting the notation

$$u = \frac{r^2}{R^2} = \frac{r^2}{\alpha\left(r^2 - a^2\right) + A^2}, \qquad x = \frac{a}{A}, \qquad \gamma = \frac{B^2}{A^2} - 1, \qquad (5.155)$$

we can rewrite

$$\left(\frac{a}{A}\frac{\ddot{a}}{A} + \frac{\dot{a}^2}{A^2}\right) \log \frac{b^2}{a^2} + \frac{\dot{a}^2}{A^2}\left(\frac{a^2}{b^2} - 1\right) = \left(\ddot{x}x + \dot{x}^2\right) \log\left(1 + \frac{\gamma}{\alpha x^2}\right) - \dot{x}^2 \frac{\frac{\gamma}{\alpha x^2}}{1 + \frac{\gamma}{\alpha x^2}}$$

$$= \frac{1}{2x}\frac{d}{dx}\left[\dot{x}^2 x^2 \log\left(1 + \frac{\gamma}{\alpha x^2}\right)\right] \qquad (5.156)$$

and

$$\int_a^b \tilde{\mu}\left(\frac{r^2}{R^2} - \frac{R^2}{\alpha^2 r^2}\right)\frac{dr}{r} = \int_a^b \tilde{\mu}\left[\frac{r^2}{\alpha\left(r^2 - a^2\right) + A^2} - \frac{\alpha\left(r^2 - a^2\right) + A^2}{\alpha^2 r^2}\right]\frac{dr}{r}$$

$$= \frac{1}{2}\int_{x^2 + \frac{\gamma}{\alpha}}^{x^2} \tilde{\mu}\frac{1 + \alpha u}{\alpha^2 u^2} du. \qquad (5.157)$$

To obtain the above identity, the following relations were used:

$$r = \left[\frac{u(A^2 - \alpha a^2)}{1 - \alpha u}\right]^{1/2}, \qquad (5.158)$$

$$\frac{dr}{du} = \frac{A^2 - \alpha a^2}{2\left(1 - \alpha u\right)^2}\left[\frac{u(A^2 - \alpha a^2)}{1 - \alpha u}\right]^{-1/2} = \frac{r}{2u(1 - \alpha u)}, \qquad (5.159)$$

$$\left(u - \frac{1}{\alpha^2 u}\right)\frac{1}{2u(1 - \alpha u)} = \frac{\alpha^2 u^2 - 1}{2\alpha^2 u^2(1 - \alpha u)} = -\frac{1 + \alpha u}{2\alpha^2 u^2}. \qquad (5.160)$$

We can now express Eq. (5.154) equivalently as follows:

$$2x\frac{T_1(t) - T_2(t)}{\rho A^2} = \frac{1}{2}\frac{d}{dx}\left[\dot{x}^2 x^2 \log\left(1 + \frac{\gamma}{\alpha x^2}\right)\right] + \frac{x}{\rho A^2}\int_{x^2 + \frac{\gamma}{\alpha}}^{x^2} \tilde{\mu}\frac{1 + \alpha u}{\alpha^2 u^2} du. \qquad (5.161)$$

Note that, when the BE inequalities (2.52) hold, $\tilde{\mu} > 0$, and the integral in (5.154), or equivalently in (5.161), is negative if $0 < u < 1/\alpha$ (i.e., if $0 < x < 1/\sqrt{\alpha}$) and positive if $u > 1/\alpha$ (i.e., if $x > 1/\sqrt{\alpha}$).

In the static case, where $\dot{a} = 0$ and $\ddot{a} = 0$, (5.154) becomes

$$T_1(t) - T_2(t) = \int_a^b \tilde{\mu}\left(\frac{r^2}{R^2} - \frac{R^2}{\alpha^2 r^2}\right)\frac{dr}{r}, \qquad (5.162)$$

and (5.161) reduces to

$$2\frac{T_1(t) - T_2(t)}{\rho A^2} = \frac{1}{\rho A^2} \int_{\frac{x^2 + \frac{\gamma}{\alpha}}{1+\gamma}}^{x^2} \tilde{\mu} \frac{1 + \alpha u}{\alpha^2 u^2} du. \tag{5.163}$$

For the cylindrical tube in finite dynamic deformation, we define the function

$$G(x, \gamma) = \frac{1}{\rho A^2} \int_{\frac{1}{\sqrt{\alpha}}}^{x} \left(\zeta \int_{\frac{\zeta^2 + \frac{\gamma}{\alpha}}{1+\gamma}}^{\zeta^2} \tilde{\mu} \frac{1 + \alpha u}{\alpha^2 u^2} du \right) d\zeta. \tag{5.164}$$

This function is useful in establishing whether the radial motion is oscillatory or not.

5.4.1   Tubes of Stochastic Mooney–Rivlin Material Under Impulse Traction

For a cylindrical tube of stochastic Mooney–Rivlin material defined by (4.49), with $\mu = \mu_1 + \mu_2 > 0$, evaluating the integral in (5.164) gives [358]

$$G(x, \gamma) = \frac{\tilde{\mu}}{2\alpha \rho A^2} \left(x^2 - \frac{1}{\alpha} \right) \log \frac{1 + \gamma}{1 + \frac{\gamma}{\alpha x^2}}, \tag{5.165}$$

where $\tilde{\mu} = \mu_1 + \mu_2 \alpha^2$. Then, $G(x, \gamma)$ is monotonically decreasing when $0 < x < 1/\sqrt{\alpha}$ and increasing when $x > 1/\sqrt{\alpha}$.

Next, we set the pressure impulse

$$2\alpha \frac{T_1(t) - T_2(t)}{\rho A^2} = \begin{cases} 0 & \text{if } t \leq 0, \\ p_0 & \text{if } t > 0, \end{cases} \tag{5.166}$$

where p_0 is constant in time. Then, integrating (5.161) once gives

$$\frac{1}{2} \dot{x}^2 x^2 \log \left(1 + \frac{\gamma}{\alpha x^2} \right) + G(x, \gamma) = \frac{p_0}{2\alpha} \left(x^2 - \frac{1}{\alpha} \right) + C_0, \tag{5.167}$$

with $G(x, \gamma)$ defined by (5.164) and

$$C_0 = \frac{1}{2} \dot{x}_0^2 x_0^2 \log \left(1 + \frac{\gamma}{\alpha x_0^2} \right) + G(x_0, \gamma) - \frac{p_0}{2\alpha} \left(x_0^2 - \frac{1}{\alpha} \right), \tag{5.168}$$

where $x(0) = x_0$ and $\dot{x}(0) = \dot{x}_0$ are the initial conditions. By (5.167), the velocity is equal to

$$\dot{x} = \pm \sqrt{\frac{\frac{p_0}{\alpha}\left(x^2 - \frac{1}{\alpha}\right) + 2C_0 - 2G(x, \gamma)}{x^2 \log\left(1 + \frac{\gamma}{\alpha x^2}\right)}}. \tag{5.169}$$

The radial motion is periodic if and only if the following equation

$$G(x, \gamma) = \frac{p_0}{2\alpha}\left(x^2 - \frac{1}{\alpha}\right) + C_0 \tag{5.170}$$

has exactly two distinct solutions, representing the amplitudes of the oscillation, $x = x_1$ and $x = x_2$, such that $0 < x_1 < x_2 < \infty$. Then, by (5.155), the minimum and maximum radii of the inner surface in the oscillation are equal to $x_1 A$ and $x_2 A$, respectively, and by (5.169), the period of oscillation is equal to

$$T = 2 \left| \int_{x_1}^{x_2} \frac{dx}{\dot{x}} \right| = 2 \left| \int_{x_1}^{x_2} \sqrt{\frac{x^2 \log\left(1 + \frac{\gamma}{\alpha x^2}\right)}{\frac{p_0}{\alpha}\left(x^2 - \frac{1}{\alpha}\right) + 2C_0 - 2G(x, \gamma)}} dx \right|. \tag{5.171}$$

For the stochastic tube, the amplitudes and period of the oscillation are random variables defined by probability distributions.

Assuming that the nonlinear shear modulus $\tilde{\mu}$ has a uniform lower bound, i.e.,

$$\tilde{\mu} > \mu_0, \tag{5.172}$$

for some constant $\mu_0 > 0$, it follows that

$$\lim_{x \to 0} G(x, \gamma) = \lim_{x \to \infty} G(x, \gamma) = \infty. \tag{5.173}$$

(i) If $p_0 = 0$ and $C_0 > 0$, then Eq. (5.170) has exactly two solutions, $x = x_1$ and $x = x_2$, satisfying $0 < x_1 < 1/\sqrt{\alpha} < x_2 < \infty$, for any positive constant C_0 (see Fig. 5.26, for example). However, these oscillations are not "free" in general, since, by (5.151), if $T_{rr}(r, t) = 0$ at $r = a$ and $r = b$, so that $T_1(t) = T_2(t) = 0$, then $T_{\theta\theta}(r, t) \neq 0$ and $T_{zz}(r, t) \neq 0$ at $r = a$ and $r = b$, unless $\alpha \to 1$ and $r^2/R^2 \to 1$ [488].

In Fig. 5.26, we represent the stochastic function $G(x, \gamma)$, defined by (5.165), intersecting the line $C = 10$, to solve Eq. (5.170) when $p_0 = 0$, and the associated velocity, given by (5.169), assuming that $\alpha = 1$, $\rho = 1$, $A = 1$, $\gamma = 1$, and μ follows the Gamma distribution with hyperparameters $\rho_1 = 405$ and $\rho_2 = 0.01$.

– For a thin-walled tube [301, 490], where $0 < \gamma \ll 1$ and $\alpha = 1$, Eq. (5.167) takes the form

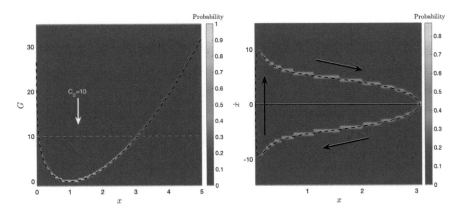

Fig. 5.26 The function $G(x, \gamma)$, defined by (5.165), intersecting the (dashed red) line $C_0 = 10$ when $p_0 = 0$ (left), and the associated velocity, given by (5.169) (right), for a cylindrical tube of stochastic Mooney–Rivlin material when $\alpha = 1$, $\rho = 1$, $A = 1$, $\gamma = 1$, and $\widetilde{\mu} = \mu = \mu_1 + \mu_2$ is drawn from the Gamma distribution with $\rho_1 = 405$ and $\rho_2 = 0.01$. The black lines correspond to the expected values based only on mean value, $\underline{\mu} = \rho_1 \rho_2 = 4.05$ [358]

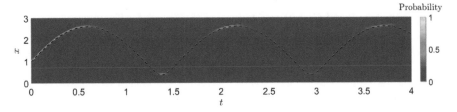

Fig. 5.27 Stochastic solution given by (5.175), with the initial conditions $x_0 = 1$ and $\dot{x}_0 = 4.5$, for a thin-walled tube, where $\rho = 1$, $A = 1$, and μ is drawn from the Gamma distribution with $\rho_1 = 405$ and $\rho_2 = 0.01$. The black line corresponds to the expected values based only on mean value, $\underline{\mu} = \rho_1 \rho_2 = 4.05$ [358]

$$\dot{x}^2 + \frac{\mu}{\rho A^2}\left(x^2 + \frac{1}{x^2}\right) = \dot{x}_0^2 + \frac{\mu}{\rho A^2}\left(x_0^2 + \frac{1}{x_0^2}\right) \tag{5.174}$$

and has the explicit solution (see Fig. 5.27) [490]

$$x = \sqrt{\left[x_0 \cos\left(\frac{t}{A}\sqrt{\frac{\mu}{\rho}}\right) + \dot{x}_0 A\sqrt{\frac{\rho}{\mu}} \sin\left(\frac{t}{A}\sqrt{\frac{\mu}{\rho}}\right)\right]^2 + \frac{1}{x_0^2}\sin^2\left(\frac{t}{A}\sqrt{\frac{\mu}{\rho}}\right)}. \tag{5.175}$$

In this case, assuming that the shear modulus, μ, has a uniform lower bound, Eq. (5.170) becomes [301]

$$x^2 + \frac{1}{x^2} = \frac{\rho A^2}{\mu}\dot{x}_0^2 + x_0^2 + \frac{1}{x_0^2}.$$

(5.176)

This equation can be solved directly to find the amplitudes

$$x_{1,2} = \sqrt{\frac{\frac{\rho A^2}{\mu}\dot{x}_0^2 + x_0^2 + \frac{1}{x_0^2} \pm \sqrt{\left(\frac{\rho A^2}{\mu}\dot{x}_0^2 + x_0^2 + \frac{1}{x_0^2}\right)^2 - 4}}{2}}.$$

(5.177)

Noting that $x_2 = 1/x_1$, the period of the oscillations can be calculated as

$$T = 2\sqrt{\frac{\rho A^2}{\mu}} \left| \int_{x_1}^{1/x_1} \frac{dx}{\sqrt{\frac{\rho A^2}{\mu}\dot{x}_0^2 + x_0^2 + \frac{1}{x_0^2} - x^2 - \frac{1}{x^2}}} \right| = \pi A\sqrt{\frac{\rho}{\mu}}.$$

(5.178)

(ii) When $p_0 \neq 0$ and $C_0 \geq 0$, substitution of (5.165) in (5.170) gives

$$p_0 = \frac{\tilde{\mu}}{\rho A^2}\log\frac{1+\gamma}{1+\frac{\gamma}{\alpha x^2}} - \frac{2\alpha C_0}{x^2 - \frac{1}{\alpha}}.$$

(5.179)

As the right-hand side of the above equation is a monotonically increasing function of x, there exists a unique positive x satisfying (5.179) if and only if the following condition holds:

$$\lim_{x\to 0}\left(\frac{\tilde{\mu}}{\rho A^2}\log\frac{1+\gamma}{1+\frac{\gamma}{\alpha x^2}} - \frac{2\alpha C_0}{x^2 - \frac{1}{\alpha}}\right) < p_0 < \lim_{x\to\infty}\left(\frac{\tilde{\mu}}{\rho A^2}\log\frac{1+\gamma}{1+\frac{\gamma}{\alpha x^2}} - \frac{2\alpha C_0}{x^2 - \frac{1}{\alpha}}\right),$$

that is,

$$-\infty < p_0 < \frac{\tilde{\mu}}{\rho A^2}\log(1+\gamma).$$

(5.180)

Then, by (5.155), (5.166), and (5.180), the necessary and sufficient condition that oscillatory motion occurs is that the nonlinear shear modulus, $\tilde{\mu}$, is uniformly bounded from below as follows:

$$\tilde{\mu} > \frac{p_0\rho A^2}{\log(1+\gamma)} = \alpha\frac{T_1(t) - T_2(t)}{\log B - \log A}.$$

(5.181)

By (5.152),

$$\tilde{\mu} = \mu_1 + \mu_2 \alpha^2 = \mu_1 + (\mu - \mu_1)\alpha^2 = \mu\alpha^2 + \mu_1\left(1 - \alpha^2\right).$$

Hence, (5.181) is equivalent to

$$\mu > \frac{p_0 \rho A^2}{\alpha^2 \log(1 + \gamma)} + \mu_1 \frac{1 - \alpha^2}{\alpha^2}. \tag{5.182}$$

Then, the probability distribution of oscillatory motion occurring is

$$P_1(p_0) = 1 - \int_0^{\frac{p_0 \rho A^2}{\alpha^2 \log(1+\gamma)} + \mu_1 \frac{1-\alpha^2}{\alpha^2}} g(u; \rho_1, \rho_2) du, \tag{5.183}$$

and that of non-oscillatory motion is

$$P_2(p_0) = 1 - P_1(p_0) = \int_0^{\frac{p_0 \rho A^2}{\alpha^2 \log(1+\gamma)} + \mu_1 \frac{1-\alpha^2}{\alpha^2}} g(u; \rho_1, \rho_2) du, \tag{5.184}$$

where $g(u; \rho_1, \rho_2)$ is the Gamma probability density function describing μ.

For example, when $\alpha = 1$, $\rho = 1$, $A = 1$, $\gamma = 1$, and $\tilde{\mu} = \mu = \mu_1 + \mu_2$ satisfies the Gamma distribution with $\rho_1 = 405$ and $\rho_2 = 0.01$, the probability distributions given by (5.183) and (5.184) are shown in Fig. 5.28 (blue lines for P_1 and red lines for P_2). Specifically, $(0, \mu)$, where $\mu = \rho_1\rho_2 = 4.05$ is the mean value of μ, was divided into 100 steps; then for each value of p_0, 100 random values of μ were numerically generated from the specified Gamma distribution and compared with the inequalities defining the two intervals for values of p_0. For the deterministic elastic tube, the critical value $p_0 = \mu \log 2 \approx 2.8072$ strictly divides the cases of oscillations occurring or not. For the stochastic problem, for the same critical value,

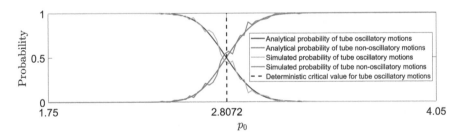

Fig. 5.28 Probability distributions of whether oscillatory motion can occur or not for a cylindrical tube of stochastic Mooney–Rivlin material, with $\alpha = 1$, $\rho = 1$, $A = 1$, $\gamma = 1$, and the shear modulus, μ, following the Gamma distribution with $\rho_1 = 405$ and $\rho_2 = 0.01$. Dark coloured lines represent analytically derived solutions, given by Eq. (5.183) and (5.184), whereas the lighter versions represent stochastically generated data. The vertical black line at the critical value $p_0 = 2.8072$ separates the expected regions based only on mean value, $\mu = \rho_1\rho_2 = 4.05$ [358]

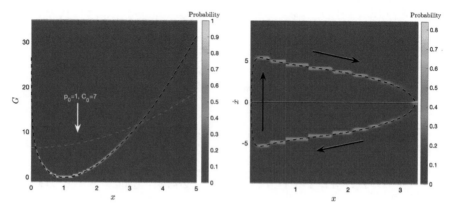

Fig. 5.29 The function $G(x, \gamma)$, defined by (5.165), intersecting the (dashed red) curve $\frac{p_0}{2\alpha}\left(x^2 - \frac{1}{\alpha}\right) + C_0$, with $p_0 = 1$ and $C_0 = 7$, (left), and the associated velocity, given by (5.169) (right), for a cylindrical tube of stochastic Mooney–Rivlin material when $\alpha = 1$, $\rho = 1$, $A = 1$, $\gamma = 1$, and $\widetilde{\mu} = \mu = \mu_1 + \mu_2$ is drawn from the Gamma distribution with $\rho_1 = 405$ and $\rho_2 = 0.01$. The black lines correspond to the expected values based only on mean value, $\underline{\mu} = \rho_1 \rho_2 = 4.05$ [358]

there is, by definition, exactly 50% chance of that the motion is oscillatory, and 50% chance that is not. To increase the probability of oscillatory motion ($P_1 \approx 1$), one must apply a sufficiently small impulse, p_0, below the expected critical point, whereas a non-oscillatory motion is certain to occur ($P_2 \approx 1$) if p_0 is sufficiently large. However, the inherent variability in the probabilistic system means that there will also exist events where there is competition between the two cases.

In Fig. 5.29, we illustrate the stochastic function $G(x, \gamma)$, defined by (5.165), intersecting the curve $p_0 \left(x^2 - 1/\alpha\right)/(2\alpha) + C$, with $p_0 = 1$ and $C = 7$, to find the solutions of Eq. (5.170), and the associated velocity, given by (5.169), assuming that $\alpha = 1$, $\rho = 1$, $A = 1$, $\gamma = 1$, and μ satisfies the Gamma distribution with $\rho_1 = 405$ and $\rho_2 = 0.01$.

When $C_0 = 0$, Eq. (5.179) can be solved explicitly to find the amplitude

$$x_1 = \sqrt{\frac{\gamma/\alpha}{(1 + \gamma)\exp\left[-\left(p_0 \rho A^2\right)/(\widetilde{\mu})\right] - 1}} = \sqrt{\frac{\left(B^2 - A^2\right)/\alpha}{B^2 \exp\left[-2\alpha\left(P_1 - P_2\right)/\widetilde{\mu}\right] - A^2}}. \tag{5.185}$$

Note that, in the static case, by (5.163) and (5.166), at $x = x_1$, the required pressure takes the form

$$p_0^{(s)} = \frac{\widetilde{\mu}}{\alpha x^2 \rho A^2} \frac{\gamma - \frac{\gamma}{\alpha x^2}}{1 + \frac{\gamma}{\alpha x^2}} + \frac{\widetilde{\mu}}{\rho A^2} \log \frac{1 + \gamma}{1 + \frac{\gamma}{\alpha x^2}}. \tag{5.186}$$

Thus, the difference between the applied pressure in the static and dynamic case, given by (5.186) and (5.179), with $C = 0$, respectively, is

$$p_0^{(s)} - p_0 = \frac{\tilde{\mu}}{\alpha x^2 \rho A^2} \frac{\gamma - \frac{\gamma}{\alpha x^2}}{1 + \frac{\gamma}{\alpha x^2}}. \tag{5.187}$$

Hence, $p_0^{(s)} < p_0$ if $0 < x_1 < \sqrt{\alpha}$, and $p_0^{(s)} > p_0$ if $x_1 > \sqrt{\alpha}$.

– If the tube wall is thin [302, 490], then $0 < \gamma \ll 1$ and $\alpha = 1$, and (5.179) becomes

$$\frac{p_0}{\gamma} = \frac{\mu}{\rho A^2}\left(1 - \frac{1}{x^2}\right) - \frac{2C_0}{x^2 - 1}. \tag{5.188}$$

In this case, the necessary and sufficient condition that oscillatory motion occurs is that

$$-\infty = \lim_{x \to 0}\left[\frac{\mu}{\rho A^2}\left(1 - \frac{1}{x^2}\right) - \frac{2C_0}{x^2 - 1}\right] < \frac{p_0}{\gamma} < \lim_{x \to \infty}\left[\frac{\mu}{\rho A^2}\left(1 - \frac{1}{x^2}\right) - \frac{2C_0}{x^2 - 1}\right] = \frac{\mu}{\rho A^2}. \tag{5.189}$$

Thus, for the motion to be oscillatory, the shear modulus must be bounded from below as follows:

$$\mu > \frac{p_0}{\gamma}\rho A^2 = \frac{2}{\gamma}(T_1(t) - T_2(t)). \tag{5.190}$$

Then, the probability distribution of oscillatory motion occurring is

$$P_1(p_0/\gamma) = 1 - \int_0^{\frac{p_0}{\gamma}\rho A^2} g(u; \rho_1, \rho_2)du, \tag{5.191}$$

and that of non-oscillatory motion is

$$P_2(p_0/\gamma) = 1 - P_1(p_0/\gamma) = \int_0^{\frac{p_0}{\gamma}\rho A^2} g(u; \rho_1, \rho_2)du. \tag{5.192}$$

For $\rho = 1$, $A = 1$, and $\tilde{\mu} = \mu = \mu_1 + \mu_2$ drawn from the Gamma distribution with $\rho_1 = 405$ and $\rho_2 = 0.01$, the probability distributions given by (5.191) and (5.192) are shown in Fig. 5.30 (blue lines for P_1 and red lines for P_2). For the deterministic thin-walled tube, the critical value $p_0/\gamma = \mu = 4.05$ strictly separates the cases of oscillations occurring or not. However, for the stochastic tube, the two cases compete.

If $C_0 = 0$, then setting $x_0 = 1$ and $\dot{x}_0 = 0$, the equation of motion has the explicit solution [490]

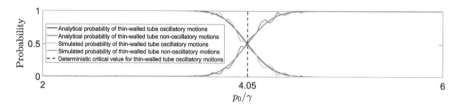

Fig. 5.30 Probability distributions of whether oscillatory motion can occur or not for a thin-walled cylindrical tube of stochastic Mooney–Rivlin material, with $\rho = 1$, $A = 1$, and the shear modulus, μ, following the Gamma distribution with $\rho_1 = 405$ and $\rho_2 = 0.01$. Dark coloured lines represent analytically derived solutions, given by Eqs. (5.183)–(5.184), whereas the lighter versions represent stochastically generated data. The vertical black line at the critical value $p_0/\gamma = 4.05$ separates the expected regions based only on mean value, $\underline{\mu} = \rho_1\rho_2 = 4.05$ [358]

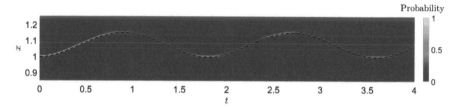

Fig. 5.31 Stochastic solution given by (5.193), with $p_0/\gamma = 1$, for a thin-walled tube, where $\rho = 1$, $A = 1$, and μ is drawn from the Gamma distribution with $\rho_1 = 405$ and $\rho_2 = 0.01$. The black line corresponds to the expected values based only on mean value, $\underline{\mu} = \rho_1\rho_2 = 4.05$ [358]

$$x = \sqrt{\frac{\frac{\mu}{\rho A^2} - \frac{p_0}{2\gamma}}{\frac{\mu}{\rho A^2} - \frac{p_0}{\gamma}} - \frac{\frac{p_0}{2\gamma}}{\frac{\mu}{\rho A^2} - \frac{p_0}{\gamma}} \cos\left(2t\sqrt{\frac{\mu}{\rho A^2} - \frac{p_0}{\gamma}}\right)}. \qquad (5.193)$$

In Fig. 5.31, we illustrate the stochastic solution given by (5.193), with $p_0/\gamma = 1$, assuming that $\rho = 1$, $A = 1$, and μ satisfies the Gamma distribution with hyperparameters $\rho_1 = 405$ and $\rho_2 = 0.01$.

Notably, the absence of the limit point instability does not imply non-oscillatory quasi-equilibrated motion, since circular cylindrical tubes of both neo-Hookean (with $\mu_2 = 0$) and Mooney–Rivlin material can oscillate, even though the inflation of neo-Hookean tubes is stable and that of Mooney–Rivlin tubes is unstable [357].

– If the tube wall is infinitely thick [488], then $\gamma \to \infty$, and assuming that the nonlinear shear modulus, $\tilde{\mu}$, has a uniform lower bound, (5.180) becomes

$$-\infty = \lim_{x \to 0}\left[\frac{\tilde{\mu}}{\rho A^2}\log\left(\alpha x^2\right) - \frac{2\alpha C_0}{x^2 - \frac{1}{\alpha}}\right] < p_0 < \lim_{x \to \infty}\left[\frac{\tilde{\mu}}{\rho A^2}\log\left(\alpha x^2\right) - \frac{2\alpha C_0}{x^2 - \frac{1}{\alpha}}\right] = \infty. \qquad (5.194)$$

Hence, the motion is always oscillatory for any value of the applied impulse.

5.4.2 Tubes with Two Stochastic Neo-Hookean Phases Under Impulse Traction

To study the behaviour under the quasi-equilibrated radial motion of a composite formed from two concentric homogeneous cylindrical tubes (see Fig. 5.32), we focus on composite tubes with two stochastic neo-Hookean phases [354], for which we define the following strain-energy function:

$$\mathcal{W}(\alpha_1, \alpha_2, \alpha_3) = \begin{cases} \frac{\mu^{(1)}}{2} \left(\alpha_1^2 + \alpha_2^2 + \alpha_3^2 - 3 \right), & C < R < A, \\ \frac{\mu^{(2)}}{2} \left(\alpha_1^2 + \alpha_2^2 + \alpha_3^2 - 3 \right), & A < R < B, \end{cases} \tag{5.195}$$

where $C < R < A$ and $A < R < B$ are the radii of the inner and outer tubes in the reference configuration, and the associated shear moduli $\mu^{(1)}$ and $\mu^{(2)}$ are spatially independent random variables characterised by the Gamma distributions $g(u; \rho_1^{(1)}, \rho_2^{(1)})$ and $g(u; \rho_1^{(2)}, \rho_2^{(2)})$, respectively.

For the deformed composite tube, we denote the radial pressures acting on the curvilinear surfaces $r = c(t)$ and $r = b(t)$ at time t as $T_1(t)$ and $T_2(t)$, respectively, and impose the continuity condition for the stress components across their interface, $r = a(t)$. Evaluating $T_1(t) = -T_{rr}(c, t)$ and $T_2(t) = -T_{rr}(b, t)$ at $r = c$ and $r = b$, respectively, and then subtracting the results, gives

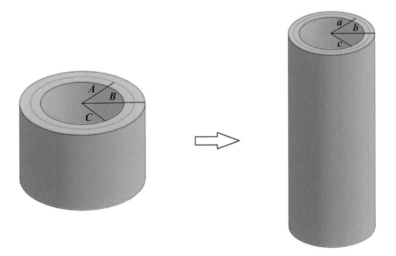

Fig. 5.32 Schematic of a composite cylindrical tube made of two concentric homogeneous tubes, with undeformed outer radii A and B, respectively, showing the reference state (left), and the deformed state, with outer radii a and b of the concentric tubes, respectively (right)

$$T_1(t) - T_2(t) = \frac{\rho}{2}\left[c\ddot{c}\log\frac{b^2}{c^2} + \dot{c}^2\log\frac{b^2}{c^2} + \dot{c}^2\left(\frac{c^2}{b^2} - 1\right)\right]$$
$$+ \int_c^a \mu^{(1)}\left(\frac{r^2}{R^2} - \frac{R^2}{\alpha^2 r^2}\right)\frac{dr}{r} + \int_a^b \mu^{(2)}\left(\frac{r^2}{R^2} - \frac{R^2}{\alpha^2 r^2}\right)\frac{dr}{r}.$$
$$= \frac{\rho C^2}{2}\left[\left(\frac{c}{C}\frac{\ddot{c}}{C} + \frac{\dot{c}^2}{C^2}\right)\log\frac{b^2}{c^2} + \frac{\dot{c}^2}{C^2}\left(\frac{c^2}{b^2} - 1\right)\right]$$
$$+ \int_c^a \mu^{(1)}\left(\frac{r^2}{R^2} - \frac{R^2}{\alpha^2 r^2}\right)\frac{dr}{r} + \int_a^b \mu^{(2)}\left(\frac{r^2}{R^2} - \frac{R^2}{\alpha^2 r^2}\right)\frac{dr}{r}.$$
$$\tag{5.196}$$

We set the notation

$$u = \frac{r^2}{R^2} = \frac{r^2}{\alpha\left(r^2 - c^2\right) + C^2}, \qquad x = \frac{c}{C}, \qquad \gamma = \frac{B^2}{C^2} - 1 \tag{5.197}$$

and rewrite

$$\left(\frac{c}{C}\frac{\ddot{c}}{C} + \frac{\dot{c}^2}{C^2}\right)\log\frac{b^2}{c^2} + \frac{\dot{c}^2}{C^2}\left(\frac{c^2}{b^2} - 1\right) = (\ddot{x}x + \dot{x}^2)\log\left(1 + \frac{\gamma}{\alpha x^2}\right) - \dot{x}^2\frac{\frac{\gamma}{\alpha x^2}}{1 + \frac{\gamma}{\alpha x^2}}$$
$$= \frac{1}{2}\frac{d}{dx}\left[\dot{x}^2 x \log\left(1 + \frac{\gamma}{\alpha x^2}\right)\right].$$
$$\tag{5.198}$$

By (5.141) and (5.197), we obtain

$$\int_c^a \mu^{(1)}\left(\frac{r^2}{R^2} - \frac{R^2}{\alpha^2 r^2}\right)\frac{dr}{r} + \int_a^b \mu^{(2)}\left(\frac{r^2}{R^2} - \frac{R^2}{\alpha^2 r^2}\right)\frac{dr}{r}$$
$$= \int_c^a \mu^{(1)}\left[\frac{r^2}{\alpha(r^2 - c^2) + C^2} - \frac{\alpha(r^2 - c^2) + C^2}{\alpha^2 r^2}\right]\frac{dr}{r}$$
$$+ \int_a^b \mu^{(2)}\left[\frac{r^2}{\alpha(r^2 - c^2) + C^2} - \frac{\alpha(r^2 - c^2) + C^2}{\alpha^2 r^2}\right]\frac{dr}{r} \tag{5.199}$$
$$= -\frac{1}{2}\int_{c^2/C^2}^{a^2/A^2} \mu^{(1)}\frac{1 + \alpha u}{\alpha^2 u^2}du - \frac{1}{2}\int_{a^2/A^2}^{b^2/B^2} \mu^{(2)}\frac{1 + \alpha u}{\alpha^2 u^2}du$$
$$= \frac{1}{2}\int_{\frac{C^2}{A^2}\left(x^2 - \frac{1}{\alpha}\right) + \frac{1}{\alpha}}^{x^2} \mu^{(1)}\frac{1 + \alpha u}{\alpha^2 u^2}du + \frac{1}{2}\int_{\frac{x^2 + \frac{\gamma}{\alpha}}{1 + \gamma}}^{\frac{C^2}{A^2}\left(x^2 - \frac{1}{\alpha}\right) + \frac{1}{\alpha}} \mu^{(2)}\frac{1 + \alpha u}{\alpha^2 u^2}du.$$

In the above calculations, the following relations were used,

$$r = \left[\frac{u(C^2 - \alpha c^2)}{1 - \alpha u}\right]^{1/2}, \tag{5.200}$$

$$\frac{dr}{du} = \frac{C^2 - \alpha c^2}{2\left(1 - \alpha u\right)^2} \left[\frac{u(C^2 - \alpha c^2)}{1 - \alpha u}\right]^{-1/2} = \frac{r}{2u(1 - \alpha u)}, \tag{5.201}$$

$$\left(u - \frac{1}{\alpha^2 u}\right) \frac{1}{2u(1 - \alpha u)} = \frac{\alpha^2 u^2 - 1}{2\alpha^2 u^2 (1 - \alpha u)} = -\frac{1 + \alpha u}{2\alpha^2 u^2}. \tag{5.202}$$

We then express Eq. (5.196) equivalently as follows:

$$2x \frac{T_1(t) - T_2(t)}{\rho C^2} = \frac{1}{2} \frac{d}{dx}\left[\dot{x}^2 x^2 \log\left(1 + \frac{\gamma}{\alpha x^2}\right)\right]$$

$$+ \frac{x}{\rho C^2} \int_{\frac{C^2}{A^2}\left(x^2 - \frac{1}{\alpha}\right) + \frac{1}{\alpha}}^{x^2} \mu^{(1)} \frac{1 + \alpha u}{\alpha^2 u^2} du \tag{5.203}$$

$$+ \frac{x}{\rho C^2} \int_{\frac{x^2 + \frac{\gamma}{\alpha}}{1 + \gamma}}^{\frac{C^2}{A^2}\left(x^2 - \frac{1}{\alpha}\right) + \frac{1}{\alpha}} \mu^{(2)} \frac{1 + \alpha u}{\alpha^2 u^2} du.$$

For the dynamic tube, we define

$$G(x, \gamma) = \frac{1}{\rho C^2} \int_{\frac{1}{\sqrt{\alpha}}}^{x} \zeta \left[\int_{\zeta^2 \frac{C^2}{A^2} + \frac{1}{\alpha}\left(1 - \frac{C^2}{A^2}\right)}^{\zeta^2} \mu^{(1)} \frac{1 + \alpha u}{\alpha^2 u^2} du\right] d\zeta$$

$$+ \frac{1}{\rho C^2} \int_{\frac{1}{\sqrt{\alpha}}}^{x} \zeta \left[\int_{\frac{\zeta^2 + \frac{\gamma}{\alpha}}{1 + \gamma}}^{\zeta^2 \frac{C^2}{A^2} + \frac{1}{\alpha}\left(1 - \frac{C^2}{A^2}\right)} \mu^{(2)} \frac{1 + \alpha u}{\alpha^2 u^2} du\right] d\zeta. \tag{5.204}$$

This function will be used to establish whether the radial motion of the tube is oscillatory or not.

Next, setting the pressure impulse (suddenly applied pressure difference)

$$2\alpha \frac{T_1(t) - T_2(t)}{\rho C^2} = \begin{cases} 0 & \text{if } t \le 0, \\ p_0 & \text{if } t > 0, \end{cases} \tag{5.205}$$

with p_0 constant in time, and integrating (5.203) imply

$$\frac{1}{2}\dot{x}^2 x^2 \log\left(1 + \frac{\gamma}{\alpha x^2}\right) + G(x, \gamma) = \frac{p_0}{2\alpha}\left(x^2 - \frac{1}{\alpha}\right) + C_0, \tag{5.206}$$

where $G(x, \gamma)$ is defined by (5.204) and

$$C_0 = \frac{1}{2}\dot{x}_0^2 x_0^2 \log\left(1 + \frac{\gamma}{\alpha x_0^2}\right) + G(x_0, \gamma) - \frac{p_0}{2\alpha}\left(x_0^2 - \frac{1}{\alpha}\right), \tag{5.207}$$

with the initial conditions $x(0) = x_0$ and $\dot{x}(0) = \dot{x}_0$. By (5.206), the velocity is

$$\dot{x} = \pm \sqrt{\frac{\frac{p_0}{\alpha}(x^2 - \frac{1}{\alpha}) + 2C_0 - 2G(x, \gamma)}{x^2 \log(1 + \frac{\gamma}{\alpha x^2})}}. \tag{5.208}$$

Then, the radial motion is periodic if and only if the equation

$$G(x, \gamma) = \frac{p_0}{2\alpha}\left(x^2 - \frac{1}{\alpha}\right) + C_0 \tag{5.209}$$

has exactly two distinct solutions, representing the amplitudes of the oscillation, $x = x_1$ and $x = x_2$, such that $0 < x_1 < x_2 < \infty$. By (5.197), the minimum and maximum radii of the inner surface in the oscillation are equal to $x_1 C$ and $x_2 C$, respectively, and by (5.208), the period of oscillation is equal to

$$T = 2\left|\int_{x_1}^{x_2} \frac{dx}{\dot{x}}\right| = 2\left|\int_{x_1}^{x_2} \sqrt{\frac{x^2 \log\left(1 + \frac{\gamma}{\alpha x^2}\right)}{\frac{p_0}{\alpha}\left(x^2 - \frac{1}{\alpha}\right) + 2C_0 - 2G(x, \gamma)}} \, dx\right|. \tag{5.210}$$

For the stochastic composite tube, the amplitude and period of the oscillation are random variables described by probability distributions.

To examine $G(x, \gamma)$ defined by (5.204), we rewrite this function equivalently as

$$G(x, \gamma) = G_1(x, \gamma) + G_2(x, \gamma), \tag{5.211}$$

where

$$G_1(x, \gamma) = \frac{1}{\rho C^2} \int_{\frac{1}{\sqrt{\alpha}}}^{x} \zeta \left[\int_{\zeta^2 \frac{C^2}{A^2} + \frac{1}{\alpha}\left(1 - \frac{C^2}{A^2}\right)}^{\zeta^2} \mu^{(1)} \frac{1 + \alpha u}{\alpha^2 u^2} \, du\right] d\zeta \tag{5.212}$$

and

$$G_2(x, \gamma) = \frac{1}{\rho C^2} \int_{\frac{1}{\sqrt{\alpha}}}^{x} \zeta \left[\int_{\zeta^2}^{\zeta^2 \frac{C^2}{A^2} + \frac{1}{\alpha}\left(1 - \frac{C^2}{A^2}\right)} \mu^{(2)} \frac{1 + \alpha u}{\alpha^2 u^2} \, du + \int_{\frac{\zeta^2 + \frac{\gamma}{\alpha}}{1+\gamma}}^{\zeta^2} \mu^{(2)} \frac{1 + \alpha u}{\alpha^2 u^2} \, du\right] d\zeta. \tag{5.213}$$

Proceeding as in [358], we obtain

$$G_1(x, \gamma) = \frac{\mu^{(1)}}{2\alpha \rho C^2}\left(\frac{1}{\alpha} - x^2\right)\log\left[\frac{C^2}{A^2} + \frac{1}{\alpha x^2}\left(1 - \frac{C^2}{A^2}\right)\right] \tag{5.214}$$

and

$$G_2(x, \gamma) = \frac{\mu^{(2)}}{2\alpha\rho C^2} \left(x^2 - \frac{1}{\alpha} \right) \left\{ \log \left[\frac{C^2}{A^2} + \frac{1}{\alpha x^2} \left(1 - \frac{C^2}{A^2} \right) \right] + \log \frac{1+\gamma}{1 + \frac{\gamma}{\alpha x^2}} \right\}.$$
$$(5.215)$$

Note that the double integral in (5.212) and the first double integral in (5.213) can be evaluated by simply replacing γ with $A^2/C^2 - 1$ in the calculations for $G(x, \gamma)$ in the Appendix of [358], while the second double integral in (5.213) is similar to that for $G(x, \gamma)$ in the Appendix of [358].

By (5.214) and (5.215), the function $G(x, \gamma)$ defined by (5.211) takes the form

$$G(x, \gamma) = \frac{1}{2\alpha\rho C^2} \left(x^2 - \frac{1}{\alpha} \right) \left\{ \left(\mu^{(2)} - \mu^{(1)} \right) \log \left[\frac{C^2}{A^2} + \frac{1}{\alpha x^2} \left(1 - \frac{C^2}{A^2} \right) \right] + \mu^{(2)} \log \frac{1+\gamma}{1 + \frac{\gamma}{\alpha x^2}} \right\}.$$
$$(5.216)$$

This function is monotonically decreasing for $0 < x < 1/\sqrt{\alpha}$ and increasing for $x > 1/\sqrt{\alpha}$. When $\mu^{(1)} = \mu^{(2)}$, the function for a homogeneous tube is recovered (see Appendix of [358]).

Assuming that the shear moduli $\mu^{(1)}$ and $\mu^{(2)}$ have a lower bound

$$\mu^{(1)} > \mu_0, \qquad \mu^{(2)} > \mu_0, \qquad\qquad (5.217)$$

for some constant $\mu_0 > 0$, it follows that

$$\lim_{x \to 0} G(x, \gamma) = \lim_{x \to \infty} G(x, \gamma) = \infty. \qquad\qquad (5.218)$$

We consider the following two cases:

(i) If $p_0 = 0$ and $C_0 > 0$, then Eq. (5.209) has exactly two solutions, $x = x_1$ and $x = x_2$, satisfying $0 < x_1 < 1/\sqrt{\alpha} < x_2 < \infty$, for any positive constant C_0. Note that these oscillations are not "free" in general, since, by (5.151), if $T_{rr}(r, t) = 0$ at $r = c$ and $r = b$, so that $T_1(t) = T_2(t) = 0$, then $T_{\theta\theta}(r, t) \neq 0$ and $T_{zz}(r, t) \neq 0$ at $r = c$ and $r = b$, unless $\alpha \to 1$ and $r^2/R^2 \to 1$ [488]. In Fig. 5.33, we represent the stochastic function $G(x, \gamma)$, defined by (5.216), intersecting the line $C_0 = 7$, and the associated velocity, given by (5.208), when $\alpha = 1$, $\rho = 1$, and the shear modulus of the inner phase, $\mu^{(1)}$, follows a Gamma distribution with $\rho_1^{(1)} = 405$ and $\rho_2^{(1)} = 4.05/\rho_1^{(1)} = 0.01$, while the shear modulus of the outer phase, $\mu^{(2)}$, is drawn from a Gamma distribution with $\rho_1^{(2)} = 405$ and $\rho_2^{(2)} = 4.2/\rho_1^{(2)}$.

(ii) When $p_0 \neq 0$ and $C_0 \geq 0$, substitution of (5.216) in (5.209) gives

$$p_0 = \frac{2\alpha (G - C_0)}{x^2 - \frac{1}{\alpha}}. \qquad\qquad (5.219)$$

The right-hand side of the above equation is a monotonically increasing function of x, implying that there exists a unique positive x satisfying (5.219)

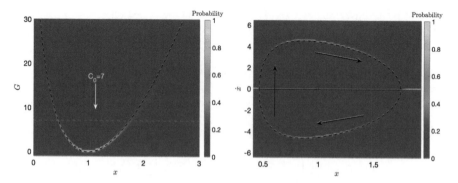

Fig. 5.33 The function $G(x, \gamma)$, defined by (5.216), intersecting the (dashed red) line $C_0 = 7$ when $p_0 = 0$ (left), and the associated velocity, given by (5.208) (right), for a dynamic composite tube with two concentric stochastic neo-Hookean phases, with inner radii $A = 1$ and $C = 1/2$, respectively, under dead-load traction, assuming that $\alpha = 1$, $\rho = 1$, and $\mu^{(1)}$ follows a Gamma distribution with $\rho_1^{(1)} = 405$ and $\rho_2^{(1)} = 4.05/\rho_1^{(1)} = 0.01$, while $\mu^{(2)}$ is drawn from a Gamma distribution with $\rho_1^{(2)} = 405$ and $\rho_2^{(2)} = 4.2/\rho_1^{(2)}$. The dashed black lines correspond to the expected values based only on mean values, $\underline{\mu}^{(1)} = 4.05$ and $\underline{\mu}^{(2)} = 4.2$. Each distribution was calculated from the average of 1000 stochastic simulations

if and only if the following condition holds:

$$-\infty = \lim_{x \to 0} \frac{2\alpha (G - C_0)}{x^2 - \frac{1}{\alpha}} < p_0 < \lim_{x \to \infty} \frac{2\alpha (G - C_0)}{x^2 - \frac{1}{\alpha}} = \frac{\mu^{(2)} - \mu^{(1)}}{\rho C^2} \log \frac{C^2}{A^2} + \frac{\mu^{(2)}}{\rho C^2} \log (1 + \gamma).$$

$$(5.220)$$

By (5.197) and (5.220), the necessary and sufficient condition that oscillatory motion occurs is that

$$-\infty < p_0 < \frac{\mu^{(1)}}{\rho C^2} \log \frac{A^2}{C^2} + \frac{\mu^{(2)}}{\rho C^2} \log \frac{B^2}{A^2}, \qquad (5.221)$$

where $\mu^{(1)}$ and $\mu^{(2)}$ are described by the Gamma probability density functions $g(u; \rho_1^{(1)}, \rho_2^{(1)})$ and $g(u; \rho_1^{(2)}, \rho_2^{(2)})$, respectively. After rescaling, $\mu^{(1)} \log \frac{A^2}{C^2}$ follows the Gamma distribution with shape parameter $\rho_1^{(1)}$ and scale parameter $\rho_2^{(1)} \log \frac{A^2}{C^2}$ and $\mu^{(2)} \log \frac{B^2}{A^2}$ follows the Gamma distribution with shape parameter $\rho_1^{(2)}$ and scale parameter $\rho_2^{(2)} \log \frac{B^2}{A^2}$. Thus, $\mu^{(1)} \log \frac{A^2}{C^2} + \mu^{(2)} \log \frac{B^2}{A^2}$ is a random variable characterised by the sum of the two rescaled Gamma distributions [346, 387] (see Appendix B.5). An example is shown in Fig. 5.34, where $p_0 = 1$ and $C_0 = 2$, while the geometric and material parameters for the composite tube are as in the previous case.

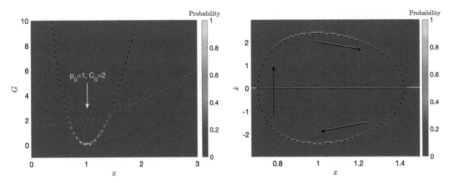

Fig. 5.34 The function $G(x, \gamma)$, defined by (5.216), intersecting the (dashed red) line $\frac{p_0}{2\alpha}\left(x^2 - \frac{1}{\alpha}\right) + C_0$ when $p_0 = 1$ and $C_0 = 2$ (left), and the associated velocity, given by (5.208) (right), for a dynamic composite tube with two concentric stochastic neo-Hookean phases, with inner radii $A = 1$ and $C = 1/2$, respectively, under dead-load traction, assuming that $\alpha = 1$, $\rho = 1$, and $\mu^{(1)}$ follows a Gamma distribution with $\rho_1^{(1)} = 405$ and $\rho_2^{(1)} = 4.05/\rho_1^{(1)} = 0.01$, while $\mu^{(2)}$ is drawn from a Gamma distribution with $\rho_1^{(2)} = 405$ and $\rho_2^{(2)} = 4.2/\rho_1^{(2)}$. The dashed black lines correspond to the expected values based only on mean values, $\underline{\mu}^{(1)} = 4.05$ and $\underline{\mu}^{(2)} = 4.2$. Each distribution was calculated from the average of 1000 stochastic simulations

5.4.3 Stochastic Radially Inhomogeneous Tubes Under Impulse Traction

We further consider the inflation of a radially inhomogeneous tube of stochastic neo-Hookean-like hyperelastic material characterised by the strain-energy function (4.49). We can regard the radially inhomogeneous tube as an extension of the composite with two concentric tubes to the case with infinitely many concentric layers and continuous inhomogeneity. Our inhomogeneous model, originally presented in [354], is similar to those proposed in [158] where the dynamic inflation of cylindrical tubes and spherical shells was treated explicitly.

Specifically, we assume a class of stochastic inhomogeneous hyperelastic models (4.49) where $\mu = \mu(R)$ takes the form

$$\mu(R) = C_1 + C_2 \frac{R^2}{C^2}, \tag{5.222}$$

such that $\mu(R) > 0$, for all $C \leq R \leq B$, $C_1 > 0$ is a single-valued (deterministic) constant, and C_2 is a random value defined by a given probability distribution.

When the mean value of the shear modulus $\mu(R)$, described by (5.222), does not depend on R, it follows that, for any fixed R,

$$\underline{\mu} = C_1, \qquad Var[\mu] = Var[C_2]\frac{R^4}{C^4}, \tag{5.223}$$

Fig. 5.35 Examples of Gamma distribution, with hyperparameters $\rho_1 = 405 \cdot B^4/R^4$ and $\rho_2 = 0.01 \cdot R^4/B^4$, for the nonlinear shear modulus $\mu(R)$ given by (5.222)

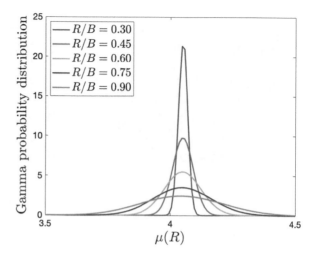

where Var$[C_2]$ is the variance of C_2.

By (5.223), the hyperparameters of the corresponding Gamma distribution take the form

$$\rho_1 = \frac{C_1}{\rho_2}, \qquad \rho_2 = \frac{\mathrm{Var}[\mu]}{C_1} = \frac{\mathrm{Var}[C_2]}{C_1}\frac{R^4}{C^4}. \tag{5.224}$$

For example, we can choose two constant values, $C_0 > 0$ and $C_1 > 0$, and set the hyperparameters for the Gamma distribution at any given R as follows:

$$\rho_1 = \frac{C_1}{C_0}\frac{C^4}{R^4}, \qquad \rho_2 = C_0\frac{R^4}{C^4}. \tag{5.225}$$

By (5.222), $C_2 = (\mu(R) - C_1)\,C^2/R^2$ is the shifted Gamma-distributed random variable with mean value $\underline{C}_2 = 0$ and variance $\mathrm{Var}[C_2] = \rho_1\rho_2^2 C^4/R^4 = C_0 C_1$.

In Fig. 5.35, we show Gamma distributions with $\rho_1 = 405 \cdot B^4/R^4$ and $\rho_2 = 0.01 \cdot R^4/B^4$. By (5.222) and (5.225), $C_0 = 0.01 \cdot C^4/B^4$, $C_1 = \mu = \rho_1\rho_2 = 4.05$, and $C_2 = \mu(C) - C_1$. In particular, for a tube with infinitely thick wall, as R decreases to C, ρ_1 increases, while ρ_2 decreases, and the Gamma distribution converges to a normal distribution [167, 357].

The shear modulus given by (5.222) takes the equivalent form

$$\mu = C_1 + C_2\frac{x^2 - \frac{1}{\alpha}}{u - \frac{1}{\alpha}}, \tag{5.226}$$

where $u = r^2/R^2$ and $x = c/C$, as denoted in (5.197).

Next, writing the invariants given by (5.146) in the equivalent form

$$I_1 = \frac{1}{\alpha^2 u} + u + \alpha^2, \qquad I_2 = \alpha^2 u + \frac{1}{u} + \frac{1}{\alpha^2}, \qquad I_3 = 1, \tag{5.227}$$

and substituting these in (5.148), gives

$$\beta_1 = 2\frac{\partial W}{\partial I_1} = \mu(u) + \frac{d\mu}{du}\frac{du}{dI_1}(I_1 - 3),$$

$$\beta_{-1} = -2\frac{\partial W}{\partial I_2} = -\frac{d\mu}{du}\frac{du}{dI_2}(I_1 - 3). \tag{5.228}$$

Therefore,

$$\beta_1 = C_1 + C_2 \frac{x^2 - \frac{1}{\alpha}}{u - \frac{1}{\alpha}}\left[1 - \frac{u^3 + u^2(\alpha^2 - 3) + \frac{u}{\alpha^2}}{\left(u - \frac{1}{\alpha}\right)^2\left(u + \frac{1}{\alpha}\right)}\right],$$

$$\beta_{-1} = C_2 \frac{x^2 - \frac{1}{\alpha}}{u - \frac{1}{\alpha}} \frac{\frac{u^3}{\alpha^2} + u^2\left(1 - \frac{3}{\alpha^2}\right) + \frac{u}{\alpha^4}}{\left(u - \frac{1}{\alpha}\right)^2\left(u + \frac{1}{\alpha}\right)}, \tag{5.229}$$

and

$$\beta_1 - \beta_{-1}\alpha^2 = C_1 + 2C_2 \frac{x^2 - \frac{1}{\alpha}}{u - \frac{1}{\alpha}}\left[\frac{1}{2} - \frac{u^3 + u^2(\alpha^2 - 3) + \frac{u}{\alpha^2}}{\left(u - \frac{1}{\alpha}\right)^2\left(u + \frac{1}{\alpha}\right)}\right]. \tag{5.230}$$

Recalling that the stress components are described by (5.151), and following a similar procedure as in the previous section, we set the pressure impulse as in (5.205). Then, similarly to (5.204), using (5.230), we define the function

$$G(x, \gamma) = \frac{C_1}{\rho C^2}\int_{\frac{1}{\sqrt{\alpha}}}^{x}\left(\zeta\int_{\frac{\zeta^2 + \frac{\gamma}{\alpha}}{1 + \gamma}}^{\zeta^2}\frac{1 + \alpha u}{\alpha^2 u^2}du\right)d\zeta$$

$$+ \frac{2C_2}{\rho C^2}\int_{\frac{1}{\sqrt{\alpha}}}^{x}\left\{\left(\zeta^3 - \frac{\zeta}{\alpha}\right)\int_{\frac{\zeta^2 + \frac{\gamma}{\alpha}}{1 + \gamma}}^{\zeta^2}\frac{1 + \alpha u}{\alpha^2 u^2\left(u - \frac{1}{\alpha}\right)}\left[\frac{1}{2} - \frac{u^3 + u^2(\alpha^2 - 3) + \frac{u}{\alpha^2}}{\left(u - \frac{1}{\alpha}\right)^2\left(u + \frac{1}{\alpha}\right)}\right]du\right\}d\zeta. \tag{5.231}$$

- If the tube wall is infinitely thick wall, such that $\gamma \to \infty$ and $\alpha = 1$, then (5.231) takes the form

$$G(x) = \frac{C_1}{\rho C^2}\left(x^2 - 1\right)\log x - \frac{C_2}{4\rho C^2}\left(x^4 - 4x^2 + 4\log x + 3\right). \tag{5.232}$$

Examples for this case are presented in Figs. 5.36 and 5.37, where $C = 1$, $\rho = 1$, and μ follows a Gamma distribution with $\rho_1 = 405/R^4$ and $\rho_2 = 0.01 \cdot R^4$.

- If the tube wall is thin, such that $0 < \gamma \ll 1$ and $\alpha = 1$, then the shear modulus defined by (5.222) takes the form $\mu = C_1 + C_2$, and the problem reduces to that of a homogeneous tube with thin wall [358].

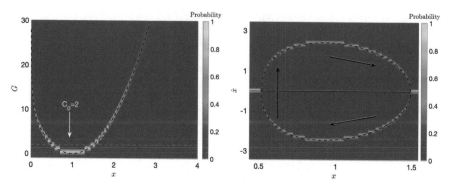

Fig. 5.36 The function $G(x, \gamma)$, defined by (5.232), intersecting the (dashed red) line $C_0 = 2$ when $p_0 = 0$ (left), and the associated velocity, given by (5.208) (right), for a dynamic radially inhomogeneous tube with infinitely thick wall having inner radius $C = 1$, under dead-load traction, assuming that $\alpha = 1$, $\rho = 1$, and μ follows a Gamma distribution with $\rho_1 = 405/R^4$ and $\rho_2 = 0.01 \cdot R^4$. The dashed black lines correspond to the expected values based only on mean values. Each distribution was calculated from the average of 1000 stochastic simulations

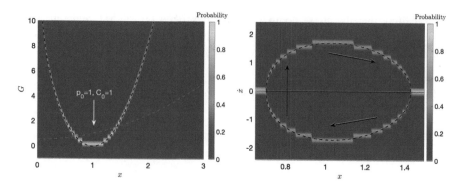

Fig. 5.37 The function $G(x, \gamma)$, defined by (5.232), intersecting the (dashed red) line $\frac{p_0}{2\alpha}\left(x^2 - \frac{1}{\alpha}\right) + C_0$ when $p_0 = 1$ and $C_0 = 1$ (left), and the associated velocity, given by (5.208) (right), for a dynamic radially inhomogeneous tube with infinitely thick wall having inner radius $C = 1$, under dead-load traction, assuming that $\alpha = 1$, $\rho = 1$, and μ follows a Gamma distribution with $\rho_1 = 405/R^4$ and $\rho_2 = 0.01 \cdot R^4$. The dashed black lines correspond to the expected values based only on mean values. Each distribution was calculated from the average of 1000 stochastic simulations

5.5   Generalised Shear Motion of a Stochastic Cuboid

We finally examine a stochastic hyperelastic cuboid subject to dynamic generalised shear [358]. The generalised shear motion of an elastic body is described by Destrade et al. [135]

$$x = \frac{X}{\sqrt{\alpha}}, \qquad y = \frac{Y}{\sqrt{\alpha}}, \qquad z = \alpha Z + u(X, Y, t), \tag{5.233}$$

where (X, Y, Z) and (x, y, z) are the Cartesian coordinates for the reference (Lagrangian, material) and current (Eulerian, spatial) configurations, respectively, $\alpha > 0$ is a given constant, and $u = z - Z$, representing the displacement in the third direction, is a time-dependent function to be determined. Here, we assume that the edges of the cuboid are aligned with the directions of the Cartesian axes in the undeformed state (see Fig. 5.38).

By the governing equations (5.233), the condition (2.168) is valid for $\mathbf{x} = (x, y, z)^T$ if and only if

$$\mathbf{0} = \text{curl } \ddot{\mathbf{x}} = \begin{bmatrix} \partial \ddot{z}/\partial y - \partial \ddot{y}/\partial z \\ \partial \ddot{x}/\partial z - \partial \ddot{z}/\partial x \\ \partial \ddot{y}/\partial x - \partial \ddot{x}/\partial y \end{bmatrix} = \begin{bmatrix} \partial \ddot{u}/\partial Y \\ -\partial \ddot{u}/\partial X \\ 0 \end{bmatrix}. \tag{5.234}$$

This condition imposes very strict constraints on the motion. Yet, we will see that even though the generalised shear motion (5.233) is not quasi-equilibrated, exact solutions are still available, although these solutions are not universal [404, 578].

For the deformation (5.233), the gradient tensor is equal to

$$\mathbf{A} = \begin{bmatrix} 1/\sqrt{\alpha} & 0 & 0 \\ 0 & 1/\sqrt{\alpha} & 0 \\ u_X & u_Y & \alpha \end{bmatrix},$$

where u_X and u_Y denote the partial first derivatives of u with respect to X and Y, respectively. The corresponding left Cauchy–Green tensor is

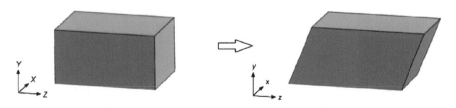

Fig. 5.38 Schematic of generalised shear of a cuboid, showing the reference state (left) and the deformed state (right), respectively

$$\mathbf{B} = \mathbf{A}\mathbf{A}^T = \begin{bmatrix} 1/\alpha & 0 & u_X/\sqrt{\alpha} \\ 0 & 1/\alpha & u_Y/\sqrt{\alpha} \\ u_X/\sqrt{\alpha} & u_Y/\sqrt{\alpha} & u_X^2 + u_Y^2 + \alpha^2 \end{bmatrix} \tag{5.235}$$

and has the principal invariants

$$
\begin{aligned}
I_1 &= \mathrm{tr}\,(\mathbf{B}) = u_X^2 + u_Y^2 + \frac{2}{\alpha} + \alpha^2, \\
I_2 &= \frac{1}{2}\left[(\mathrm{tr}\,\mathbf{B})^2 - \mathrm{tr}\left(\mathbf{B}^2\right)\right] = \frac{u_X^2}{\alpha} + \frac{u_Y^2}{\alpha} + \frac{1}{\alpha^2} + 2\alpha, \\
I_3 &= \det \mathbf{B} = 1.
\end{aligned}
\tag{5.236}
$$

The associated Cauchy stress tensor takes the form [216, pp. 87–91]

$$\mathbf{T} = -p\mathbf{I} + \beta_1 \mathbf{B} + \beta_{-1} \mathbf{B}^{-1}, \tag{5.237}$$

where p is the Lagrange multiplier for the incompressibility constraint ($I_3 = 1$), and

$$\beta_1 = 2\frac{\partial W}{\partial I_1}, \qquad \beta_{-1} = -2\frac{\partial W}{\partial I_2} \tag{5.238}$$

are the nonlinear material parameters, with I_1, I_2 given by (5.236).

5.5.1   Shear Oscillations of a Cuboid of Stochastic Neo-Hookean Material

We specialise our analysis to the case of a cuboid of stochastic neo-Hookean material, with $\mu_1 = \mu > 0$ and $\mu_2 = 0$ in (4.49), where the non-zero components of the Cauchy stress tensor given by (5.237) are as follows:

$$T_{xx} = T_{yy} = -p + \frac{\mu}{\alpha}, \quad T_{zz} = -p + \mu\left(u_X^2 + u_Y^2 + \alpha^2\right), \quad T_{xz} = \frac{\mu}{\sqrt{\alpha}}u_X, \quad T_{yz} = \frac{\mu}{\sqrt{\alpha}}u_Y. \tag{5.239}$$

Then, by the equation of motion (2.165),

$$\frac{\partial p}{\partial x} = 0, \qquad \frac{\partial p}{\partial y} = 0, \qquad \frac{\partial p}{\partial z} = -\rho\ddot{u} + \mu\left(u_{XX} + u_{YY}\right), \tag{5.240}$$

where u_{XX} and u_{YY} represent the second derivatives of u with respect to X and Y, respectively. Hence, p is independent of x and y.

We consider the undeformed cuboid to be long in the Z-direction and impose an initial displacement $u_0(X, Y) = u(X, Y, 0)$ and velocity $\dot{u}_0(X, Y) = \dot{u}(X, Y, 0)$. For the boundary condition, we distinguish the following two cases:

(i) If we impose null normal Cauchy stresses, $T_{xx} = T_{yy} = 0$, on the faces perpendicular to the X- and Y-directions, at all times, then $p = \mu/\alpha$ is constant and $T_{zz} = \mu \left(u_X^2 + u_Y^2 + \alpha^2 - 1/\alpha\right)$.

(ii) If $T_{xx} = T_{yy} \neq 0$, as T_{zz} cannot be made pointwise zero, we denote the normal force acting on the cross-sections of area $\mathcal{A}$ in the z-direction at time t by

$$N_z(t) = \int_{\mathcal{A}} T_{zz} d\mathcal{A} \qquad (5.241)$$

and consider this force to be zero, i.e., $N_z(t) = 0$ at all times. Then, p is independent of z, and, by (5.240), it is also independent of x and y; hence, $p = p(t)$.

In both the above cases, (i) and (ii), respectively, by (5.240),

$$\ddot{u} = \frac{\mu}{\rho} (u_{XX} + u_{YY}). \qquad (5.242)$$

It remains to solve, by standard procedures, the linear wave equation (5.242), describing the propagation of waves, subject to the given initial and boundary conditions. To solve this equation, we let the shear stresses T_{xz} and T_{yz}, defined by (5.239), vanish at the sides, i.e.,

$$T_{xz}(0, Y, Z, t) = T_{xz}(1, Y, Z, t) = 0 \quad \Longleftrightarrow \quad u_X(0, Y, t) = u_X(1, Y, t) = 0,$$

$$T_{yz}(X, 0, Z, t) = T_{yz}(X, 1, Z, t) = 0 \quad \Longleftrightarrow \quad u_Y(X, 0, t) = u_Y(X, 1, t) = 0.$$
$$(5.243)$$

In this case, the general solution takes the form

$$u(X, Y, t) = \sum_{m=1}^{\infty} \sum_{n=1}^{\infty} [A_{mn} \cos(\omega_{mn} t) + B_{mn} \sin(\omega_{mn} t)] \cos(\pi m X) \cos(\pi n Y),$$
$$(5.244)$$

where

$$\omega_{mn} = \pi \sqrt{(m^2 + n^2) \frac{\mu}{\rho}}, \qquad (5.245)$$

and

$$A_{mn} = 4 \int_0^1 \left[\int_0^1 u_0(X, Y) \cos(\pi m X) \, dX \right] \cos(\pi n Y) \, dY, \qquad (5.246)$$

$$B_{mn} = \frac{4}{\omega_{mn}} \int_0^1 \left[\int_0^1 \dot{u}_0(X, Y) \cos(\pi m X) \, dX \right] \cos(\pi n Y) \, dY. \quad (5.247)$$

These oscillations under the generalised shear motion (5.233) cannot be completely "free", due to the non-zero tractions corresponding to the cases (i) and (ii), respectively. Note that the condition (5.234) is not satisfied.

As μ is a random variable, it follows that the speed of wave propagation, $\sqrt{\mu/\rho}$, is stochastic. As an example, we consider the initial data $u_0(X, Y) = \cos(\pi X) \cos(\pi Y)$ and $\dot{u}_0(X, Y) = 0$ leading to $A_{11} = 1$ and $B_{11} = 0$. In Fig. 5.39, we illustrate the stochastic dynamic displacement on the edges $(X, Y, Z) \in \{(0, 0, Z), (1, 1, , Z)\}$ when $m = n = 1$, $A_{11} = 1$, $B_{11} = 0$, $\rho = 1$, and μ is drawn from the Gamma distribution with hyperparameters $\rho_1 = 405$ and $\rho_2 = 0.01$. The top plot of Fig. 5.39 represents two single simulations, with two different values of μ drawn from the distribution, illustrating the variety of outcomes that can be obtained. The middle plot of Fig. 5.39 then represents histograms of the ensemble data. Namely, since not all material parameters are equally likely, not all outcomes are equally likely. Specifically, the values of $u(0, 0, t)$ are most likely going to be near the mean value (dashed line) with the probability of observing alternative values of u decreasing as we tend away from the mean. We note from Fig. 5.39 that, as we might expect, extremal probabilities always occur at the extreme displacement of the oscillations, i.e., when the cuboid is slowest. However, in between these probability maxima, the variance grows over time. Thus, although the displacements

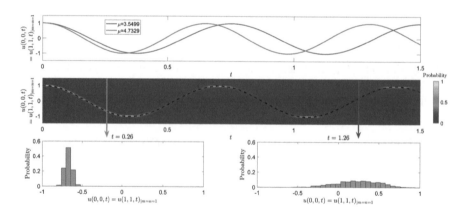

Fig. 5.39 Stochastic displacement $u(X, Y, t)$ of the edges $(X, Y, Z) \in \{(0, 0, Z), (1, 1, Z)\}$ of the cuboid in dynamic generalised shear, when $m = n = 1$, $A_{11} = 1$, $B_{11} = 0$, $\rho = 1$, and μ is drawn from the Gamma distribution with $\rho_1 = 405$ and $\rho_2 = 0.01$. The top figure shows the displacement over time of two cuboids, with randomly chosen values of μ, derived from the specified Gamma distribution. The middle figure illustrates a probability histogram at each time instant. The histogram comprises 1000 stochastic simulations, and the colour bar defines the probability of finding a given displacement at a given time. The black line corresponds to the expected values based only on mean value, $\mu = \rho_1 \rho_2 = 4.05$. The bottom two figures represent histogram distributions at two specified times [358]

are initially close (seen explicitly in the top of Fig. 5.39 and by the tight distribution around the mean in the bottom left of Fig. 5.39), eventually, the phase difference dominates causing the displacements to diverge (top of Fig. 5.39), and an increase in the variance of the distribution (bottom right of Fig. 5.39).

Chapter 6
Liquid Crystal Elastomers

*what a theorist can and should systematically introduce is
comparisons with other fields*

—P.G. de Gennes [123]

Liquid crystal elastomers (LCEs) are cross-linked networks of polymeric chains containing liquid crystal mesogens [124, 164, 594]. The molecular order in a solid crystal, a liquid crystal, and a liquid is compared schematically in Fig. 6.1. Computer simulations of a nematic[1] LCE are illustrated in Fig. 6.2.

Because of their molecular architecture, LCEs are capable of large reversible deformations and are highly responsive to external stimuli such as heat[2] [5, 44, 73, 122, 314, 491, 531, 588], light [9, 36, 115, 121, 166, 255, 313, 434, 486, 491, 545, 586, 587, 596, 598, 623, 626–628], solvents [67, 588, 626], and electric [562, 563, 624], or magnetic fields [291, 609]. Recyclable and self-healing liquid crystalline solids can also be synthesised [243, 478, 586]. Such extraordinary properties can be harnessed for a variety of technological applications, including soft actuators or sensors [83, 119, 131, 171, 189, 222, 231, 322, 401, 530, 546, 568, 583, 585, 606, 618] and soft tissue engineering [386, 447, 483, 540, 547, 582]. However, a better understanding of these materials is needed before they can be exploited on an industrial scale [15, 126, 272, 286, 309, 336, 348, 380, 422, 475, 537, 560, 561, 577, 589, 601, 603, 604, 611].

Despite their difficult synthetic processes, the intriguing mechanical behaviour of LCEs has been probed extensively in laboratories around the world thanks to recent developments in additive manufacturing (3D printing). Nevertheless, their accurate description can only be useful if fully integrated in a multiphysics framework combining elasticity and liquid crystal theories. Due to many similarities with conventional rubber, considerable progress can be made in their description by

---

[1] From the Ancient Greek νῆμα (*néma*, thread). The word *nematic* was coined by G. Friedel (1922) [181] (see also [125]).

[2] The tendency of stretched elastomers to contract when heated, or vice versa, the heating of an elastomer when stretched, is known as the Gough-Joule effect [212, 352], [558, p. 360].

© Springer Nature Switzerland AG 2022
L. A. Mihai, *Stochastic Elasticity*, Interdisciplinary Applied Mathematics 55,
https://doi.org/10.1007/978-3-031-06692-4_6

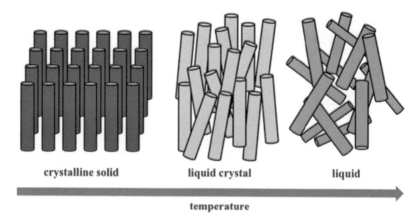

Fig. 6.1 Comparison of molecular order in a solid crystal, a liquid crystal, and a liquid

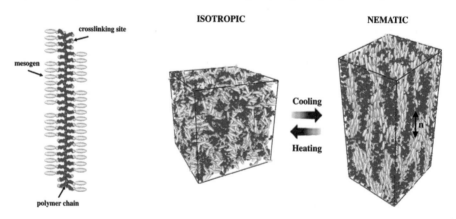

Fig. 6.2 Molecular dynamics simulation of structure and physical responses in a thermotropic nematic elastomer: the material sample extends on cooling as the liquid crystal mesogens align in the nematic phase at low temperature, with the orientation of nematic director **n** shown, and contracts on heating as the mesogens are unaligned in the isotropic phase at high temperatures [370]

adapting existing hyperelastic models, but differences from classical rubber models are also significant and need to be examined carefully.

Many nematic elastomers are synthesised as polydomains, where the liquid crystal mesogens are separated into different domains, such that, in every domain, they are aligned in a preferred direction [106, 107, 246, 313, 479, 548, 564]. Depending on the fabrication process, polydomains may have very different material properties and behaviours. Monodomains, where mesogenic molecules are uniformly aligned throughout the material, can be formed from polydomains through mechanical stretching or by cooling from the isotropic to the nematic phase under the external stress field.

A continuum model for ideal monodomains is described by the so-called neoclassical strain-energy function proposed in [63, 591, 595]. This is a phenomenological model based on the molecular network theory of rubber [550] where the parameters of the neo-Hookean-type strain-energy density can be derived from macroscopic shape changes at small strain or through statistical averaging at microscopic scale [593, 594]. Extensions to polydomains where every domain has the same strain-energy density as a monodomain are provided in [58, 59]. Typically, elastic stresses dominate over Frank elasticity induced by the distortion of mesogens alignment [125, 174], known as the *director*, and therefore, Frank effects are generally neglected [36, 381, 382]. These descriptions have been generalised by employing other hyperelastic models (e.g., Mooney–Rivlin, Gent, Ogden) that are known to capture the nonlinear elastic behaviour at large strains [3, 4, 132] (molecular interpretations of the Mooney–Rivlin and Gent constitutive models in rubber elasticity are presented in [177, 266]). Further generalisations are proposed in [17, 94, 625]. In particular, when Frank energy also plays an important role, it needs to be taken into account as well.

We are interested in developing tools for a consistent stochastic theory to analyse the mechanical behaviour of nematic LCEs. First, we generalise the continuum model equations and then present the stochastic description. Following the classical work of Flory [169] on polymer elasticity, we use the stress-free state of a virtual isotropic phase at high temperature as the reference configuration [36, 104, 128–130, 132], rather than the nematic phase in which the cross-linking might have been produced [17, 63, 572, 590, 591, 595, 625]. Within this theoretical framework, the material deformation due to the interaction between external stimuli and mechanical loads can be expressed as a composite deformation from a reference configuration to the current configuration, via an elastic distortion followed by a natural (stress-free) shape change. The multiplicative decomposition of the associated gradient tensor is similar in some respects to those found in the constitutive theories of thermoelasticity, elastoplasticity, and growth [205, 331] (see also [204, 477]) but is different on one major aspect, namely that the stress-free geometrical change is superposed on the elastic deformation, which is applied directly to the reference state [207, 208, 364–366]. This difference is important since, although the elastic configuration obtained by this deformation may not be observed in practice, it may still be possible for the nematic body to assume such a configuration under suitable external stimuli. The resulting elastic stresses can then be used to analyse the final deformation where the particular geometry also plays a role. Moreover, in liquid crystalline materials, asymmetric Cauchy stresses generally occur, unlike in purely elastic materials [594, p. 80].

The focus here is on the stability of LCE bodies acted upon by external loads (see also [363, 367]). In addition to the *soft elasticity* phenomenon where alternating shear stripes develop at very low stress if a nematic body is stretched, it is instructive to explore theoretically a set of classical instabilities inherited from the parent polymeric networks, namely, *necking* under uniaxial tension and *inflation instability* of an internally pressurised spherical shell. The aim is to determine conditions for the onset of instability and to show how nematic materials perform compared to their purely elastic analogue. Experimental observations on necking of LCE material under tensile loading are presented in [244, 250]. Pressurised nematic spherical shells are examined in [168, 277, 278, 493]. Inflation tests of elongated nematic balloons are also reported in [245]. The inflation of nematic cylindrical balloons is treated theoretically in [94, 199, 318]. Other intriguing phenomena inherited from rubber elasticity, such as the *Poynting effect* [198] (see Chap. 2), can also be studied using our theoretical framework.

We further present a continuum model that captures the auxeticity observed in a novel nematic LCE where the thickness of a stretched material sample increases at sufficiently large strain. The auxetic response in this material occurs at a molecular level, while volume is conserved and there is no evidence of porosity emerging as the system is strained [377–379, 453]. When viewed as a two-dimensional system, this auxetic effect (from the Greek word $\alpha\nu\xi\eta\sigma\iota\varsigma$ for "growth" or "increase") coincides with an apparent sharp rotation, by $\pi/2$, of the average nematic alignment direction. However, in three dimensions, the sharp director rotation is accompanied by a gradual decrease and then increase in uniaxial orientational order coupled with the emergence and subsequent loss of biaxial symmetry [453]. This is different from the mechanical response of most other nematic LCEs, with a continuous rotation of the director and a constant uniaxial order parameter, where shear striping patterns can develop.

As in the previous chapters, the LCE problems included here are all amenable to analytical treatment. Computational studies of LCEs can be found, for example, in [113, 114, 625] where the finite element method is applied [96, 370, 445, 605] where molecular dynamics simulations are used, and [501, 502] where molecular Monte Carlo is employed. Numerical methods for LC and LCE modeling are reviewed in [621] and [512], respectively.

6.1   Uniaxial Elastic Models

For an ideal uniaxial nematic liquid crystalline solid, the neoclassical strain-energy density function takes the form

$$W^{(nc)}(\mathbf{F}, \mathbf{n}) = W(\mathbf{A}), \tag{6.1}$$

where $\mathbf{F}$ represents the deformation gradient from the reference isotropic state, $\mathbf{n}$ is the nematic director, i.e., a unit vector for the localised direction of uniaxial nematic alignment, in the present configuration, and $W(\mathbf{A})$ is the strain-energy density of the isotropic polymer network, depending only on the (local) elastic deformation tensor $\mathbf{A}$. The tensors $\mathbf{F}$ and $\mathbf{A}$ satisfy the relation (see Fig. 6.3 and also Figure 1 of [132])

$$\mathbf{F} = \mathbf{G}\mathbf{A}, \tag{6.2}$$

Fig. 6.3 Multiplicative decomposition of the deformation gradient for a nematic solid, whereby the elastic deformation is applied to the reference configuration and is followed by an anelastic strain deformation

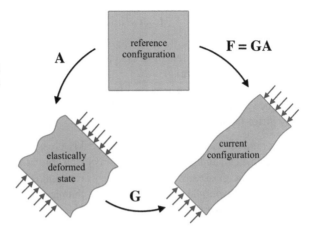

where $\mathbf{G}$ is the "spontaneous" (or "natural") deformation tensor defining a change of frame of reference from the isotropic phase to a nematic phase and is equal to

$$\mathbf{G} = a^{1/3}\mathbf{n} \otimes \mathbf{n} + a^{-\nu/3}(\mathbf{I} - \mathbf{n} \otimes \mathbf{n}) = a^{-\nu/3}\mathbf{I} + \left(a^{1/3} - a^{-\nu/3}\right)\mathbf{n} \otimes \mathbf{n}. \quad (6.3)$$

In (6.3), $a > 0$ is a temperature-dependent shape parameter, ν denotes the optothermal analogue to the Poisson ratio [383] relating responses in directions parallel or perpendicular to the director $\mathbf{n}$, $\otimes$ stands for the tensor product of two vectors, and $\mathbf{I} = \mathrm{diag}(1, 1, 1)$ is the identity tensor. We assume that a and ν are spatially independent (i.e., no differential swelling). These parameters can be estimated independently (e.g., by examining the thermal or light-induced response of the nematic elastomer with uniform planar alignment [304]). The ratio $r = a^{1/3}/a^{-\nu/3} = a^{(\nu+1)/3}$ represents the anisotropy parameter, which, in an ideal nematic solid, is the same in all directions. In the nematic phase, both the cases with $r > 1$ (prolate molecules) and $r < 1$ (oblate molecules) are possible, while when $r = 1$, the energy function reduces to that of an isotropic hyperelastic material. Monodomains can be synthesised with the parameter a taking values from 1.05 for liquid crystalline glasses to 60 for liquid crystal elastomers. This would correspond to changes in natural length between 7 and 400% (spontaneous extension ratio $a^{1/3}$ between 1.02 and 4) [105]. For nematic elastomers, the Poisson's ratio is $\nu = 1/2$, i.e., their volume remains unchanged during deformation, while for glasses, $\nu \in (1/2, 2)$ [568, 589].

In (6.1), the elastic strain-energy function W is minimised by any deformation satisfying $\mathbf{A}\mathbf{A}^T = \mathbf{I}$ [415, 558], and the nematic strain-energy $W^{(nc)}$ is minimised by any deformation satisfying $\mathbf{F}\mathbf{F}^T = \mathbf{G}^2$. Hence, every pair $(\mathbf{G}\mathbf{R}, \mathbf{n})$, where $\mathbf{R}$ is an arbitrary rigid-body rotation (i.e., $\mathbf{R}^{-1} = \mathbf{R}^T$ and $\det \mathbf{R} = 1$), is a natural (i.e., stress-free) state for this material model. By (6.3), if $\mathbf{R}$ is a rigid-body rotation, i.e., $\mathbf{R}^{-1} = \mathbf{R}^T$ and $\det \mathbf{R} = 1$, then the following identity holds:

$$\mathbf{R}^T \mathbf{G} \mathbf{R} = a^{-\nu/3}\mathbf{I} + \left(a^{1/3} - a^{-\nu/3}\right)\left(\mathbf{R}^T \mathbf{n}\right) \otimes \left(\mathbf{R}^T \mathbf{n}\right). \tag{6.4}$$

Note that the nematic tensor $\mathbf{G}$ is symmetric, whereas the elastic tensor $\mathbf{A}$ may not be symmetric in general.

The nematic director $\mathbf{n}$ is an observable (spatial) quantity. Denoting by $\mathbf{n}_0$ the reference orientation of the local director corresponding to the cross-linking state, $\mathbf{n}$ may differ from $\mathbf{n}_0$ both by a rotation and by a change in r.

- For nematic elastomers, which are weakly cross-linked, the nematic director can rotate freely, and the material displays isotropic mechanical properties.
- In nematic glasses, which are densely cross-linked, the nematic director $\mathbf{n}$ cannot rotate relative to the elastic matrix but changes through convection due to elastic strain and satisfies [104, 381–383, 388]

$$\mathbf{n} = \frac{\mathbf{F}\mathbf{n}_0}{|\mathbf{F}\mathbf{n}_0|}. \tag{6.5}$$

This constraint enables patterning of the director field at cross-linking and guarantees that the "written-in" pattern remains virtually the same during natural shape changes [381, 382, 592]. The elastic anisotropy of NLC materials where the director cannot rotate was investigated experimentally in [163, 378, 379].

The strain energy given by (6.1) satisfies the following conditions inherited from isotropic finite elasticity [363]:

(C1) *Material objectivity*, stating that constitutive equations must be unaffected by a superimposed rigid-body transformation (which involves a change of position after deformation). Note that, as $\mathbf{n}$ is defined with respect to the deformed configuration, it transforms when this configuration is rotated, whereas $\mathbf{n}_0$ does not [178]. Material objectivity is guaranteed by defining strain-energy functions in terms of the scalar invariants.

Indeed, by the material frame indifference of W,

$$W(\mathbf{R}^T \mathbf{A}) = W(\mathbf{A}), \tag{6.6}$$

and by (6.2),

$$\mathbf{R}^T \mathbf{F} = \left(\mathbf{R}^T \mathbf{G} \mathbf{R}\right)\left(\mathbf{R}^T \mathbf{A}\right). \tag{6.7}$$

Then, (6.1), (6.4), (6.6), and (6.7) imply

$$W^{(nc)}(\mathbf{R}^T \mathbf{F}, \mathbf{R}^T \mathbf{n}) = W(\mathbf{R}^T \mathbf{A}) = W(\mathbf{A}) = W^{(nc)}(\mathbf{F}, \mathbf{n}). \tag{6.8}$$

(C2) *Material isotropy*, requiring that the strain-energy density is unaffected by a rigid-body transformation prior to deformation. Note that, as $\mathbf{n}$ is defined with respect to the deformed configuration, it does not change when the reference configuration is rotated, whereas $\mathbf{n}_0$ does [178]. For isotropic materials, the strain-energy function is a symmetric function of the principal stretch ratios.

This is because, as W is isotropic, i.e.,

$$W(\mathbf{A}) = W(\mathbf{AR}), \tag{6.9}$$

and (6.2) holds, it follows that

$$\mathbf{FR} = \mathbf{G}\,(\mathbf{AR})\,. \tag{6.10}$$

Hence, by (6.1), (6.9), and (6.10),

$$W^{(nc)}(\mathbf{FR}, \mathbf{n}) = W(\mathbf{AR}) = W(\mathbf{A}) = W^{(nc)}(\mathbf{F}, \mathbf{n}). \tag{6.11}$$

(C3) For any given finite deformation, at any point in the material, the shear modulus $\mu > 0$, the shape parameter $a > 0$, and their inverse, $1/\mu$ and $1/a$, respectively, have finite mean value and finite variance.

Under condition (C1), the neoclassical model defined by (6.1) takes the equivalent form

$$W^{(nc)}(I_1^{(nc)}, I_2^{(nc)}, I_3^{(nc)}, I_4^{(nc)}, I_5^{(nc)}) = W(I_1, I_2, I_3), \tag{6.12}$$

where

$$I_1^{(nc)} = \text{tr}\left(\mathbf{FF}^T\right) = \text{tr}\left(\mathbf{F}^T\mathbf{F}\right), \tag{6.13}$$

$$I_2^{(nc)} = \text{tr}\left[\text{Cof}\left(\mathbf{FF}^T\right)\right] = \text{tr}\left[\text{Cof}\left(\mathbf{F}^T\mathbf{F}\right)\right], \tag{6.14}$$

$$I_3^{(nc)} = \det\left(\mathbf{FF}^T\right) = \det\left(\mathbf{F}^T\mathbf{F}\right), \tag{6.15}$$

$$I_4^{(nc)} = \mathbf{n} \cdot \mathbf{FF}^T \mathbf{n}, \tag{6.16}$$

$$I_5^{(nc)} = \mathbf{n} \cdot \left(\mathbf{FF}^T\right)^2 \mathbf{n} \tag{6.17}$$

are the scalar invariants. Then, the following relations hold:

$$I_1 = a^{1/3}\left[I_1^{(nc)} - \left(1 - a^{-1}\right)I_4^{(nc)}\right], \tag{6.18}$$

$$I_2 = a^{-1/3}\left[I_2^{(nc)} - (1-a)\left(I_3^{(nc)}\right)^{-1}\left(I_5^{(nc)} - I_1^{(nc)}I_4^{(nc)} + I_2^{(nc)}\right)\right], \tag{6.19}$$

$$I_3 = I_3^{(nc)}. \tag{6.20}$$

Indeed, identity (6.20) is valid since $\det \mathbf{G} = 1$, and therefore

$$I_3 = (\det \mathbf{A})^2 = \left(\det \mathbf{G}^{-1} \det \mathbf{F} \right)^2 = (\det \mathbf{F})^2 = I_3^{(nc)}. \tag{6.21}$$

Identity (6.18) holds since

$$\begin{aligned}
I_1 &= \operatorname{tr}\left(\mathbf{A}^T \mathbf{A} \right) \\
&= \operatorname{tr}\left[\mathbf{F}^T \mathbf{G}^{-2} \mathbf{F} \right] \\
&= a^{1/3} \left[\operatorname{tr}\left(\mathbf{F}^T \mathbf{F} \right) - \left(1 - a^{-1} \right) \mathbf{n} \cdot \mathbf{F} \mathbf{F}^T \mathbf{n} \right] \\
&= a^{1/3} \left[I_1^{(nc)} - \left(1 - a^{-1} \right) I_4^{(nc)} \right].
\end{aligned} \tag{6.22}$$

To show identity (6.19), we recall the Cayley–Hamilton theorem that

$$\left(\mathbf{F}\mathbf{F}^T \right)^3 - I_1^{(nc)} \left(\mathbf{F}\mathbf{F}^T \right)^2 + I_2^{(nc)} \left(\mathbf{F}\mathbf{F}^T \right) - I_3^{(nc)} \mathbf{I} = 0. \tag{6.23}$$

Multiplying the above equation by $\left(\mathbf{F}\mathbf{F}^T \right)^{-1}$ implies

$$I_3^{(nc)} \left(\mathbf{F}\mathbf{F}^T \right)^{-1} = \left(\mathbf{F}\mathbf{F}^T \right)^2 - I_1^{(nc)} \left(\mathbf{F}\mathbf{F}^T \right) + I_2^{(nc)} \mathbf{I}. \tag{6.24}$$

Hence,

$$\begin{aligned}
\mathbf{n} \cdot \mathbf{F}^{-T} \mathbf{F}^{-1} \mathbf{n} &= \mathbf{n} \cdot \left(I_3^{(nc)} \right)^{-1} \left[\left(\mathbf{F}\mathbf{F}^T \right)^2 - I_1^{(nc)} \left(\mathbf{F}\mathbf{F}^T \right) + I_2^{(nc)} \mathbf{I} \right] \mathbf{n} \\
&= \left(I_3^{(nc)} \right)^{-1} \left(I_5^{(nc)} - I_1^{(nc)} I_4^{(nc)} + I_2^{(nc)} \right).
\end{aligned} \tag{6.25}$$

Finally, by calculations analogous to those in (6.22), together with (6.20) and (6.25), it follows that

$$\begin{aligned}
I_2 &= I_3 \operatorname{tr}\left(\mathbf{A}^T \mathbf{A} \right)^{-T} \\
&= I_3^{(nc)} \operatorname{tr}\left[\mathbf{F}^{-1} \mathbf{G}^2 \mathbf{F}^{-T} \right] \\
&= I_3^{(nc)} a^{-1/3} \left[\operatorname{tr}\left(\mathbf{F}^{-1} \mathbf{F}^{-T} \right) - (1 - a) \mathbf{n} \cdot \mathbf{F}^{-T} \mathbf{F}^{-1} \mathbf{n} \right] \\
&= a^{-1/3} \left[I_2^{(nc)} - (1 - a) \left(I_5^{(nc)} - I_1^{(nc)} I_4^{(nc)} + I_2^{(nc)} \right) \right].
\end{aligned} \tag{6.26}$$

By condition (C2), the nematic model given by (6.1) can be written equivalently in the form

$$W^{(nc)}(\lambda_1, \lambda_2, \lambda_3, \mathbf{n}) = W(\alpha_1, \alpha_2, \alpha_3), \qquad (6.27)$$

where $\lambda_1, \lambda_2, \lambda_3$ are the principal stretches of the LCE, and $\alpha_1, \alpha_2, \alpha_3$ are the principal stretches for the polymeric network. The usual relations between the principal invariants and the principal stretches hold, i.e.,

$$I_1^{(nc)} = \lambda_1^2 + \lambda_2^2 + \lambda_3^2, \qquad I_2^{(nc)} = \lambda_1^2 \lambda_2^2 + \lambda_2^2 \lambda_3^2 + \lambda_3^2 \lambda_1^2, \qquad I_3^{(nc)} = \lambda_1^2 \lambda_2^2 \lambda_3^2,$$
$$(6.28)$$

and

$$I_1 = \alpha_1^2 + \alpha_2^2 + \alpha_3^2, \qquad I_2 = \alpha_1^2 \alpha_2^2 + \alpha_2^2 \alpha_3^2 + \alpha_3^2 \alpha_3^2 \qquad I_3 = \alpha_1^2 \alpha_2^2 \alpha_3^2. \quad (6.29)$$

6.2   Stresses in Liquid Crystal Elastomers

We can exploit the multiplicative decomposition (6.2) of the deformation gradient $\mathbf{F}$ to formulate the stress tensors of the nematic material in terms of the stresses of the elastically deformed polymeric network [364]. We distinguish two cases: first, when the nematic director is "free" to rotate relative to the elastic matrix, in which case it tends to align in the direction of the maximum principal stretch ratio, and second, when the nematic director is "frozen" and satisfies condition (6.5).

6.2.1   The Case of Free Nematic Director

When the nematic director is "free," $\mathbf{F}$ and $\mathbf{n}$ are independent variables. Since $\mathbf{G}$ is symmetric, the Cauchy stress tensor for the nematic material with the strain-energy function described by (6.1) is calculated as follows:

$$\begin{aligned}
\mathbf{T}^{(nc)} &= J^{-1} \frac{\partial W^{(nc)}}{\partial \mathbf{F}} \mathbf{F}^T - p^{(nc)} \mathbf{I} \\
&= J^{-1} \mathbf{G}^{-1} \frac{\partial W}{\partial \mathbf{A}} \mathbf{A}^T \mathbf{G} - p^{(nc)} \mathbf{I} \qquad (6.30) \\
&= J^{-1} \mathbf{G}^{-1} \mathbf{T} \mathbf{G},
\end{aligned}$$

where $\mathbf{T}$ is the Cauchy stress tensor defined by (2.24), $J = \det \mathbf{F} = \det \mathbf{G} \det \mathbf{A}$, and the scalar $p^{(nc)}$ represents the Lagrange multiplier for the internal constraint $J = \det \mathbf{G} = a^{(1-2v)/3}$.

The principal components $\left(T_1^{(nc)}, T_2^{(nc)}, T_3^{(nc)} \right)$ of the Cauchy stress defined by (6.30) are the solutions to the characteristic equation

$$\det \left(\mathbf{T}^{(nc)} - \Lambda \mathbf{I} \right) = 0. \tag{6.31}$$

Since

$$\det \left(\mathbf{T}^{(nc)} - \Lambda \mathbf{I} \right) = \det \left[\mathbf{G}^{-1} \left(J^{-1}\mathbf{T} - \Lambda \mathbf{I} \right) \mathbf{G} \right] = J^{-1} \det \left(\mathbf{T} - J\Lambda \mathbf{I} \right), \tag{6.32}$$

it follows that the principal Cauchy stresses for the underlying hyperelastic model satisfy

$$(T_1, T_2, T_3) = J \left(T_1^{(nc)}, T_2^{(nc)}, T_3^{(nc)} \right). \tag{6.33}$$

Therefore, if the Baker–Ericksen inequalities (2.52) hold for the hyperelastic model, then the greater principal Cauchy stress occurs in the direction of the greater principal elastic stretch for the nematic model.

In the presence of a nematic field, the total Cauchy stress tensor $\mathbf{T}^{(nc)}$ given by (6.30) is not symmetric in general [17, 515, 625]. In addition, the following condition is required [17, 625]:

$$\frac{\partial W^{(nc)}}{\partial \mathbf{n}} = \mathbf{0}. \tag{6.34}$$

Equivalently, by the principle of material objectivity stating that constitutive equations must be invariant under changes of frame of reference (see [17] for details),

$$\frac{1}{2} \left(\mathbf{T}^{(nc)} - \mathbf{T}^{(nc)^T} \right) \mathbf{n} = \mathbf{0}, \tag{6.35}$$

where $\left(\mathbf{T}^{(nc)} - \mathbf{T}^{(nc)^T} \right) / 2$ represents the skew-symmetric part of the Cauchy stress tensor.

The first Piola–Kirchhoff stress tensor for the nematic material is equal to

$$\mathbf{P}^{(nc)} = \mathbf{T}^{(nc)}\mathrm{Cof}\mathbf{F} = \mathbf{G}^{-1}\mathbf{T}\mathbf{A}^{-T} = \mathbf{G}^{-1}\mathbf{P}, \tag{6.36}$$

where $\mathbf{P}$ is the first Piola–Kirchhoff stress given by (2.34).

The corresponding second Piola–Kirchhoff stress tensor is

$$\mathbf{S}^{(nc)} = \mathbf{F}^{-1}\mathbf{P}^{(nc)} = \mathbf{A}^{-1}\mathbf{G}^{-2}\mathbf{P} = \mathbf{A}^{-1}\mathbf{G}^{-2}\mathbf{A}\mathbf{S}, \tag{6.37}$$

where $\mathbf{S}$ is the elastic second Piola–Kirchhoff stress tensor given by (2.41).

6.2.2   The Case of Frozen Nematic Director

When the nematic director is "frozen," first, we define a modified strain-energy function with independent variables $\mathbf{F}$ and $\mathbf{n}$,

$$\widehat{W}^{(nc)}(\mathbf{F}, \mathbf{n}) = W^{(nc)}(\mathbf{F}, \mathbf{n}) - q\left(\mathbf{n} \cdot \frac{\mathbf{Fn}_0}{|\mathbf{Fn}_0|} - 1\right), \tag{6.38}$$

where the scalar q is the Lagrange multiplier for the constraint (6.5).

Then, the Cauchy stress tensor for the nematic material takes the form

$$\widehat{\mathbf{T}}^{(nc)} = J^{-1}\frac{\partial \widehat{W}^{(nc)}}{\partial \mathbf{F}}\mathbf{F}^T - p^{(nc)}\mathbf{I}$$
$$= J^{-1}\mathbf{G}^{-1}\mathbf{TG} - J^{-1}q\left(\mathbf{I} - \frac{\mathbf{Fn}_0 \otimes \mathbf{Fn}_0}{|\mathbf{Fn}_0|^2}\right)\mathbf{n} \otimes \frac{\mathbf{Fn}_0}{|\mathbf{Fn}_0|}, \tag{6.39}$$

where $\mathbf{T}$ is the Cauchy stress defined by (2.24) and $p^{(nc)}$ is the Lagrange multiplier for the volume constraint $J = a^{(1-2\nu)/3}$.

As the Cauchy stress tensor given by (6.39) is not symmetric in general, the following additional condition must hold:

$$\frac{\partial \widehat{W}^{(nc)}}{\partial \mathbf{n}} = \mathbf{0}, \tag{6.40}$$

or equivalently,

$$\frac{1}{2}\left(\widehat{\mathbf{T}}^{(nc)} - \widehat{\mathbf{T}}^{(nc)T}\right)\mathbf{n} = \mathbf{0}. \tag{6.41}$$

The corresponding first Piola–Kirchhoff stress tensor for the nematic material is equal to

$$\widehat{\mathbf{P}}^{(nc)} = \widehat{\mathbf{T}}^{(nc)}\mathrm{CofF}. \tag{6.42}$$

The associated second Piola–Kirchhoff stress tensor is

$$\widehat{\mathbf{S}}^{(nc)} = \mathbf{F}^{-1}\widehat{\mathbf{P}}^{(nc)}. \tag{6.43}$$

6.3   A Continuum Elastic–Nematic Model

To describe the nematic material, we combine isotropic hyperelastic and neoclassical strain-energy densities as follows [363]:

$$W^{(lce)}(\mathbf{F}, \mathbf{n}) = W^{(1)}\left(\mathbf{FG}_0^{-1}\right) + W^{(2)}(\mathbf{F}, \mathbf{n}), \tag{6.44}$$

where, on the right-hand side, the first term represents the hyperelastic strain-energy density associated with the overall macroscopic deformation [113, 178], and the second term is the strain-energy density for the polymer microstructure described by (6.1). Specifically, $\mathbf{F}$ is the deformation gradient tensor, $\mathbf{n}$ is a unit vector for the director orientation in the present configuration, and $\mathbf{G}_0$ takes the form (6.3) with the director orientation in the reference configuration, $\mathbf{n}_0$, which may be spatially varying, instead of $\mathbf{n}$. For LCE, $\nu = 1/2$.

Because the classical Mooney–Rivlin (MR) hyperelastic model [385, 465] captures well the elastic behaviour of polymeric networks at small and medium strains [362, 417, 450], we consider the MR hyperelastic strain-energy function

$$W(\mathbf{A}) = \frac{\mu_1}{2}\left[\mathrm{tr}\left(\mathbf{AA}^T\right) - 3\right] + \frac{\mu_2}{2}\left\{\mathrm{tr}\left[\mathrm{Cof}\left(\mathbf{AA}^T\right)\right] - 3\right\}, \tag{6.45}$$

where $\mu_1 + \mu_2 > 0$ is the shear modulus for infinitesimal deformations, with $\mu_1 \geq 0$ and $\mu_2 \geq 0$ material constants. When $\mu_2 = 0$, this model reduces to the classical neo-Hookean (NH) model [549].

The associated MR-type neoclassical strain-energy function, defined by (6.1), is equal to [180, 496]

$$W^{(nc)}(\mathbf{F}, \mathbf{n}) = \frac{\mu_1}{2}\left\{a^{1/3}\left[\mathrm{tr}\left(\mathbf{FF}^T\right) - \left(1 - a^{-1}\right)\mathbf{n}\cdot\mathbf{FF}^T\mathbf{n}\right] - 3\right\}$$
$$+ \frac{\mu_2}{2}\left\{a^{-1/3}\left[\mathrm{tr}\left(\mathbf{F}^{-T}\mathbf{F}^{-1}\right) - (1-a)\mathbf{n}\cdot\mathbf{F}^{-T}\mathbf{F}^{-1}\mathbf{n}\right] - 3\right\}. \tag{6.46}$$

This function takes the following equivalent form in terms of the invariants defined by (6.13)–(6.17), with $I_3^{(nc)} = 1$,

$$W^{(nc)}(I_1^{(nc)}, I_2^{(nc)}, I_4^{(nc)}, I_5^{(nc)}) = \frac{\mu_1}{2}\left\{a^{1/3}\left[I_1^{(nc)} - \left(1 - a^{-1}\right)I_4^{(nc)}\right] - 3\right\}$$
$$+ \frac{\mu_2}{2}\left\{a^{-1/3}\left[I_2^{(nc)} - (1-a)\left(I_5^{(nc)} - I_1^{(nc)}I_4^{(nc)} + I_2^{(nc)}\right)\right] - 3\right\}. \tag{6.47}$$

Equivalently, in terms of the principal stretches, this strain-energy function is given by

$$W^{(nc)}(\lambda_1, \lambda_2, \lambda_3, \mathbf{n}) = \frac{\mu_1}{2}a^{1/3}\left\{\left[\sum_{i=1}^{3}\lambda_i^2 - \left(1 - a^{-1}\right)\sum_{i=1}^{3}\lambda_i^2(\mathbf{e}_i\cdot\mathbf{n})^2\right] - 3\right\}$$
$$+ \frac{\mu_2}{2}\left\{a^{-1/3}\left[\sum_{i=1}^{3}\lambda_i^{-2} - (1-a)\sum_{i=1}^{3}\lambda_i^{-2}(\mathbf{e}_i\cdot\mathbf{n})^2\right] - 3\right\}, \tag{6.48}$$

where $\{\lambda_i^2\}_{i=1,2,3}$ are the principal eigenvalues and $\{\mathbf{e}_i\}_{i=1,2,3}$ are the principal eigenvectors of the tensor $\mathbf{B} = \mathbf{F}\mathbf{F}^T$.

Taking $W^{(1)}$ and $W^{(2)}$ in (6.44) to be a classical and a neoclassical MR-type strain-energy density, respectively, we define the following LCE composite strain-energy density:

$$
\begin{aligned}
W^{(lce)}(\mathbf{F}, \mathbf{n}) = {} & \frac{\mu_1^{(1)}}{2}\left[\mathrm{tr}\left(\mathbf{F}\mathbf{G}_0^{-2}\mathbf{F}^T\right) - 3\right] \\
& + \frac{\mu_2^{(1)}}{2}\left[\mathrm{tr}\left(\mathbf{F}^{-T}\mathbf{G}_0^2\mathbf{F}^{-1}\right) - 3\right] \\
& + \frac{\mu_1^{(2)}}{2}\left\{a^{1/3}\left[\mathrm{tr}\left(\mathbf{F}\mathbf{F}^T\right) - \left(1 - a^{-1}\right)\mathbf{n}\cdot\mathbf{F}\mathbf{F}^T\mathbf{n}\right] - 3\right\} \\
& + \frac{\mu_2^{(2)}}{2}\left\{a^{-1/3}\left[\mathrm{tr}\left(\mathbf{F}^{-T}\mathbf{F}^{-1}\right) - (1 - a)\mathbf{n}\cdot\mathbf{F}^{-T}\mathbf{F}^{-1}\mathbf{n}\right] - 3\right\},
\end{aligned}
$$
(6.49)

where the shear modulus at infinitesimal strain is equal to $\mu = \mu^{(1)} + \mu^{(2)} = \mu_1^{(1)} + \mu_2^{(1)} + \mu_1^{(2)} + \mu_2^{(2)} > 0$, with $\mu^{(1)} = \mu_1^{(1)} + \mu_2^{(1)} > 0$ and $\mu^{(2)} = \mu_1^{(2)} + \mu_2^{(2)} > 0$.

6.4 Shear Striping Instability

Many macroscopic deformations of nematic liquid crystal elastomers (LCEs) induce a re-orientation of the director with a general tendency for the director to become parallel to the direction of the largest principal stretch. This re-orientation is typically uniform across the material. However, non-uniform behaviours are also possible. In particular, under appropriate uniaxial tension or biaxial stretch, bifurcation to a pattern of stripe domains is generated, where adjacent stripes deform by the same shear but in opposite directions. Early experimental investigations of this phenomenon, known as *soft elasticity* [419, 571–573, 590], were reported in [165, 311, 312, 533, 630] (see also [432, 433]). Its theoretical explanation is that, for these materials, the energy is minimised by passing through a state exhibiting a microstructure of many homogeneously deformed parts [113, 129, 132, 178–180, 311] (Fig. 6.4).

We analyse shear striping under biaxial stretch and assume that the nematic director can only rotate in the biaxial plane. In a Cartesian system of rectangular coordinates, we set [132]

$$
\mathbf{n}_0 = \begin{bmatrix} 0 \\ 0 \\ 1 \end{bmatrix}, \qquad \mathbf{n} = \begin{bmatrix} 0 \\ \sin\theta \\ \cos\theta \end{bmatrix}, \tag{6.50}
$$

Fig. 6.4 Schematics of a monodomain nematic solid, with uniform director field aligned vertically, developing periodic shear stripes under biaxial stretch before full rotation of the nematic director is achieved. In practice, the material appears opaque during director rotation and translucent when the director is uniformly aligned [311, 435]

where $\theta \in [0, \pi/2]$. We examine small shear perturbations of biaxial stretch deformations, with gradient tensor [132]

$$\mathbf{F} = \begin{bmatrix} a^{-1/6} & 0 & 0 \\ 0 & \lambda & \varepsilon \\ 0 & 0 & a^{1/6}\lambda^{-1} \end{bmatrix}, \tag{6.51}$$

where $a > 1$ is the nematic parameter, $\lambda > 0$ is the stretch ratio, and $\varepsilon > 0$ is the small perturbation. The corresponding left Cauchy–Green tensor and its cofactor are, respectively,

$$\mathbf{B} = \mathbf{FF}^T = \begin{bmatrix} a^{-1/3} & 0 & 0 \\ 0 & \lambda^2 + \varepsilon^2 & \varepsilon a^{1/6}\lambda^{-1} \\ 0 & \varepsilon a^{1/6}\lambda^{-1} & a^{1/3}\lambda^{-2} \end{bmatrix}, \tag{6.52}$$

$$\text{Cof}\mathbf{B} = \mathbf{F}^{-T}\mathbf{F}^{-1} = \begin{bmatrix} a^{1/3} & 0 & 0 \\ 0 & \lambda^{-2} & -\varepsilon a^{-1/6}\lambda^{-1} \\ 0 & -\varepsilon a^{-1/6}\lambda^{-1} & a^{-1/3}\left(\lambda^2 + \varepsilon^2\right) \end{bmatrix}. \tag{6.53}$$

It follows that

$$\text{tr}\left(\mathbf{FF}^T\right) = a^{-1/3} + \lambda^2 + \varepsilon^2 + a^{1/3}\lambda^{-2}, \tag{6.54}$$

$$\mathbf{n} \cdot \mathbf{FF}^T\mathbf{n} = \left(\lambda^2 + \varepsilon^2\right)\sin^2\theta + \varepsilon a^{1/6}\lambda^{-1}\sin\left(2\theta\right) + a^{1/3}\lambda^{-2}\cos^2\theta, \tag{6.55}$$

$$\text{tr}\left(\mathbf{F}^{-T}\mathbf{F}^{-1}\right) = a^{1/3} + \lambda^{-2} + a^{-1/3}\left(\lambda^2 + \varepsilon^2\right), \tag{6.56}$$

$$\mathbf{n} \cdot \mathbf{F}^{-T}\mathbf{F}^{-1}\mathbf{n} = \lambda^{-2}\sin^2\theta - \varepsilon a^{-1/6}\lambda^{-1}\sin\left(2\theta\right) + a^{-1/3}\left(\lambda^2 + \varepsilon^2\right)\cos^2\theta. \tag{6.57}$$

We also have

$$\mathbf{FG}_0^{-1} = \begin{bmatrix} 1 & 0 & 0 \\ 0 & a^{1/6}\lambda & a^{-1/3}\varepsilon \\ 0 & 0 & a^{-1/6}\lambda^{-1} \end{bmatrix}. \tag{6.58}$$

Denoting $w(\lambda, \varepsilon, \theta) = W^{(lce)}(\mathbf{F}, \mathbf{n})$, we obtain

$$
\begin{aligned}
w(\lambda, \varepsilon, \theta) =\ & \frac{\mu_1^{(1)} + \mu_2^{(1)}}{2} \left(a^{1/3}\lambda^2 + a^{-2/3}\varepsilon^2 + a^{-1/3}\lambda^{-2} - 2 \right) \\
& + \frac{\mu_1^{(2)}}{2} a^{1/3} \Big\{ a^{-1/3} + \lambda^2 + \varepsilon^2 + a^{1/3}\lambda^{-2} \\
& \quad - \left(1 - a^{-1} \right) \left[\left(\lambda^2 + \varepsilon^2 \right) \sin^2\theta + \varepsilon a^{1/6}\lambda^{-1}\sin(2\theta) + a^{1/3}\lambda^{-2}\cos^2\theta \right] \Big\} \\
& + \frac{\mu_2^{(2)}}{2} a^{-1/3} \Big\{ a^{1/3} + \lambda^{-2} + a^{-1/3}\left(\lambda^2 + \varepsilon^2 \right) \\
& \quad - (1 - a)\left[\lambda^{-2}\sin^2\theta - \varepsilon a^{-1/6}\lambda^{-1}\sin(2\theta) + a^{-1/3}\left(\lambda^2 + \varepsilon^2 \right)\cos^2\theta \right] \Big\} \\
& - 3\frac{\mu_1^{(2)} + \mu_2^{(2)}}{2}.
\end{aligned}
\tag{6.59}
$$

Differentiating the above function with respect to ε and θ, respectively, gives

$$
\begin{aligned}
\frac{\partial w(\lambda, \varepsilon, \theta)}{\partial \varepsilon} =\ & \left(\mu_1^{(1)} + \mu_2^{(1)} \right) a^{-2/3}\varepsilon \\
& + \frac{\mu_1^{(2)}}{2} a^{1/3} \left\{ 2\varepsilon - \left(1 - a^{-1} \right)\left[2\varepsilon\sin^2\theta + a^{1/6}\lambda^{-1}\sin(2\theta) \right] \right\} \\
& + \frac{\mu_2^{(2)}}{2} a^{-1/3} \left\{ 2\varepsilon a^{-1/3} - (1 - a)\left[2\varepsilon a^{-1/3}\cos^2\theta - a^{-1/6}\lambda^{-1}\sin(2\theta) \right] \right\},
\end{aligned}
\tag{6.60}
$$

$$
\begin{aligned}
\frac{\partial w(\lambda, \varepsilon, \theta)}{\partial \theta} =\ & \frac{\mu_1^{(2)}}{2} \left(a^{-2/3} - a^{1/3} \right)\left[\left(\lambda^2 + \varepsilon^2 - a^{1/3}\lambda^{-2} \right)\sin(2\theta) + 2\varepsilon a^{1/6}\lambda^{-1}\cos(2\theta) \right] \\
& + \frac{\mu_2^{(2)}}{2} \left(a^{2/3} - a^{-1/3} \right)\left\{ \left[\lambda^{-2} - a^{-1/3}\left(\lambda^2 + \varepsilon^2 \right) \right]\sin(2\theta) - 2\varepsilon a^{-1/6}\lambda^{-1}\cos(2\theta) \right\}.
\end{aligned}
\tag{6.61}
$$

The equilibrium solution minimises the energy and hence satisfies

$$\frac{\partial w(\lambda, \varepsilon, \theta)}{\partial \varepsilon} = 0, \qquad \frac{\partial w(\lambda, \varepsilon, \theta)}{\partial \theta} = 0. \tag{6.62}$$

At $\varepsilon = 0$ and $\theta = 0$, both the partial derivatives defined by (6.60)–(6.61) are equal to zero, i.e.,

$$\frac{\partial w}{\partial \varepsilon}(\lambda, 0, 0) = \frac{\partial w}{\partial \theta}(\lambda, 0, 0) = 0. \tag{6.63}$$

Therefore, this trivial solution is always an equilibrium state, and for sufficiently small values of ε and θ, we have the second-order approximation [132]

$$w(\lambda, \varepsilon, \theta) \approx w(\lambda, 0, 0) + \frac{1}{2}\left(\varepsilon^2 \frac{\partial^2 w}{\partial \varepsilon^2}(\lambda, 0, 0) + 2\varepsilon\theta \frac{\partial^2 w}{\partial \varepsilon \partial \theta}(\lambda, 0, 0) + \theta^2 \frac{\partial^2 w}{\partial \theta^2}(\lambda, 0, 0)\right), \tag{6.64}$$

where

$$\frac{\partial^2 w}{\partial \varepsilon^2}(\lambda, 0, 0) = \mu^{(1)}a^{-2/3} + \mu^{(2)}a^{1/3}, \tag{6.65}$$

$$\frac{\partial^2 w}{\partial \varepsilon \partial \theta}(\lambda, 0, 0) = \mu^{(2)}\lambda^{-1}\left(a^{-1/2} - a^{1/2}\right), \tag{6.66}$$

$$\frac{\partial^2 w}{\partial \theta^2}(\lambda, 0, 0) = \mu^{(2)}\left(a^{-2/3} - a^{1/3}\right)\left(\lambda^2 - a^{1/3}\lambda^{-2}\right), \tag{6.67}$$

with $\mu^{(1)} = \mu_1^{(1)} + \mu_2^{(1)}$ and $\mu^{(2)} = \mu_1^{(2)} + \mu_2^{(2)}$. First, we find the equilibrium value θ_0 for θ as a function of ε by solving the second equation in (6.62), which, after the approximation (6.64), takes the form

$$\varepsilon \frac{\partial^2 w}{\partial \varepsilon \partial \theta}(\lambda, 0, 0) + \theta \frac{\partial^2 w}{\partial \theta^2}(\lambda, 0, 0) = 0. \tag{6.68}$$

This implies

$$\theta_0(\varepsilon) = -\varepsilon \frac{\partial^2 w}{\partial \varepsilon \partial \theta}(\lambda, 0, 0) \Big/ \frac{\partial^2 w}{\partial \theta^2}(\lambda, 0, 0), \tag{6.69}$$

and substituting $\theta = \theta_0(\varepsilon)$ in (6.64) gives the following function of ε:

$$w(\lambda, \varepsilon, \theta_0(\varepsilon)) - w(\lambda, 0, 0) \approx \frac{\varepsilon^2}{2}\left[\frac{\partial^2 w}{\partial \varepsilon^2}(\lambda, 0, 0) - \left(\frac{\partial^2 w}{\partial \varepsilon \partial \theta}(\lambda, 0, 0)\right)^2 \Big/ \frac{\partial^2 w}{\partial \theta^2}(\lambda, 0, 0)\right]. \tag{6.70}$$

Depending on whether the expression on the right-hand side in (6.70) is positive, zero, or negative, the equilibrium state with $\varepsilon = 0$ and $\theta = 0$ is stable, neutrally stable, or unstable [132]. Clearly, if $\mu^{(1)} > 0$ and $\mu^{(2)} = 0$ (purely elastic case), then the equilibrium state is stable. If $\mu^{(2)} > 0$, then, assuming $a > 1$, this equilibrium state is unstable for [363]

$$a^{-1/6} \leq a^{1/12}\left(\frac{\eta + 1}{\eta + a}\right)^{1/4} < \lambda < a^{1/12}, \tag{6.71}$$

where $\eta = \mu^{(1)}/\mu^{(2)}$ is the parameter ratio of the elastic and neoclassical contributions.

There is also an equilibrium state with $\varepsilon = 0$ and $\theta = \pi/2$, satisfying (6.62), where the nematic director is fully rotated so that it aligns uniformly with the direction of macroscopic extension. By symmetry arguments, this state is unstable for

$$a^{1/12} < \lambda < a^{1/12} \left(\frac{\eta + a}{\eta + 1}\right)^{1/4} \leq a^{1/3}. \tag{6.72}$$

When λ satisfies (6.71) or (6.72), solving the simultaneous equations (6.62), we obtain

$$\varepsilon_0 = \pm \frac{\lambda(a-1)\sin(2\theta_0)}{2\sqrt{(\eta+1)(\eta+a)}}, \qquad \theta_0 = \pm \arccos\sqrt{\frac{a^{1/6}\sqrt{(\eta+1)(\eta+a)}}{\lambda^2(a-1)} - \frac{\eta+1}{a-1}}. \tag{6.73}$$

For the resulting strip pattern, the gradient tensors of alternating shear deformations in two adjacent stripe domains are $\mathbf{F}_\pm$ with $\varepsilon = \pm\varepsilon_0$, respectively. The two deformations are geometrically compatible in the sense that there exist two non-zero vectors $\mathbf{q}$ and $\mathbf{p}$, such that the Hadamard jump condition is satisfied,

$$\mathbf{F}_+ - \mathbf{F}_- = \mathbf{q} \otimes \mathbf{p}, \tag{6.74}$$

where $\mathbf{p}$ is the normal vector to the interface between the two phases corresponding to the deformation gradients $\mathbf{F}_+$ and $\mathbf{F}_-$ [40, 41]. In other words, $\mathbf{F}_+$ and $\mathbf{F}_-$ are rank-one connected,[3] i.e.,

$$\text{rank}(\mathbf{F}_+ - \mathbf{F}_-) = 1. \tag{6.75}$$

If $\eta = 0$, then the above equilibrium states are unstable for $\lambda \in (a^{-1/6}, a^{1/12})$ and $\lambda \in (a^{1/12}, a^{1/3})$, respectively. Thus, soft elasticity is always presented by the purely neoclassical model [113, 132]. When $\eta \to \infty$, there is no shear striping since the material is practically elastic.

The strain-energy function $w(\lambda, \varepsilon, \theta)$, defined by (6.59), is illustrated in Fig. 6.5. For λ with values between the lower and upper bounds given by (6.71) and (6.72), respectively, the minimum energy is attained when $(\varepsilon, \theta) = (\varepsilon_0, \theta_0)$. Figure 6.5b then suggests that, when $\eta = 0$, i.e., for the purely neoclassical form, the director rotates and alternating shear stripes develop when $\lambda \in (a^{-1/6}, a^{1/3})$, at zero load, as the slope of the curve is equal to zero within this interval.[4] In contrast, if $\eta > 0$, then from Fig. 6.5a, we infer that the applied load increases with deformation.

---

[3] For homogeneous isotropic hyperelastic bodies, the possibility of a homogeneous Cauchy stress induced by an inhomogeneous finite deformation was demonstrated in [368, 369, 398, 484].

[4] A similar physical behaviour can be found when cellular solids, such as cork, are compressed. Namely, the stress first increases as the deformation increases, then exhibits an almost constant plateau where it increases or decreases slightly, and then monotonically increases again [597].

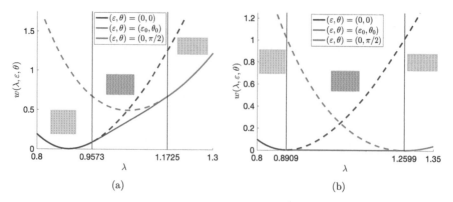

Fig. 6.5 The strain-energy function $w(\lambda, \varepsilon, \theta)$, defined by equation (6.59), for $(\varepsilon, \theta) = (0, 0)$, $(\varepsilon, \theta) = (\varepsilon_0, \theta_0)$, with ε_0 and θ_0 given by (6.73), and $(\varepsilon, \theta) = (0, \pi/2)$, when $a = 2$ and **(a)** $\mu^{(1)} = 4.05$ or **(b)** $\mu^{(1)} = 0$, while $\mu_1^{(2)} = 2.05$ and $\mu_2^{(2)} = 2$. The two vertical lines correspond to the lower and upper bounds on λ, given by equations (6.71) and (6.72), respectively. Between these bounds, the second solution, with $(\varepsilon, \theta) = (\varepsilon_0, \theta_0)$, minimises the energy

Due to the geometric compatibility, and since the intervals for stretch ratios λ where shear striping occurs are at a maximum length when $\mathbf{n}_0 = [0, 0, 1]^T$, the bounds (6.71)–(6.72) give also the maximum interval for shear striping when $\mathbf{n}_0$ is not uniformly aligned. The minimum length of those intervals is attained for monodomains with $\mathbf{n}_0 = [0, 1, 0]^T$. Experimental results for monodomains where the tensile load forms different angles with the initial nematic director are reported in [378, 418]. If $\mathbf{G}_0 = \mathbf{I}$ (see [58]), then the solution with $\varepsilon = 0$ and $\theta = 0$ is unstable for

$$a^{-1/6} \le a^{1/12} \left(\frac{\eta + a^{-1}}{\eta + 1} \right)^{1/4} < \lambda < a^{1/12} \tag{6.76}$$

and that with $\varepsilon = 0$ and $\theta = \pi/2$ is unstable for

$$a^{1/12} < \lambda < a^{1/12} \left(\frac{\eta + 1}{\eta + a^{-1}} \right)^{1/4} \le a^{1/3}. \tag{6.77}$$

For example, when $a = 2$, the bounds given by (6.71)–(6.72) and by (6.76)–(6.77), respectively, are represented as functions of the parameter ratio η in Fig. 6.6.

Next, we examine the shear striping problem when the model parameters are random variables following spatially independent probability distributions [363]. One can now regard the material sample of stochastic nematic elastomer as an ensemble of samples with the same geometry, such that each sample is made from a homogeneous nematic material with the model parameters not known with certainty, but drawn from known probability distributions. Then, for each individual sample, the finite strain theory for homogeneous materials applies. The question is: *What*

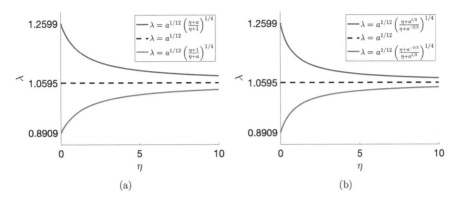

Fig. 6.6 The lower and upper bounds on λ, given by (**a**) (6.71)–(6.72) and (**b**) (6.76)–(6.77), respectively, as functions of the parameter ratio η when $a = 2$. For $\eta \to \infty$, the model approaches a purely elastic form where the homogeneous deformation is always stable. For $\eta = 0$, the bounds are the same and correspond to the neoclassical form

is the probability that shear striping occurs under a given stretch? By treating the shear modulus $\mu > 0$ and the shape parameter $a > 0$ as independent input parameters, we answer this question by showing the effects of fluctuations in these physical quantities separately. We first set μ as a random variable, while a is a single-valued constant, then keep μ constant, and let a fluctuate.

(I) Stochastic Shear Modulus and Deterministic Shape Parameter First, we assume that the shear modulus $\mu = \mu^{(1)} + \mu^{(2)} = \mu_1^{(1)} + \mu_1^{(1)} + \mu_1^{(2)} + \mu_2^{(2)}$ is a random variable, such that $\mu^{(1)} = \mu_1^{(1)} + \mu_2^{(1)} > 0$ and $\mu^{(2)} = \mu_1^{(2)} + \mu_2^{(2)} > 0$, while a is deterministic, such that $a > 1$. For purely elastic models, probability distributions for the individual model coefficients contributing to the shear parameter μ are discussed in [357, 359, 371, 372]. Here, we only require the probability density function of μ.

Substituting $\mu^{(2)} = \mu - \mu^{(1)}$ in the inequality involving the lower bound on λ in (6.71) implies

$$\frac{\mu}{\mu^{(1)}} > \left(1 - \frac{1}{a}\right) \frac{\lambda^4}{\lambda^4 - a^{-2/3}}. \tag{6.78}$$

Hence, given $a > 1$ and $\lambda > 0$, for any $\mu^{(1)} > 0$, the probability that the equilibrium state at $\varepsilon = 0$ and $\theta = 0$ is unstable, such that shear striping develops, is

$$P_1(\mu^{(1)}) = 1 - \int_0^{\mu^{(1)}\left(1 - \frac{1}{a}\right)\frac{\lambda^4}{\lambda^4 - a^{-2/3}}} g(u; \rho_1, \rho_2)\mathrm{d}u, \tag{6.79}$$

while the probability of stable equilibrium at $\varepsilon = 0$ and $\theta = 0$ is

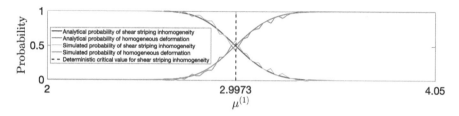

Fig. 6.7 Probability distributions of whether shear striping can occur or not for a monodomain nematic elastomer where $a = 2$ and $\lambda = 1$, and the shear modulus μ is drawn from the Gamma distribution with $\rho_1 = 405$ and $\rho_2 = 0.01$. Dark coloured lines represent analytically derived solutions, given by Eqs. (6.79)–(6.80), whereas the lighter versions represent stochastically generated data. The vertical line at the critical value $\mu^{(1)} = 2.9973$ corresponds to the deterministic solution based only on the mean value $\underline{\mu} = \rho_1\rho_2 = 4.05$. The probabilities were calculated from the average of 100 stochastic simulations

$$P_2(\mu^{(1)}) = 1 - P_1(\mu^{(1)}) = \int_0^{\mu^{(1)}\left(1-\frac{1}{a}\right)\frac{\lambda^4}{\lambda^4-a^{-2/3}}} g(u; \rho_1, \rho_2)du. \qquad (6.80)$$

For instance, setting $a = 2$ and $\lambda = 1$, the probability distributions given by equations (6.79)–(6.80) are illustrated both analytically and numerically in Fig. 6.7 (blue lines for P_1 and red lines for P_2). For the numerical approximation, the interval $(2, \underline{\mu})$ for $\mu^{(1)}$ was discretised into 100 points, and then for each value of $\mu^{(1)}$, 100 random values of μ were numerically generated from a specified Gamma distribution, with hyperparameters $\rho_1 = 405$ and $\rho_2 = 0.01$, and compared with the inequalities defining the two intervals for values of $\mu^{(1)}$. Note that, although only 100 random numbers were generated, the numerical approximation captures well the analytical solution.

In the deterministic case based on the mean value of the shear modulus, $\underline{\mu} = \rho_1\rho_2 = 4.05$, the critical value of $\mu^{(1)} = \underline{\mu}/1.3512 = 2.9973$ strictly divides the regions where striping inhomogeneity can occur or not. For the stochastic problem, to increase the probability of homogeneous deformations ($P_2 \approx 1$), one must consider sufficiently large values of $\mu^{(1)}$, whereas shear striping is certain ($P_1 \approx 1$) only if the model reduces to the neoclassical one (i.e., when $\mu^{(1)} = 0$). However, while $\mu^{(1)} > 0$, the inherent variability in the probabilistic system means that there will always be competition between the homogeneous and inhomogeneous deformations.

In Fig. 6.8, the stochastic shear parameter ε_0 and director angle θ_0, given by (6.73), are represented when $a = 2$ is deterministic and the shear modulus $\mu = \mu^{(1)} + \mu^{(2)}$ follows a Gamma distribution with shape and scale parameters $\rho_1 = 405$ and $\rho_2 = 0.01$, respectively, while $R^{(1)} = \mu^{(1)}/\mu$ is drawn from a Beta distribution with hyperparameters $\xi_1 = 100$ and $\xi_2 = 100$. Then the mean value of μ is $\underline{\mu} = \rho_1\rho_2 = 4.05$, and the mean value of $R^{(1)}$ is $\underline{R}^{(1)} = \xi_1/(\xi_1 + \xi_2) = 0.5$. To compare directly our stochastic results with the deterministic ones, we sampled

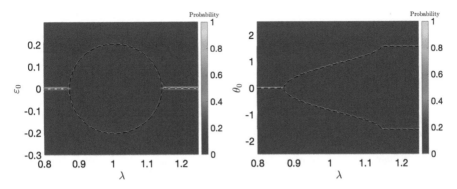

Fig. 6.8 The stochastic shear parameter ε_0 and director angle θ_0, given by (6.73), when $a = 2$, while the shear modulus μ follows a Gamma distribution with $\rho_1 = 405$ and $\rho_2 = 0.01$, and $R^{(1)} = \mu^{(1)}/\mu$ is drawn from a Beta distribution with $\xi_1 = 100$ and $\xi_2 = 100$. The dashed black lines correspond to the deterministic solutions based only on the mean values $\underline{\mu} = \rho_1\rho_2 = 4.05$ and $\underline{R}_1 = \xi_1/(\xi_1 + \xi_2) = 0.5$, whereas the red versions show the arithmetic mean value solutions

from a distribution where the shear modulus was set to have the mean value corresponding to the deterministic system. Due to the linear form in which this modulus is computed from the model coefficients, the mean value solutions are guaranteed to converge to the deterministic solutions.

(II) Stochastic Shape Parameter and Deterministic Shear Modulus Next, we assume that the value of the nematic shape parameter a "fluctuates," while the director remains uniformly oriented [571, 572] (see also the discussion in [180]). Given that a takes finite positive values, it is reasonable to require that assumption (C3) also holds for this parameter. Then, a also follows a Gamma distribution. To probe the independent effect of fluctuating a, we take the coefficients $\mu^{(1)}$ and $\mu^{(2)}$ to be deterministic constants. Hence $\mu = \mu^{(1)} + \mu^{(2)}$ is deterministic.

For example, setting $\lambda = 1$, the inequality involving the lower bound on λ in (6.71) is equivalent to

$$
a > a_c = \left[\sqrt{\left(\frac{\mu}{\mu^{(1)}} - 1 \right)\left(\frac{\mu}{\mu^{(1)}} + 3 \right)} - \frac{\mu}{\mu^{(1)}} + 1 \right]^3 \bigg/ \left[8\left(\frac{\mu}{\mu^{(1)}} - 1 \right)^3 \right].
$$
(6.81)

Hence, for any $\mu/\mu^{(1)} > 1$, the probability that the equilibrium state at $\varepsilon = 0$ and $\theta = 0$ is unstable, such that shear stripes form, is

$$
P_1\left(\frac{\mu}{\mu^{(1)}} \right) = 1 - \int_0^{a_c} g(u; \rho_1, \rho_2)\mathrm{d}u,
$$
(6.82)

while the probability of stable equilibrium at $\varepsilon = 0$ and $\theta = 0$ is

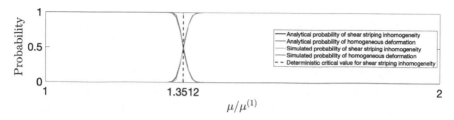

Fig. 6.9 Probability distributions of whether shear striping can occur or not for a monodomain nematic elastomer where $\lambda = 1$ and the shape parameter a is drawn from the Gamma distribution with $\rho_1 = 200$ and $\rho_2 = 0.01$. Dark coloured lines represent analytically derived solutions, given by Eqs. (6.82)–(6.83), while the lighter colours represent stochastically generated data. The vertical line at the critical value $\mu/\mu^{(1)} = 1.3512$ corresponds to the deterministic solution based only on the mean value $\underline{a} = \rho_1\rho_2 = 2$. The probabilities were calculated from the average of 100 stochastic simulations

$$P_2\left(\frac{\mu}{\mu^{(1)}}\right) = 1 - P_1\left(\frac{\mu}{\mu^{(1)}}\right) = \int_0^{a_c} g(u; \rho_1, \rho_2)du, \tag{6.83}$$

where $g(u; \rho_1, \rho_2)$ now represents the probability density function for the Gamma-distributed shape parameter a.

The probability distributions given by equations (6.82)–(6.80) are illustrated in Fig. 6.9 (blue lines for P_1 and red lines for P_2). For the numerical realisations in these plots, the interval $(1, 2)$ $\mu/\mu^{(1)}$ was discretised into 100 representative points, then for each value of $\mu/\mu^{(1)}$, 100 random values of a were numerically generated from the specified Gamma distribution and compared with the inequalities defining the two intervals for values of $\mu/\mu^{(1)}$.

For the deterministic solution based on the mean value of the nematic parameter, $\underline{a} = \rho_1\rho_2 = 2$, the critical value of $\mu/\mu^{(1)} = 1.3512$ strictly separates the regions where striping inhomogeneity occurs or not. For the stochastic version, to increase the probability of homogeneous deformations ($P_2 \approx 1$), one must decrease the values of $\mu/\mu^{(1)}$, whereas shear striping is certain ($P_1 \approx 1$) only if the model reduces to the neoclassical one (i.e., when $\mu^{(1)} = 0$). When $\mu^{(1)} > 0$, there will always be competition between the homogeneous and striping deformations.

To illustrate the effect of the probabilistic shape parameter a, in Fig. 6.10, we represent the stochastic shear parameter ε_0 and director angle θ_0, given by (6.73), when $\mu = 4.05$ and $R^{(1)} = \mu^{(1)}/\mu = 0.5$ are deterministic, while the shape parameter a follows a Gamma distribution with shape and scale parameters $\rho_1 = 200$ and $\rho_2 = 0.01$, respectively. Then the mean value of a is $\underline{a} = \rho_1\rho_2 = 2$. In this case, there is a significant difference between the mean value solutions and the deterministic solutions.

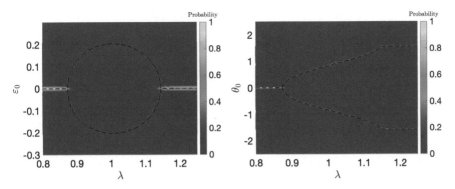

Fig. 6.10 The stochastic shear parameter ε_0 and director angle θ_0, given by (6.73), when $\mu = 4.05$ and $R^{(1)} = \mu^{(1)}/\mu = 0.5$, while the shape parameter a follows a Gamma distribution with $\rho_1 = 200$ and $\rho_2 = 0.01$. The dashed black lines correspond to the deterministic solutions based only on the mean value $\underline{a} = \rho_1\rho_2 = 2$, whereas the red versions show the arithmetic mean value solutions

6.5 Inflation of a Nematic Spherical Shell

Next, we analyse *inflation instabilities* in nematic spherical shells of Mooney–Rivlin-type nematic material given by (6.46), with the nematic director aligned in the radial direction [277, 278].

Taking a spherical system of reference with coordinates (R, Θ, Φ), the sphere is deformed by radially symmetric inflation with deformation gradient $\mathbf{F} = \text{diag}\left(\lambda^{-2}, \lambda, \lambda\right)$, and the natural deformation tensor is equal to $\mathbf{G} = \text{diag}\left(a^{-1/3}, a^{1/6}, a^{1/6}\right)$, such that $\lambda > a^{1/6} > 1$. We denote by $\mathcal{W}^{(nc)}(\lambda, \mathbf{n}) = W^{(nc)}(\mathbf{F}, \mathbf{n})$ the material strain-energy function and assume that the shell is thin, i.e., $0 < \epsilon = (B - A)/A \ll 1$, where A and B represent the inner and outer radii of the reference shell, respectively. If the deformation is caused by a radial pressure applied uniformly on the inner surface in the present configuration, then the radial Cauchy stress at the inner surface takes the form $T^{(nc)} = \epsilon a^{1/2}\lambda^{-2}d\mathcal{W}^{(nc)}/d\lambda$. The relation between the Cauchy stress at the inner surface in the nematic and a purely elastic shell with the same shear modulus is $T^{(nc)} = T$. In Fig. 6.11a–b, we show that, if the parameter ratio μ_1/μ is sufficiently small, then the required stress changes from increasing to decreasing, i.e., the material displays inflation instability. Moreover, for the nematic sphere, instability is expected at larger deformation than for the purely elastic sphere. However, this value decreases if the constitutive model includes also an additional elastic energy as in (6.44), while the shear modulus μ remains the same. For the nematic sphere, in Fig. 6.12, either the shear modulus $\mu = \mu_1+\mu_2$ or the shape parameter a is a random variable. In both cases, the critical load for instability resides in a probabilistic interval where the stable and unstable states compete. To decrease the chance of stable inflation, one must increase the value of μ_1/μ, and only when $\mu_1 = \mu$ unstable inflation is guaranteed.

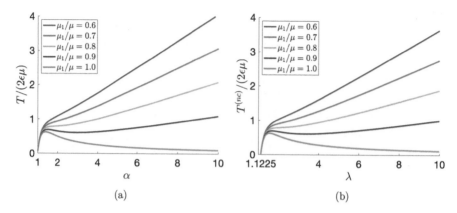

(a)   (b)

Fig. 6.11 The effect of changing the value of the parameter ratio μ_1/μ on the normalised internal pressure for a spherical shell of (**a**) MR hyperelastic material, defined by (6.45), when $\mathbf{A} = \text{diag}\left(\alpha^{-2}, \alpha, \alpha\right)$, with $\alpha > 1$, and (**b**) MR-type LCE material, defined by (6.46), when $\mathbf{F} = \text{diag}\left(\lambda^{-2}, \lambda, \lambda\right)$ and $\mathbf{G} = \text{diag}\left(a^{-1/3}, a^{1/6}, a^{1/6}\right)$, with $a = 2$ and $\lambda > a^{1/6} \approx 1.1225$. Note that, for the nematic model, instability is expected at larger deformation than for the underlying hyperelastic model

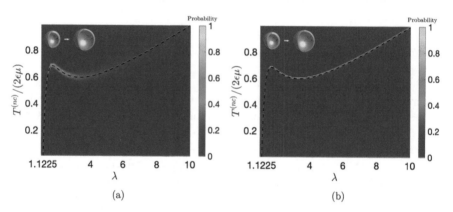

(a)   (b)

Fig. 6.12 Probability distribution of internal pressure for a spherical shell of (**a**) MR-type LCE material with $a = 2$, while the shear modulus μ is drawn from a Gamma distribution with $\rho_1 = 405$ and $\rho_2 = 0.01$, and $R_1 = \mu_1/\mu$ follows a Beta distribution with $\xi_1 = 900$ and $\xi_2 = 100$, and (**b**) MR-type LCE material with $\mu = 4.05$ and $R_1 = \mu_1/\mu = 0.9$, while the shape parameter a follows a Gamma distribution with $\rho_1 = 200$ and $\rho_2 = 0.01$. The dashed black lines correspond to the deterministic solutions based only on the mean values of the parameters, whereas the red versions show the arithmetic mean value solutions

6.6 Necking of a Nematic Elastomer

Like other rubber-like materials, when stretched, LCEs may suffer from *necking instability* [106, 107]. However, neo-Hookean or Mooney–Rivlin-type hyperelastic models are incapable of capturing necking (see Chap. 4), and this property is inherited by the associated nematic LCE models. For example, in [244], a necking instability observed experimentally under uniaxial tension could not be captured by the neoclassical LCE model based on the neo-Hookean strain-energy function alone. As necking was initiated during director rotation, a composite model consisting of a purely neoclassical form and an additional elastic form, as presented in Sect. 6.4, should be useful in predicting the non-zero stress plateau associated with the neck formation.

To explore necking instability that may occur when the director is parallel to the applied tensile force and compare that with the same behaviour in a purely elastic material, we consider the hyperelastic Gent-Thomas (GT) model [194] defined by (see Table 2.5)

$$W(\mathbf{A}) = \frac{\mu_1}{2}\left[\operatorname{tr}\left(\mathbf{A}\mathbf{A}^T\right) - 3\right] + \frac{3\mu_2}{2}\ln\frac{\operatorname{tr}\left[\operatorname{Cof}\left(\mathbf{A}\mathbf{A}^T\right)\right]}{3}, \tag{6.84}$$

where $\mu = \mu_1 + \mu_2 > 0$ is the shear modulus at small strain. For this model, if μ_1/μ is sufficiently small, then necking instability is observed under uniaxial tension, with deformation gradient tensor $\mathbf{A} = \operatorname{diag}\left(\alpha^{-1/2}, \alpha, \alpha^{-1/2}\right)$. This is demonstrated by the change in monotonicity of the function represented in Fig. 6.13a. In this case, denoting $\mathcal{W}(\alpha) = W(\alpha^{-1/2}, \alpha, \alpha^{-1/2})$, the uniaxial tensile load is equal to $P_2 = d\mathcal{W}/d\alpha$.

A Gent–Thomas-type neoclassical strain-energy function for the nematic material then reads

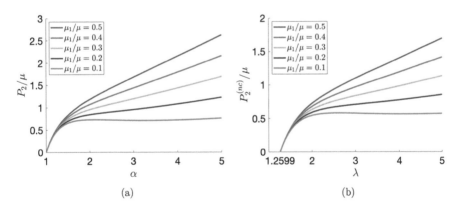

(a)   (b)

Fig. 6.13 The effect of changing the deterministic value of parameter ratio μ_1/μ on the normalised tensile first Piola–Kirchhoff stress for (**a**) the Gent–Thomas model, defined by (6.84), when $\mathbf{A} = \operatorname{diag}\left(\alpha^{-1/2}, \alpha, \alpha^{-1/2}\right)$ with $\alpha > 1$, and (**b**) the Gent–Thomas-type LCE model, defined by (6.85), when $\mathbf{F} = \operatorname{diag}\left(\lambda^{-1/2}, \lambda, \lambda^{-1/2}\right)$ and $\mathbf{G} = \operatorname{diag}\left(a^{-1/6}, a^{1/3}, a^{-1/6}\right)$, with $a = 2$ and $\lambda > a^{1/3} \approx 1.2599$. If the parameter ratio is sufficiently small, then the required dead load changes from increasing to decreasing, and the material displays necking instability. Note that, for the nematic model, necking is expected at larger deformation and lower maximum load than for the underlying hyperelastic model

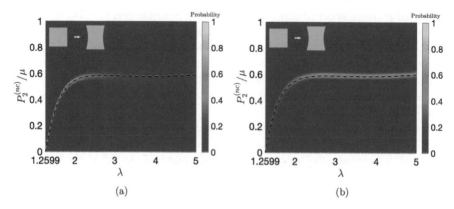

Fig. 6.14 Probability distribution of the normalised tensile load $P_2^{(nc)}$ for the Gent–Thomas-type LCE model when (**a**) $a = 2$, while the shear modulus μ is drawn from a Gamma distribution with $\rho_1 = 405$ and $\rho_2 = 0.01$, and $R_1 = \mu_1/\mu$ follows a Beta distribution with $\xi_1 = 12$ and $\xi_2 = 100$, and (**b**) $\mu = 4.05$ and $R_1 = \mu_1/\mu = 0.11$, while the shape parameter a follows a Gamma distribution with $\rho_1 = 200$ and $\rho_2 = 0.01$. The dashed black lines correspond to the deterministic solutions based only on the mean values of the parameters, whereas the red versions show the arithmetic mean value solutions

$$W^{(nc)}(\mathbf{F}, \mathbf{n}) = W(\mathbf{G}^{-1}\mathbf{F}).\qquad(6.85)$$

We take $\mathbf{F} = \mathrm{diag}\left(\lambda^{-1/2}, \lambda, \lambda^{-1/2}\right)$, while $\mathbf{G} = \mathrm{diag}\left(a^{-1/6}, a^{1/3}, a^{-1/6}\right)$ and $\lambda > a^{1/3} > 1$. Denoting $\mathcal{W}^{(nc)}(\lambda, \mathbf{n}) = W^{(nc)}(\mathbf{F}, \mathbf{n})$, the uniaxial tensile load is given by the first Piola–Kirchhoff stress $P_2^{(nc)} = d\mathcal{W}^{(nc)}/d\lambda$, and necking occurs when the ratio μ_1/μ is sufficiently small. The relation between the first Piola–Kirchhoff tensile stress in the nematic and purely elastic case with the same shear modulus is $P_2^{(nc)} = a^{-1/3} P_2$. As shown in Fig. 6.13, for the nematic model, necking is expected at larger deformation and lower maximum dead load than for the hyperelastic model. However, the maximum load will increase if the model is modified to include an additional elastic energy as in (6.44), while the shear modulus μ remains the same. In Fig. 6.14, the stochastic tensile load in the deformed LCE is illustrated.

6.7 Auxeticity and Biaxiality

The nematic LCEs studied so far have uniaxial symmetry, given by the nematic director $\mathbf{n}$. However, strain-induced biaxial nematic order, with a secondary axis of symmetry in a plane orthogonal to $\mathbf{n}$, is also possible [594, Sec 6.6] (for a summary of nematic order, see also [594, Sec. 2.2]). In [377–379, 453], experimental observations for a nematic LCE exhibiting auxetic effects when subject to a uniaxial tensile force were reported where biaxial symmetry emerges. We choose a Cartesian system of coordinates (X_1, X_2, X_3) in which the second direction is along the reference

Fig. 6.15 When the reference auxetic LCE sample (**a**) is stretched horizontally (in the X_1 direction), its volume remains unchanged, while its thickness h_a first decreases, $h_b < h_a$ (**b**), then remains almost constant $h_c \approx h_b$ (**c**), and then increases again $h_d > h_c$ (**d**). In this LCE, the nematic director $\mathbf{n_0}$, which is initially aligned in the second direction, rotates by $\pi/2$, along $\mathbf{n}$ to become parallel to the applied force

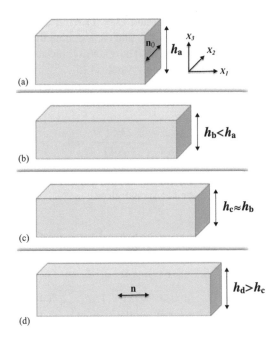

director $\mathbf{n_0}$, while the tensile force is applied in the first direction, and designate the third direction as the direction of thickness (see Fig. 6.15). Then, according to the experimental results, if the material sample is stretched longitudinally, its volume remains unchanged, but its thickness first decreases, then remains almost constant for a tensile range, and then increases again, while the director suddenly rotates to align in the direction of the applied force.

To model the auxetic LCE, we consider a composite strain-energy function of the form

$$W^{(lce)}(\mathbf{F}, \mathbf{Q}, \mathbf{n}) = W^{(1)}(\mathbf{F}) + W^{(2)}\left(\mathbf{G}^{-1}\mathbf{F}\mathbf{G_0}\right), \qquad (6.86)$$

where $\mathbf{F}$ denotes the deformation gradient from the reference cross-linking state, satisfying $\det \mathbf{F} = 1$, and $\mathbf{n}$ is the nematic director in the present configuration. We denote by $\mathbf{n_0}$ the reference orientation of the local director corresponding to the cross-linking state. On the right-hand side of Eq. (6.86), the first term is the strain-energy density associated with the overall macroscopic deformation, while the second term represents the strain-energy density of the polymer microstructure, with $\mathbf{G_0}$ and $\mathbf{G}$ denoting the "natural" deformation tensor in the reference and current configuration, respectively. In our model, these tensors satisfy the following relations [163] (see also [594, Chap. 3] and [370, 430]):

$$\mathbf{G_0^2} = c_0(\mathbf{I} + 2\mathbf{Q_0}), \qquad \mathbf{G^2} = c(\mathbf{I} + 2\mathbf{Q}), \qquad (6.87)$$

where $\mathbf{Q}_0 = \mathbf{Q}_0(\mathbf{n}_0)$ and $\mathbf{Q} = \mathbf{Q}(\mathbf{n})$ are the associated symmetric traceless order parameter tensors [125], and c_0, c denote the effective step length of the polymeric chain [594, pp.48-49]. The macroscopic tensor parameter describes orientational order in nematic liquid crystals [125].

For the above function, we assume the following conditions to be satisfied [17, 625]:

$$\frac{\partial W^{(lce)}}{\partial \mathbf{Q}} = \mathbf{0}, \qquad \frac{\partial W^{(lce)}}{\partial \mathbf{n}} = \mathbf{0}. \tag{6.88}$$

For the components of the LCE model, we use an Ogden-type strain-energy density function [412], as follows:

$$W^{(1)}(\lambda_1, \lambda_2, \lambda_3) = \sum_{j=1}^{m} \frac{c_j^{(1)}}{2\left(p_j^{(1)}\right)^2} \left(\lambda_1^{2p_j^{(1)}} + \lambda_2^{2p_j^{(1)}} + \lambda_3^{2p_j^{(1)}} - 3\right), \tag{6.89}$$

where $\{c_j^{(1)}\}_{j=1,\cdots,m}$ and $\{p_j^{(1)}\}_{j=1,\cdots,m}$ are constants independent of the deformation, and $\{\lambda_1^2, \lambda_2^2, \lambda_3^2\}$ are the eigenvalues of the tensor $\mathbf{F}^T\mathbf{F}$, such that $\lambda_1\lambda_2\lambda_3 = 1$, and

$$W^{(2)}(\alpha_1, \alpha_2, \alpha_3) = \sum_{j=1}^{n} \frac{c_j^{(2)}}{2\left(p_j^{(2)}\right)^2} \left(\alpha_1^{2p_j^{(2)}} + \alpha_2^{2p_j^{(2)}} + \alpha_3^{2p_j^{(2)}} - 3\right), \tag{6.90}$$

where $\{c_j^{(2)}\}_{j=1,\cdots,n}$ and $\{p_j^{(2)}\}_{j=1,\cdots,n}$ are constants independent of the deformation, and $\{\alpha_1^2, \alpha_2^2, \alpha_3^2\}$ are the eigenvalues of the elastic Cauchy–Green tensor $\mathbf{A}^T\mathbf{A}$, such that $\alpha_1\alpha_2\alpha_3 = 1$, with the local elastic deformation tensor $\mathbf{A} = \mathbf{G}^{-1}\mathbf{F}\mathbf{G}_0$. Hence,

$$W^{(1)}(\mathbf{F}) = W^{(1)}(\lambda_1, \lambda_2, \lambda_3) \qquad \text{and} \qquad W^{(2)}(\mathbf{A}) = W^{(2)}(\alpha_1, \alpha_2, \alpha_3). \tag{6.91}$$

The composite model defined by (6.86) then takes the form

$$W^{(lce)}(\lambda_1, \lambda_2, \lambda_3, \mathbf{Q}, \mathbf{n}) = W^{(1)}(\lambda_1, \lambda_2, \lambda_3) + W^{(2)}(\alpha_1, \alpha_2, \alpha_3). \tag{6.92}$$

For this model, the principal Cauchy stresses are equal to

$$T_i^{(lce)} = \frac{\partial W^{(lce)}}{\partial \lambda_i} \lambda_i - p, \qquad i = 1, 2, 3. \tag{6.93}$$

The associated first Piola–Kirchhoff stresses are

$$P_i^{(lce)} = T_i^{(lce)} \lambda_i^{-1}, \qquad i = 1, 2, 3. \tag{6.94}$$

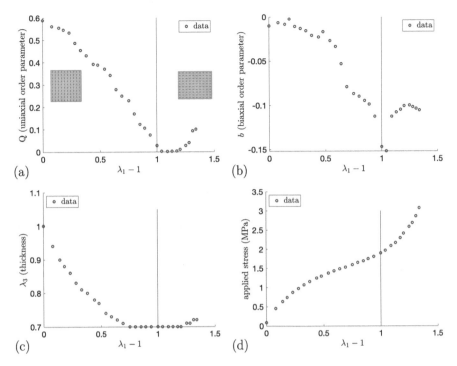

Fig. 6.16 Experimental data for (**a**) the uniaxial scalar order parameter Q, (**b**) the biaxial scalar order parameter b, (**c**) the sample thickness λ_3, and (**d**) the applied first Piola–Kirchhoff stress $P_1^{(Ice)}$ vs. the longitudinal strain $\lambda_1 - 1$. The nematic director is oriented in the second direction until a critical strain is reached and the director suddenly aligns in the first direction, i.e., parallel to the applied tensile load. In each plot, the vertical line is drawn at the critical strain where the director suddenly rotates by $\pi/2$ [367]

Following traditional deterministic approaches, in [367], the model function described by (6.92) was calibrated to the experimental data for $(\lambda_1, \theta, Q, b, \lambda_3,$ $P_1^{(Ice)})$ represented in Fig. 6.16. These data values are slightly idealised compared to those reported in [453], in the sense that the angle θ for the director orientation remains equal to $\pi/2$ until a critical extension is reached, and then it becomes 0, i.e., it is assumed that the director rotates by $\pi/2$ instantly. Note that, in Fig. 6.16c, the sample thickness first decreases and then increases at a critical large strain. Simultaneously, in Fig. 6.16a, the magnitude of the uniaxial order parameter decreases until the same critical strain is reached and then increases, while in Fig. 6.16b, the magnitude of the biaxial order parameter is larger around the critical strain. We set $Q = P_{200}$ and $b = 6P_{220}$, where P_{200} and P_{220} are determined via Raman spectroscopy in [453]. A schematic of the molecular frame with respect to the director frame used in the derivation of order parameters is shown in Fig. 6.17. For these parameters, $Q = 1$ corresponds to perfect nematic order, while $Q = 0$ is when mesogens are randomly oriented, and if $b = 0$, then the system reduces to the uniaxial case.

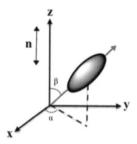

Fig. 6.17 Schematic of the molecular frame with respect to the director frame used in the derivation of uniaxial order parameter $Q = \langle (3/2) \cos^2 \beta - (1/2) \rangle$ and biaxial order parameter $b = (3/2) \langle \sin^2 \beta \cos(2\alpha) \rangle$, where $\langle \cdot \rangle$ denotes average value over mesogen angles α and β [367]

To obtain a model function with stochastic coefficients, we assume a small error in the experimental data for the measured stress and apply the calibration method described in Chap. 3.

Setting the nematic director in the reference and current configuration as $\mathbf{n}_0 = [0, 1, 0]^T$ and $\mathbf{n} = [\cos \theta, \sin \theta, 0]^T$, respectively, where $\theta \in [0, \pi/2]$, the deformation gradient takes the form

$$\mathbf{F} = \mathrm{diag}(\lambda_1, \lambda_2, \lambda_3), \tag{6.95}$$

where $\lambda_1 \lambda_2 \lambda_3 = 1$. Note that, for the uniaxial deformation under consideration, all tensors involved share the same principal directions and thus are all diagonal.

In the reference configuration, the LCE is uniaxial, and the order parameter tensor is equal to [594, p. 14]

$$\mathbf{Q}_0 = \mathrm{diag}\left(-\frac{Q_0}{2}, Q_0, -\frac{Q_0}{2} \right), \tag{6.96}$$

where Q_0 is the scalar order parameter. In this case, $\mathbf{G}_0 = \mathrm{diag}\left(a_0^{-1/6}, a_0^{1/3}, a_0^{-1/6} \right)$, where $a_0 = (1 + 2Q_0)/(1 - Q_0)$.

In the deformed configuration, when biaxiality emerges [594, p. 15]

- If $\theta = \pi/2$, then the order parameter tensor takes the form

$$\mathbf{Q} = \mathrm{diag}\left(-\frac{Q - b}{2}, Q, -\frac{Q + b}{2} \right). \tag{6.97}$$

- If $\theta = 0$, then

$$\mathbf{Q} = \mathrm{diag}\left(Q, -\frac{Q - b}{2}, -\frac{Q + b}{2} \right), \tag{6.98}$$

where Q and b are the uniaxial and biaxial scalar order parameters, respectively.

For the elastic Cauchy–Green tensor $\mathbf{A}^T \mathbf{A}$, by (6.87):

- If $\theta = \pi/2$, then

$$\alpha_1^2 = \left[\frac{\det (\mathbf{I} + 2\mathbf{Q})}{\det (\mathbf{I} + 2\mathbf{Q}_0)} \right]^{1/3} \frac{1 - Q_0}{1 - (Q - b)} \lambda_1^2,$$

$$\alpha_2^2 = \left[\frac{\det (\mathbf{I} + 2\mathbf{Q})}{\det (\mathbf{I} + 2\mathbf{Q}_0)} \right]^{1/3} \frac{1 + 2Q_0}{1 + 2Q} \lambda_2^2,$$

$$\alpha_3^2 = \left[\frac{\det (\mathbf{I} + 2\mathbf{Q})}{\det (\mathbf{I} + 2\mathbf{Q}_0)} \right]^{1/3} \frac{1 - Q_0}{1 - (Q + b)} \lambda_3^2. \tag{6.99}$$

- If $\theta = 0$, then

$$\alpha_1^2 = \left[\frac{\det (\mathbf{I} + 2\mathbf{Q})}{\det (\mathbf{I} + 2\mathbf{Q}_0)} \right]^{1/3} \frac{1 - Q_0}{1 + 2Q} \lambda_1^2,$$

$$\alpha_2^2 = \left[\frac{\det (\mathbf{I} + 2\mathbf{Q})}{\det (\mathbf{I} + 2\mathbf{Q}_0)} \right]^{1/3} \frac{1 + 2Q_0}{1 - (Q - b)} \lambda_2^2,$$

$$\alpha_3^2 = \left[\frac{\det (\mathbf{I} + 2\mathbf{Q})}{\det (\mathbf{I} + 2\mathbf{Q}_0)} \right]^{1/3} \frac{1 - Q_0}{1 - (Q + b)} \lambda_3^2. \tag{6.100}$$

The calibrated model function defined by (6.86) is equal to

$$\mathcal{W}^{(lce)} = \frac{c_1^{(1)}}{2 \left(p_1^{(1)} \right)^2} \left(\lambda_1^{2p_1^{(1)}} + \lambda_2^{2p_1^{(1)}} + \lambda_3^{2p_1^{(1)}} - 3 \right)$$

$$+ \frac{c_2^{(1)}}{2 \left(p_2^{(1)} \right)^2} \left(\lambda_1^{2p_2^{(1)}} + \lambda_2^{2p_2^{(1)}} + \lambda_3^{2p_2^{(1)}} - 3 \right) \tag{6.101}$$

$$+ \frac{c_1^{(2)}}{2 \left(p_1^{(2)} \right)^2} \left(\alpha_1^{2p_1^{(2)}} + \alpha_2^{2p_1^{(2)}} + \alpha_3^{2p_1^{(2)}} - 3 \right),$$

with the parameter values listed in Table 6.1. For this strain-energy function, the coefficients are stochastic, while the exponents are deterministic constants. At infinitesimal strain, the stochastic shear modulus is equal to $\overline{\mu} = c_1^{(1)} + c_2^{(1)} + c_1^{(2)}$ and follows a Gamma probability distribution with shape and scale parameters $\rho_1 = 399.9924$ and $\rho_2 = 0.0069$. The corresponding stretch modulus is equal to $\overline{E} = 3\overline{\mu} = 8.4427 \pm 0.4221$ (MPa).

Table 6.1 Parameters of the strain-energy function described by (6.101) calibrated to data. Numerical and experimental results are compared in Fig. 6.18

Model function	Calibrated parameters (mean value $\pm$ standard deviation)
$\mathcal{W}^{(1)}(\lambda_1, \lambda_2, \lambda_3)$ given by (6.89)	$c_1^{(1)} = -0.0895 \pm 0.0045,\ c_2^{(1)} = 2.8993 \pm 0.1450,$ $p_1^{(1)} = 2.6478,\ p_2^{(1)} = -1.6629$
$\mathcal{W}^{(2)}(\alpha_1, \alpha_2, \alpha_3)$ given by (6.90)	$c_1^{(2)} = 0.0044 \pm 0.0002,$ $p_1^{(2)} = 5.4845$

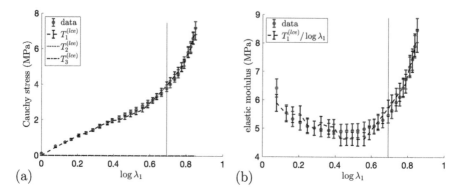

Fig. 6.18 (a) The principal Cauchy stresses given by (6.102) and (b) the nonlinear elastic modulus $T_1^{(lce)}/\log \lambda_1$ vs. the logarithmic longitudinal strain $\log \lambda_1$. The model function is described by (6.101) with parameter values recorded in Table 6.1. In each plot, the vertical line is drawn at the critical strain where the director suddenly rotates by $\pi/2$

As the second and third directions are stress-free, the principal Cauchy stresses can be written as follows:

$$
\begin{aligned}
T_i^{(lce)} &= \frac{c_1^{(1)}}{p_1^{(1)}} \left(\lambda_i^{2p_1^{(1)}} - \lambda_2^{2p_1^{(1)}} \right) + \frac{c_2^{(1)}}{p_2^{(1)}} \left(\lambda_i^{2p_2^{(1)}} - \lambda_2^{2p_2^{(1)}} \right) \\
&\quad + \frac{c_1^{(2)}}{p_1^{(2)}} \left(\alpha_i^{2p_1^{(2)}} - \alpha_2^{2p_1^{(1)}} \right) \\
&= \frac{c_1^{(1)}}{p_1^{(1)}} \left(\lambda_i^{2p_1^{(1)}} - \lambda_3^{2p_1^{(1)}} \right) + \frac{c_2^{(1)}}{p_2^{(1)}} \left(\lambda_i^{2p_2^{(1)}} - \lambda_3^{2p_2^{(1)}} \right) \\
&\quad + \frac{c_1^{(2)}}{p_1^{(2)}} \left(\alpha_i^{2p_1^{(2)}} - \alpha_3^{2p_1^{(1)}} \right), \qquad i = 1, 2, 3.
\end{aligned}
\tag{6.102}
$$

By (6.88),

$$\frac{\partial \mathcal{W}^{(lce)}}{\partial Q} = 0 \qquad \text{and} \qquad \frac{\partial \mathcal{W}^{(lce)}}{\partial b} = 0. \tag{6.103}$$

Figure 6.18 illustrates the principal Cauchy stresses given by (6.102) and the nonlinear stretch modulus $T_1^{(lce)} / \log \lambda_1$ *vs.* the logarithmic strain $\log \lambda_1$.

Conversely, given the model parameters listed in Table 6.1, from the expressions of the principal Cauchy stresses defined by (6.102), under the constraints (6.103), one should be able to obtain λ_2, λ_3, Q, and b as functions of the stretch ratio λ_1. Hence, the auxetic elastic response at large strain should be predicted. However, the inverse problem involved is highly nonlinear, and finding an effective predictive model for this complex material behaviour remains open to future investigation. Auxetic materials that are simple to fabricate and avoid porosity-related weakening have long been sought for applications in a range of industries including sports equipment, aerospace, biomedical materials, and architecture.

Chapter 7
Conclusion

In fact, I feel that my story here must stop just as it begins to get really interesting.

–G.A. Miller [375]

The opposition between "clarity" and "predictability" on the one hand, and "uncertainty" on the other hand, or between "knowledge" and "chance", has a long tradition but is starting to break down in many areas.

History reminds us of what has been achieved and helps us set a fairly lucid and honest standard for current and future achievements.

The triple mathematical, philosophical, and practical origin of probability theory, which has been inseparable from its applications, is both fascinating and frustrating at the same time. A detailed history of probabilistic and statistical reasoning in the nineteenth and twentieth centuries can be found in the two volumes edited by Krüger et al. [308]. The origins and development of probability and statistics from the mid-seventeenth to end of the nineteenth century are surveyed by Gigerenzer et al. [197].

As always, popular science books and science textbooks are the most effective source for learning the universal language of chance and uncertainty before applying it to new fields [65, 140, 197, 223, 234, 349, 399, 437, 440, 448, 503, 514, 532].

In the classical theories of continuum mechanics, determinism has been the tool of choice for the majority of mathematical models [91, 228]. For soft solid materials, deterministic models can often misrepresent the location and extent of the deformation or damage as they do not account for the range in experimental data, for example, in soft tissues or additive manufacturing composites.

This book shows how the field of elasticity can be reinvigorated, enhanced, and enriched by incorporating nondeterministic approaches in its mainstream. The goal is to demonstrate how probabilities can be included in the continuum theory of finite elasticity by building directly on classical achievements, without requiring their modification, that is, the deterministic problem should always be solved first. Then the main idea is general enough and quite simple: To characterise material parameters by probability distributions instead of tuning them to average experimental data. This not only brings mechanical models a little closer to

© Springer Nature Switzerland AG 2022
L. A. Mihai, *Stochastic Elasticity*, Interdisciplinary Applied Mathematics 55,
https://doi.org/10.1007/978-3-031-06692-4_7

reality, but makes us think more carefully about the consequence of treating model parameters as single-valued constants.

The instability problems collected here arise from universal deformations, which can be sustained by whole families of material models under external loads, and include random variables as basic concepts along with mechanical stresses and strains. For these problems, the propagation of stochastic variation from input physical quantities to output mechanical behaviour is mathematically traceable, and the particular values in the numerical calculations are for illustrative purposes only.

The solid mechanics literature offers many more interesting problems amenable to the stochastic framework [81, 332, 527]. As instabilities are often precursors of material failure, stochastic representations can lead to more realistic estimations of subsequent mechanical damage. Such problems are important in their own right and may inspire similar investigations of more complex systems [99, 118, 290, 319, 324, 344, 516, 536, 614]. Computational models are required to tackle the associated large-scale datasets. To compute effectively and responsibly, one needs to have a theoretical insight in both the physical problem they study and the numerical procedures they use.

Appendix A
Notation

Throughout this book, first-order tensors (vectors) and second-order tensors (matrices) are denoted by boldface letters, to distinguish them from zero-order tensors (scalars). There are no set rules as to what type of letters must be used to distinguish random variables from deterministic constants. Instead, their nature is made clear by comments and the context in which they appear. The notation convention that non-scalar quantities are denoted by boldface letters applies to random variables as well. It is also convenient to denote the mean value of a random variable X by the underline, i.e., $E[X] = \underline{X}$, and the standard deviation of X by $\|X\| = \sqrt{\mathrm{Var}[X]}$. The bar over a letter is used to denote a quantity from the theory of linear elasticity. See Tables A.1 and A.2.

© Springer Nature Switzerland AG 2022
L. A. Mihai, *Stochastic Elasticity*, Interdisciplinary Applied Mathematics 55,
https://doi.org/10.1007/978-3-031-06692-4

Table A.1 Frequently used symbols

Symbol	Name	First defined
χ	A finite elastic deformation	(2.1)
$\mathbf{A}$	The gradient tensor of a finite elastic deformation	(2.2)
$\mathbf{u}$	The displacement	(2.5)
$\mathbf{I}$	The tensor identity	(2.7)
$\mathbf{U}$	The right stretch tensor	(2.9)
$\mathbf{V}$	The left stretch tensor	(2.9)
$\mathbf{C}$	The right Cauchy–Green tensor $\mathbf{A}^T \mathbf{A}$	(2.10)
$\mathbf{B}$	The left Cauchy–Green tensor $\mathbf{A}\mathbf{A}^T$	(2.11)
$\alpha_1, \alpha_2, \alpha_3$	The singular values of the deformation tensor $\mathbf{A}$	(2.12)
I_1, I_2, I_3	The principal invariants of the Cauchy–Green tensors	(2.15)
W	A strain-energy density function of a hyperelastic material	(2.18)
$\mathbf{e}_n$	The elastic strain tensor	(2.19)
$\bar{\mathbf{e}}$	The strain tensor from linear elasticity	(2.20)
$\mathbf{T}$	The Cauchy stress tensor of an elastic solid	(2.21)
$\mathbf{P}$	The first Piola–Kirchhoff stress tensor of an elastic solid	(2.34)
$\mathbf{S}$	The second Piola–Kirchhoff stress tensor of an elastic solid	(2.41)
$\iota_1, \iota_2, \iota_3$	The principal invariants of the logarithmic stretch tensors $\ln \mathbf{U}$, $\ln \mathbf{V}$	(2.28)
i_1, i_2, i_3	The principal invariants of the stretch tensors $\mathbf{U}$, $\mathbf{V}$	(2.37)
E	The nonlinear stretch modulus	(2.79)
$\bar{E}$	Young's modulus from linear elasticity	(2.81)
ν_n	The nonlinear Poisson function	(2.92)
$\bar{\nu}$	Poisson's ratio from linear elasticity	(2.93)
κ	The nonlinear bulk modulus	(2.94)
$\bar{\kappa}$	The bulk modulus from linear elasticity	(2.96)
μ	The nonlinear shear modulus	(2.104)
$\bar{\mu}$	The shear modulus from linear elasticity	(2.115)
μ_P	The Poynting modulus	(2.137)
μ_T	The torsion modulus	(2.129)
$\bar{\mu}_T$	The torsion modulus from linear elasticity	(2.133)
$W^{(nc)}$	A strain-energy density function of a nematic elastomer	(6.1)
$\mathbf{F}$	The gradient tensor of the total deformation for a nematic solid	(6.1)
$\mathbf{n}$	The nematic director	(6.1)
$\mathbf{G}$	The spontaneous deformation tensor for a nematic solid	(6.2)
$\mathbf{n}_0$	The nematic director in the reference configuration	(6.5)
$I_1^{(nc)}, I_2^{(nc)}, I_3^{(nc)}$	The principal invariants of the deformation tensors $\mathbf{F}^T\mathbf{F}$, $\mathbf{F}\mathbf{F}^T$	(6.13)–(6.15)
$\lambda_1, \lambda_2, \lambda_3$	The singular values of the deformation tensor $\mathbf{F}$	(6.27)
$\mathbf{T}^{(nc)}$	The Cauchy stress tensor of a nematic solid	(2.21)
$\mathbf{P}^{(nc)}$	The first Piola–Kirchhoff stress tensor of a nematic solid	(2.34)
$\mathbf{S}^{(nc)}$	The second Piola–Kirchhoff stress tensor of a nematic solid	(2.41)

Table A.2 Specific probability distributions

Name	Probability density function	Hyperparameters	Random variables	Mean value	Standard deviation
Normal	$f(u; \underline{u}, \|u\|) = \dfrac{e^{-(u-\underline{u})^2/(2\|u\|^2)}}{\sqrt{2\pi}\|u\|}$	$\underline{u} \in (-\infty, \infty),\ \|u\| > 0$	$u \in (-\infty, \infty)$	$\underline{u}$	$\|u\|$
Gamma	$g(u; \rho_1, \rho_2) = \dfrac{u^{\rho_1-1}e^{-u/\rho_2}}{\rho_2^{\rho_1}\Gamma(\rho_1)}$, $\Gamma(\rho_1) = \int_0^\infty t^{\rho_1-1}e^{-t}\,dt$	$\rho_1, \rho_2 > 0$	$u > 0$	$\underline{u} = \rho_1\rho_2$	$\|u\| = \sqrt{\rho_1}\,\rho_2$
Dirichlet	$h(r_1, \ldots, r_n; \xi_1, \ldots, \xi_n) = \dfrac{1}{B(\xi_1,\ldots,\xi_n)} \prod_{i=1}^n r_i^{\xi_i-1}$, $B(\xi_1,\ldots,\xi_n) = \dfrac{\prod_{i=1}^n \Gamma(\xi_i)}{\Gamma\left(\sum_{i=1}^n \xi_i\right)}$	$\xi_1, \ldots, \xi_n > 0$	$r_1, \ldots, r_n \in (0, 1)$ $\sum_{i=1}^n r_i = 1$	$r_p = \dfrac{\xi_p}{\sum_{i=1}^n \xi_i}$	$\|r_p\| = \dfrac{1}{\sum_{i=1}^n \xi_i}\sqrt{\dfrac{\xi_p\psi_p}{\xi_p+\psi_p+1}}$ $\psi_p = \sum_{j=1,\ j\neq p}^n \xi_j$
Beta	$h(r; \xi_1, \xi_2) = \dfrac{r^{\xi_1-1}(1-r)^{\xi_2-1}}{B(\xi_1,\xi_2)}$, $B(\xi_1, \xi_2) = \int_0^1 t^{\xi_1-1}(1-t)^{\xi_2-1}\,dt$	$\xi_1, \xi_2 > 0$	$r \in (0, 1)$	$\underline{r} = \dfrac{\xi_1}{\xi_1+\xi_2}$	$\|r\| = \dfrac{1}{\xi_1+\xi_2}\sqrt{\dfrac{\xi_1\xi_2}{\xi_1+\xi_2+1}}$

Appendix B
Fundamental Concepts

B.1 Tensors and Tensor Fields

A scalar is described by a single number, a vector with respect to a given basis is described by an array, and a matrix with respect to a basis is described by a multi-dimensional array. The numbers in the array are known as the scalar components of the tensor. They are denoted by indices giving their position in the array. The total number of indices required to uniquely specify each component is equal to the dimension of the array and is called the order or the rank of the tensor. For example, in an n-dimensional space, a second-order tensor $\mathbf{A}$ is a matrix with components denoted A_{ij}, where $i, j = 1, 2, \ldots, n$, in particular, in a three-dimensional (3D) Euclidean space, $i, j = 1, 2, 3$. Since a vector is a one-dimensional array, it is also a first-order tensor, while scalars are single numbers and are thus zero-order tensors. A tensor having functions as its components is called a tensor field. Tensors provide a concise mathematical framework for formulating and solving problems in mechanics.[1]

Definition B.1.1 The principal invariants of a second-order tensor $\mathbf{A} \in \mathbb{R}^{n \times n}$, $n \geq 2$, are the coefficients of the characteristic polynomial of $\mathbf{A}$,

$$p(\lambda) = \det (\lambda \mathbf{I} - \mathbf{A}),$$

where "det" is the determinant operator, $\mathbf{I} = (\delta_{ij})_{i, j=1,\ldots,n} \in \mathbb{R}^{n \times n}$ is the identity second-order tensor, with the entries given by the Kronecker delta (i.e., $\delta_{ii} = 1$ for $i = 1, \ldots, n$, and $\delta_{ij} = 0$ for $i \neq j$, $i, j = 1, \ldots, n$), and λ is a scalar variable. The roots of the characteristic polynomial are the eigenvalues of $\mathbf{A}$.

---

[1] A useful introduction to tensor analysis is [64].

© Springer Nature Switzerland AG 2022
L. A. Mihai, *Stochastic Elasticity*, Interdisciplinary Applied Mathematics 55,
https://doi.org/10.1007/978-3-031-06692-4

- The principal invariants of the second-order tensor $\mathbf{A} \in \mathbb{R}^{3\times3}$ take the following form:

$$I_1(\mathbf{A}) = \text{tr } \mathbf{A},$$

$$I_2(\mathbf{A}) = \text{tr}\,(\text{CofA}) = \frac{1}{2}\left[(\text{tr } \mathbf{A})^2 - \text{tr}\left(\mathbf{A}^2\right)\right],$$

$$I_3(\mathbf{A}) = \det \mathbf{A} = \text{tr}\left(\mathbf{A}^3\right) - \frac{3}{2}(\text{tr } \mathbf{A})\,\text{tr}\left(\mathbf{A}^2\right) + \frac{1}{2}(\text{tr } \mathbf{A})^3,$$

where "tr" is the trace operator, $\text{CofA} = (\det \mathbf{A})\mathbf{A}^{-T}$ is the cofactor of $\mathbf{A}$, with "T" denoting the transpose and "$-T$" the inverse of the transpose (or, equivalently, the transpose of the inverse), and $\det \mathbf{A}$ is the determinant of $\mathbf{A}$. The scalar values $\overline{I}_1(\mathbf{A}) = \text{tr}(\mathbf{A})$, $\overline{I}_2(\mathbf{A}) = \text{tr}(\mathbf{A}^2)$, and $\overline{I}_3(\mathbf{A}) = \text{tr}(\mathbf{A}^3)$ represent the first three moments of the tensor $\mathbf{A}$.
- The principal invariants of $\mathbf{A} \in \mathbb{R}^{3\times3}$ can be expressed equivalently as follows:

$$I_1(\mathbf{A}) = \alpha_1 + \alpha_2 + \alpha_3, \qquad I_2(\mathbf{A}) = \alpha_1\alpha_2 + \alpha_2\alpha_3 + \alpha_3\alpha_1, \qquad I_3(\mathbf{A}) = \alpha_1\alpha_2\alpha_3,$$

where $\alpha_1, \alpha_2, \alpha_3$ denote the eigenvalues of $\mathbf{A}$.

Definition B.1.2 The set of orthogonal second-order tensors in $\mathbb{R}^{n\times n}$, $n \geq 2$, is defined by

$$O(n) = \{\mathbf{R} \in \mathbb{R}^{n\times n} \mid \mathbf{R}^T\mathbf{R} = \mathbf{R}\mathbf{R}^T = \mathbf{I}\}.$$

The corresponding set of proper orthogonal matrices (rotations) is

$$SO(n) = \{\mathbf{R} \in O(n) \mid \det \mathbf{R} = 1\}.$$

- The principal invariants of the second-order tensor $\mathbf{A} \in \mathbb{R}^{3\times3}$ are independent of rotation, i.e., if $\mathbf{R} \in SO(3)$, then:
  - (a) $I_1(\mathbf{A}) = \text{tr}\left(\mathbf{R}^T\mathbf{A}\mathbf{R}\right)$.
  - (b) $I_2(\mathbf{A}) = \text{tr}\left(\text{Cof}\left(\mathbf{R}^T\mathbf{A}\mathbf{R}\right)\right)$.
  - (c) $I_3(\mathbf{A}) = \det\left(\mathbf{R}^T\mathbf{A}\mathbf{R}\right)$.

Theorem B.1.3 (Cayley–Hamilton) *For any invertible matrix* $\mathbf{A} \in \mathbb{R}^{3\times3}$, *the following relationship between the powers of* $\mathbf{A}$ *holds*

$$\mathbf{A}^3 - I_1\mathbf{A}^2 + I_2\mathbf{A} - I_3\mathbf{I} = \mathbf{0},$$

where $I_1 = I_1(\mathbf{A})$, $I_2 = I_2(\mathbf{A})$, $I_3 = I_3(\mathbf{A})$ *are the principal invariants of* $\mathbf{A}$, $\mathbf{I}$ *is the identity matrix, and* $\mathbf{0}$ *is the matrix with all entries equal to zero.*

Proof For a non-singular matrix $\mathbf{A}$, the cofactor is equal to $\mathrm{Cof}\mathbf{A} = J\mathbf{A}^{-T}$, where $\mathbf{A}^{-T}$ denotes the transpose of the inverse matrix $\mathbf{A}^{-1}$ and $J = \det \mathbf{A}$ is the determinant. Hence, $\mathbf{A}\,(\mathrm{Cof}\mathbf{A})^T = J\mathbf{I}$. Similarly,

$$(\lambda\mathbf{I} - \mathbf{A})\,[\mathrm{Cof}\,(\lambda\mathbf{I} - \mathbf{A})]^T = \det\,(\lambda\mathbf{I} - \mathbf{A})\,\mathbf{I}.$$

This can be written equivalently as follows:

$$(\lambda\mathbf{I} - \mathbf{A})\left(\lambda^2\mathbf{M}_2 + \lambda\mathbf{M}_1 + \mathbf{M}_0\right) = \left(\lambda^3 + a_2\lambda^2 + a_1\lambda + a_0\right)\mathbf{I},$$

where $\mathbf{M}_2, \mathbf{M}_1, \mathbf{M}_0 \in \mathbb{R}^{3\times3}$ and $a_2, a_1, a_0 \in \mathbb{R}$ are independent of λ. Equating the coefficients of $\lambda^3, \lambda^2, \lambda^1$, and λ^0 gives

$$\mathbf{M}_2 = \mathbf{I},$$
$$\mathbf{M}_1 - \mathbf{A}\mathbf{M}_2 = a_2\mathbf{I},$$
$$\mathbf{M}_0 - \mathbf{A}\mathbf{M}_1 = a_1\mathbf{I},$$
$$-\mathbf{A}\mathbf{M}_0 = a_0\mathbf{I}.$$

Pre-multiplying the third equation in the above sequence by $\mathbf{A}$, the second equation by $\mathbf{A}^2$, and the first equation by $\mathbf{A}^3$, then adding the four equations yields

$$0 = \mathbf{A}^3 + a_2\mathbf{A}^2 + a_1\mathbf{A} + a_0\mathbf{I}.$$

This argument is similar to that of Buchheim [76].[2] To conclude the proof, it can be verified that $a_2 = -I_1$, $a_1 = I_2$, and $a_0 = -I_3$, as stated by Cayley [90]. □

Lemma B.1.4 *For every symmetric positive definite (SPD) second-order tensor* $\mathbf{C} \in \mathbb{R}^{n\times n}$, $n \geq 2$, *there exists a unique SPD tensor* $\mathbf{U} \in \mathbb{R}^{n\times n}$, *such that* $\mathbf{U}^2 = \mathbf{C}$.

Proof As $\mathbf{C}$ is SPD, all its eigenvalues are strictly positive, and we can denote them by $\{\alpha_i^2\}_{i=1,2,\ldots,n}$, and their respective unitary eigenvectors by $\{\mathbf{r}^{(i)}\}_{i=1,2,\ldots,n} \subset \mathbb{R}^{n\times1}$. Then, $\mathbf{C}$ has the following spectral decomposition:

$$\mathbf{C} = \mathbf{R}\mathbf{\Lambda}\mathbf{R}^T,$$

where $\mathbf{\Lambda} = \mathrm{diag}(\alpha_1^2, \alpha_2^2, \alpha_3^2)$ is the diagonal second-order tensor with the eigenvalues of $\mathbf{C}$ as its first principal diagonal entries, and $\mathbf{R} = (\mathbf{r}^{(1)}, \mathbf{r}^{(2)} \ldots, \mathbf{r}^{(n)})$, with the eigenvectors of $\mathbf{C}$ as its columns. Equivalently, in component notation,

---

[2] An alternative proof can be found in [558, pp. 25–26].

$$\mathbf{C} = \sum_{i=1}^{n} \alpha_i^2 \mathbf{r}^{(i)} \otimes \mathbf{r}^{(i)},$$

where $\mathbf{x} \otimes \mathbf{y} = (x_k y_l)_{k,l=1,2,\ldots,n}$ is the tensor product of two vectors $\mathbf{x}, \mathbf{y} \in \mathbb{R}^{n \times 1}$. Denoting

$$\mathbf{U} = \sum_{i=1}^{n} \alpha_i \mathbf{r}^{(i)} \otimes \mathbf{r}^{(i)},$$

we obtain

$$\mathbf{U}^2 \mathbf{r}^{(i)} = \mathbf{U} \left(\alpha_i \mathbf{r}^{(i)} \right) = \alpha_i^2 \mathbf{r}^{(i)} = \mathbf{C} \mathbf{r}^{(i)}, \qquad i = 1, 2, \ldots, n.$$

As $\{\mathbf{r}^{(i)}\}_{i=1,2,\ldots,n}$ form an orthonormal basis of $\mathbb{R}^{n \times 1}$, it follows that $\mathbf{U}^2 = \mathbf{C}$.

To prove uniqueness, let $\hat{\mathbf{U}}$ be any SPD matrix such that $\hat{\mathbf{U}}^2 = \mathbf{C}$. Then,

$$\hat{\mathbf{U}}^2 \mathbf{r}^{(i)} = \mathbf{C} \mathbf{r}^{(i)} = \alpha_i^2 \mathbf{r}^{(i)},$$

i.e., $\{\mathbf{r}^{(i)}\}_{i=1,2,\ldots,n}$ is an eigenbasis of $\hat{\mathbf{U}}^2$, and also an eigenbasis of $\hat{\mathbf{U}}$, and if $\{\hat{\lambda}_i\}_{i=1,2,\ldots,n}$ are the eigenvalues of $\hat{\mathbf{U}}$, then $\hat{\lambda}_i^2 = \alpha_i^2$. As $\hat{\mathbf{U}}$ is SPD, it follows that $0 < \hat{\lambda}_i = \alpha_i$, i.e., $\hat{\mathbf{U}} = \mathbf{U}$. $\qquad\qquad\qquad\qquad\qquad\qquad\qquad\qquad\qquad\qquad\square$

Theorem B.1.5 (Polar Decomposition) *Any invertible square second-order tensor* $\mathbf{A} \in \mathbb{R}^{n \times n}$, $n \geq 2$, *has two unique multiplicative decompositions:*

$$\mathbf{A} = \mathbf{R}\mathbf{U} \qquad \text{(right polar decomposition)},$$

$$\mathbf{A} = \mathbf{V}\mathbf{R} \qquad \text{(left polar decomposition)},$$

where $\mathbf{U}, \mathbf{V} \in \mathbb{R}^{n \times n}$ *are symmetric positive definite (SPD) and* $\mathbf{R} \in SO(n)$ *(i.e.,* $\mathbf{R}^{-1} = \mathbf{R}^T$ *and* $\det \mathbf{R} = 1$*).*

Proof Since $\mathbf{A}^T \mathbf{A}$ and $\mathbf{A}\mathbf{A}^T$ are SPD, there exist a unique $\mathbf{U}$ and a unique $\mathbf{V}$, which are also SPD, such that:

$$\mathbf{U}^2 = \mathbf{A}^T \mathbf{A}, \qquad \mathbf{V}^2 = \mathbf{A}\mathbf{A}^T.$$

First, we define $\mathbf{R} = \mathbf{A}\mathbf{U}^{-1}$ and note that

$$\mathbf{R}^T \mathbf{R} = \mathbf{U}^{-T} \mathbf{A}^T \mathbf{A} \mathbf{U}^{-1} = \mathbf{U}^{-1} \mathbf{U}^2 \mathbf{U}^{-1} = \mathbf{I},$$

i.e., $\mathbf{R}$ is orthogonal, and as $\mathbf{U}$ is unique, $\mathbf{R}$ is also unique.

Similarly, we define $\mathbf{S} = \mathbf{V}^{-1} \mathbf{A}$ and obtain

$$\mathbf{SS}^T = \mathbf{V}^{-1}\mathbf{AA}^T\mathbf{V}^{-T} = \mathbf{V}^{-1}\mathbf{V}^2\mathbf{V}^{-1} = \mathbf{I},$$

i.e., $\mathbf{S}$ is orthogonal, and as $\mathbf{V}$ is unique, $\mathbf{S}$ is unique.

It remains to show that $\mathbf{S} = \mathbf{R}$. Since

$$\mathbf{A} = \mathbf{RU} = \mathbf{RUR}^T\mathbf{R},$$

if we denote $\hat{\mathbf{V}} = \mathbf{RUR}^T$, then

$$\mathbf{A} = \hat{\mathbf{V}}\mathbf{R}$$

and

$$\hat{\mathbf{V}}^2 = \left(\mathbf{RUR}^T\right)\left(\mathbf{RUR}^T\right) = \mathbf{RU}^2\mathbf{R}^T = \left(\mathbf{AU}^{-1}\right)\mathbf{U}^2\left(\mathbf{AU}^{-1}\right)^T = \mathbf{AA}^T.$$

As the square root of $\mathbf{AA}^T$ is unique, it follows that $\hat{\mathbf{V}} = \mathbf{V}$; hence,

$$\mathbf{S} = \mathbf{V}^{-1}\mathbf{A} = \hat{\mathbf{V}}^{-1}\mathbf{A} = \mathbf{R}.$$

This completes the proof.   □

Proposition B.1.6 *A scalar-valued function $f(\mathbf{A})$, of a tensor variable $\mathbf{A} \in \mathbb{R}^{3\times3}$, is frame indifferent, i.e., $f(\mathbf{A}) = f(\mathbf{RA})$, for all $\mathbf{R} \in SO(3)$, if and only if $f(\mathbf{A}) = f(\mathbf{U})$, where $\mathbf{U}^2 = \mathbf{A}^T\mathbf{A}$.*

Proof If $f(\mathbf{A}) = f(\mathbf{RA})$, by the polar decomposition theorem, $\mathbf{A} = \mathbf{RU}$, where $\mathbf{R} \in SO(3)$ and $\mathbf{U}^2 = \mathbf{A}^T\mathbf{A}$. Then, $f(\mathbf{A}) = f(\mathbf{RU}) = f(\mathbf{U})$. Conversely, if $f(\mathbf{A}) = f(\mathbf{U})$, where $\mathbf{U}^2 = \mathbf{A}^T\mathbf{A}$, then $f(\overline{\mathbf{A}}) = f(\overline{\mathbf{U}})$, where $\overline{\mathbf{U}}^2 = \overline{\mathbf{A}}^T\overline{\mathbf{A}}$ and $\overline{\mathbf{A}} = \mathbf{RA}$. Since $\mathbf{U} = \overline{\mathbf{U}}$, it follows that $f(\mathbf{A}) = f(\mathbf{U}) = f(\mathbf{RA})$.   □

Proposition B.1.7 *A scalar-valued function $f(\mathbf{A})$, of a symmetric tensor variable $\mathbf{A} \in \mathbb{R}^{3\times3}$, is isotropic, i.e., $f(\mathbf{A}) = f\left(\mathbf{RAR}^T\right)$, for all $\mathbf{R} \in SO(3)$, if and only if it can be expressed as a function of the first three moments of $\mathbf{A}$, i.e., $f(\mathbf{A}) = h\left(\overline{I}_1, \overline{I}_2, \overline{I}_3\right)$, for some function h, where $\overline{I}_1(\mathbf{A}) = \mathrm{tr}(\mathbf{A})$, $\overline{I}_2(\mathbf{A}) = \mathrm{tr}(\mathbf{A}^2)$, and $\overline{I}_3(\mathbf{A}) = \mathrm{tr}(\mathbf{A}^3)$.*

Proof If f is isotropic and $\mathbf{R} \in SO(3)$ is selected such that $\mathbf{RAR}^T = \mathrm{diag}(\alpha_1, \alpha_2, \alpha_3)$, where $\mathbf{A} = \mathbf{A}^T$, then $f(\mathbf{A}) = f\left(\mathbf{RAR}^T\right) = g(\alpha_1, \alpha_2, \alpha_3)$, for some function g. Next, taking $\mathbf{R}$ as to permute the eigenvalues $\alpha_1, \alpha_2, \alpha_3$, it follows that g is symmetric with respect to the permutation of these eigenvalues. Thus, g is a function of the principal invariants $I_1(\mathbf{A})$, $I_2(\mathbf{A})$, $I_3(\mathbf{A})$, and consequently, of the first three moments, i.e., it takes the form $g(\alpha_1, \alpha_2, \alpha_3) = h\left(\overline{I}_1, \overline{I}_2, \overline{I}_3\right)$. Conversely, if $f(\mathbf{A}) = h\left(\overline{I}_1, \overline{I}_2, \overline{I}_3\right)$, then f is isotropic since $\mathrm{tr}(\mathbf{A}) = \mathrm{tr}\left(\mathbf{RAR}^T\right)$, $\mathrm{tr}(\mathbf{A}^2) = \mathrm{tr}\left(\mathbf{RAR}^T\right)^2$, and $\mathrm{tr}(\mathbf{A}^3) = \mathrm{tr}\left(\mathbf{RAR}^T\right)^3$, for all $\mathbf{R} \in SO(3)$.   □

Definition B.1.8 If $f(\mathbf{A})$ is a scalar-valued tensor function, where $\mathbf{A} \in \mathbb{R}^{3\times3}$, such that the corresponding component function $f(A_{ij})$ has continuous partial derivatives with respect to all its variables A_{ij}, $i, j = 1, 2, 3$, then its gradient is the tensor-valued tensor function with components

$$\left(\frac{\partial f}{\partial \mathbf{A}}\right)_{ij} = \frac{\partial f}{\partial A_{ij}}, \qquad i, j = 1, 2, 3.$$

Definition B.1.9 For φ, $\mathbf{v} = [v_1, v_2, v_3]^T$, and $\mathbf{A} = (A_{ij})_{i,j=1,2,3}$ representing a zero-order tensor (scalar), a first-order tensor (vector), and a second-order tensor (matrix) fields, respectively, that are differentiable functions of $\mathbf{x} = [x_1, x_2, x_3]^T \in \mathbb{R}^{3\times1}$ in a Cartesian system of coordinates with the orthonormal basis $\mathbf{e}_1 = [1, 0, 0]^T$, $\mathbf{e}_2 = [0, 1, 0]^T$, $\mathbf{e}_3 = [0, 0, 1]^T$, the following differential operators can be defined:

(i) The gradient operators, which transform a tensor into a higher order:

$$\operatorname{grad} \varphi = \nabla\varphi = \frac{\partial \varphi}{\partial x_i}\mathbf{e}_i,$$

$$\operatorname{grad} \mathbf{v} = \nabla \otimes \mathbf{v} = \frac{\partial}{\partial x_j}\mathbf{v} \otimes \mathbf{e}_j = \frac{\partial v_i}{\partial x_j}\mathbf{e}_i \otimes \mathbf{e}_j,$$

$$\operatorname{grad} \mathbf{A} = \nabla \otimes \mathbf{A} = \frac{\partial}{\partial x_k}\mathbf{A} \otimes \mathbf{e}_k = \frac{\partial A_{ij}}{\partial x_k}\mathbf{e}_i \otimes \mathbf{e}_j \otimes \mathbf{e}_k.$$

(ii) The divergence operator, transforming a tensor into a lower order one:

$$\operatorname{div} \mathbf{v} = \nabla \cdot \mathbf{v} = \frac{\partial v_j}{\partial x_j},$$

$$\operatorname{div} \mathbf{A} = \frac{\partial A_{ij}}{\partial x_j}\mathbf{e}_i.$$

In the above expressions, the Einstein summation convention that repeated indices denote summation over those indices was used.

Theorem B.1.10 (Divergence Theorem[3]) *If* $\mathbf{v} : \Omega \rightarrow \mathbb{R}^{3\times1}$ *is a continuously differentiable vector field, defined on a compact domain* $\Omega \in \mathbb{R}^3$, *with piecewise smooth boundary* $\partial\Omega$, *then*

---

[3] A detailed history of the divergence theorem can be found in [297].

$$\int_{\Omega} \mathrm{div} \, \mathbf{v} \, \mathrm{d}\mathcal{V} = \int_{\partial\Omega} \mathbf{v} \cdot \mathbf{n} \, \mathrm{d}\mathcal{A},$$

where $\mathrm{d}\mathcal{V}$ *and* $\mathbf{d}\mathcal{A}$ *are units of volume and area, respectively,* $\mathbf{n}$ *is the outward unit normal vector to the surface* $\partial\Omega$, *and* $\mathbf{x} \cdot \mathbf{y} = x_i y_i$ *represents the scalar product of two vectors* $\mathbf{x}, \mathbf{y} \in \mathbb{R}^{3\times 1}$.

If $\mathbf{A} : \Omega \to \mathbb{R}^{3\times 3}$ *is a continuously differentiable second-order tensor field, then*

$$\int_{\Omega} \mathrm{div} \, \mathbf{A} \, \mathrm{d}\mathcal{V} = \int_{\partial\Omega} \mathbf{A}\mathbf{n} \, \mathrm{d}\mathcal{A}.$$

This result is central to the equations of elastostatic equilibrium.

B.2   Polar Systems of Coordinates

B.2.1   Cylindrical Coordinate System

We denote by (r, θ, z) the cylindrical polar coordinates consisting of the radius $r \in [0, \infty]$, the azimuth $\theta \in (-\pi, \pi]$, and the height $z \in (-\infty, \infty)$, and the associated basis vectors by $(\mathbf{e}_r, \mathbf{e}_\theta, \mathbf{e}_z)$. The cylindrical coordinates are defined with respect to a set of Cartesian (rectangular) coordinates (x, y, z), with the basis vectors $(\mathbf{e}_x, \mathbf{e}_y, \mathbf{e}_z)$, and the two sets of coordinates are related as follows:

$$x = r \cos \theta, \qquad y = r \sin \theta, \qquad z = z,$$

and

$$r = \sqrt{x^2 + y^2}, \qquad \theta = \arctan\frac{y}{x}, \qquad z = z.$$

Their corresponding basis vectors are related by

$$\mathbf{e}_x = \mathbf{e}_r \cos \theta - \mathbf{e}_\theta \sin \theta, \qquad \mathbf{e}_y = \mathbf{e}_r \sin \theta + \mathbf{e}_\theta \cos \theta, \qquad \mathbf{e}_z = \mathbf{e}_z,$$

and

$$\mathbf{e}_r = \mathbf{e}_x \cos \theta + \mathbf{e}_y \sin \theta, \qquad \mathbf{e}_\theta = -\mathbf{e}_x \sin \theta + \mathbf{e}_y \cos \theta, \qquad \mathbf{e}_z = \mathbf{e}_z.$$

For time-dependent systems, we have $r = r(t)$, $\theta = \theta(t)$, and $z = z(t)$, where t denotes the time variable, and differentiating the cylindrical basis vectors with respect to time gives

$$\dot{\mathbf{e}}_r = \dot{\theta}\mathbf{e}_\theta, \qquad \dot{\mathbf{e}}_\theta = -\dot{\theta}\mathbf{e}_r, \qquad \dot{\mathbf{e}}_z = 0.$$

The position, velocity, and acceleration in cylindrical coordinates are, respectively,

$$\mathbf{p} = r\mathbf{e}_r + z\mathbf{e}_z,$$
$$\dot{\mathbf{p}} = \dot{r}\mathbf{e}_r + r\dot{\theta}\mathbf{e}_\theta + \dot{z}\mathbf{e}_z,$$
$$\ddot{\mathbf{p}} = \left(\ddot{r} - r\dot{\theta}^2\right)\mathbf{e}_r + \left(r\ddot{\theta} + 2\dot{r}\dot{\theta}\right)\mathbf{e}_\theta + \ddot{z}\mathbf{e}_z.$$

The gradient of a scalar function f in cylindrical polar coordinates is

$$\nabla f = \frac{\partial f}{\partial r}\mathbf{e}_r + \frac{1}{r}\frac{\partial f}{\partial \theta}\mathbf{e}_\theta + \frac{\partial f}{\partial z}\mathbf{e}_z.$$

The divergence of a vector field $\mathbf{f} = (f_r, f_\theta, f_z)^T$ is

$$\nabla \cdot \mathbf{f} = \frac{1}{r}\frac{\partial (rf_r)}{\partial r} + \frac{1}{r}\frac{\partial f_\theta}{\partial \theta} + \frac{\partial f_z}{\partial z}.$$

The curl of a vector field $\mathbf{f} = (f_r, f_\theta, f_z)^T$ is

$$\nabla \times \mathbf{f} = \left(\frac{1}{r}\frac{\partial f_z}{\partial \theta} - \frac{\partial f_\theta}{\partial z}\right)\mathbf{e}_r + \left(\frac{\partial f_r}{\partial z} - \frac{\partial f_z}{\partial r}\right)\mathbf{e}_\theta + \frac{1}{r}\left(\frac{\partial(rf_\theta)}{\partial r} - \frac{\partial f_r}{\partial \theta}\right)\mathbf{e}_z.$$

B.2.2   Spherical Coordinate System

We denote by (r, θ, ϕ) the spherical polar coordinates consisting of the radius $r \in [0, \infty]$, the azimuth $\theta \in (-\pi, \pi]$, and the inclination $\phi \in [0, \pi]$, and the associated basis vectors by $(\mathbf{e}_r, \mathbf{e}_\theta, \mathbf{e}_\phi)$. The cylindrical coordinates are defined with respect to a set of Cartesian coordinates (x, y, z), with the basis vectors $(\mathbf{e}_x, \mathbf{e}_y, \mathbf{e}_z)$, and the two sets of coordinates are related as follows:

$$x = r\cos\theta\sin\phi, \qquad y = r\sin\theta\sin\phi, \qquad z = r\cos\phi,$$

and

$$r = \sqrt{x^2 + y^2 + z^2}, \qquad \theta = \arctan\frac{y}{x}, \qquad \phi = \arccos\frac{z}{r}.$$

Their corresponding basis vectors are related as follows:

$$\mathbf{e}_x = \mathbf{e}_r \cos\theta\sin\phi - \mathbf{e}_\theta \sin\theta + \mathbf{e}_\phi \cos\theta\cos\phi,$$
$$\mathbf{e}_y = \mathbf{e}_r \sin\theta\sin\phi + \mathbf{e}_\theta \cos\theta + \mathbf{e}_\phi \sin\theta\cos\phi,$$
$$\mathbf{e}_z = \mathbf{e}_r \cos\phi - \mathbf{e}_\phi \sin\phi,$$

and

$$\mathbf{e}_r = \mathbf{e}_x \cos\theta \sin\phi + \mathbf{e}_y \sin\theta \sin\phi + \mathbf{e}_z \cos\phi,$$

$$\mathbf{e}_\theta = -\mathbf{e}_x \sin\theta + \mathbf{e}_y \cos\theta,$$

$$\mathbf{e}_\phi = \mathbf{e}_x \cos\theta \cos\phi + \mathbf{e}_y \sin\theta \cos\phi - \mathbf{e}_z \sin\phi.$$

For time-dependent systems, $r = r(t)$, $\theta = \theta(t)$, and $\phi = \phi(t)$, and differentiating the spherical basis vectors with respect to time gives

$$\dot{\mathbf{e}}_r = \left(\dot\theta \sin\phi\right)\mathbf{e}_\theta + \dot\phi \mathbf{e}_\phi,$$

$$\dot{\mathbf{e}}_\theta = -\left(\dot\theta \sin\phi\right)\mathbf{e}_r - \left(\dot\theta \cos\phi\right)\mathbf{e}_\phi,$$

$$\dot{\mathbf{e}}_\phi = -\dot\phi \mathbf{e}_r + \left(\dot\theta \cos\phi\right)\mathbf{e}_\theta.$$

The position, velocity, and acceleration in spherical coordinates are, respectively,

$$\mathbf{p} = r\mathbf{e}_r,$$

$$\dot{\mathbf{p}} = \dot r \mathbf{e}_r + \left(r\dot\theta \sin\phi\right)\mathbf{e}_\theta + r\dot\phi \mathbf{e}_\phi,$$

$$\ddot{\mathbf{p}} = \left(\ddot r - r\dot\theta^2 \sin^2\phi - r\dot\phi^2\right)\mathbf{e}_r$$

$$+ \left(r\ddot\theta \sin\phi + 2\dot r \dot\theta \sin\phi + 2r\dot\theta\dot\phi \cos\phi\right)\mathbf{e}_\theta$$

$$+ \left(r\ddot\phi + 2\dot r\dot\phi - r\dot\theta^2 \sin\phi \cos\phi\right)\mathbf{e}_\phi.$$

The gradient of a scalar function f in spherical polar coordinates is

$$\nabla f = \frac{\partial f}{\partial r}\mathbf{e}_r + \frac{1}{r}\frac{\partial f}{\partial\theta}\mathbf{e}_\theta + \frac{1}{r\sin\theta}\frac{\partial f}{\partial\phi}\mathbf{e}_\phi.$$

The divergence of a vector field $\mathbf{f} = \left(f_r, f_\theta, f_\phi\right)^T$ is

$$\nabla\cdot\mathbf{f} = \frac{1}{r^2}\frac{\partial\left(r^2 f_r\right)}{\partial r} + \frac{1}{r\sin\theta}\frac{\partial f_\theta}{\partial\theta} + \frac{1}{r\sin\theta}\frac{\partial\left(f_\phi\sin\theta\right)}{\partial\phi}.$$

The curl of a vector field $\mathbf{f} = \left(f_r, f_\theta, f_\phi\right)^T$ is

$$\nabla\times\mathbf{f} = \frac{1}{r\sin\theta}\left[\frac{\partial\left(f_\phi\sin\theta\right)}{\partial\theta} - \frac{\partial f_\theta}{\partial\phi}\right]\mathbf{e}_r + \frac{1}{r}\left[\frac{1}{\sin\theta}\frac{\partial f_r}{\partial\phi} - \frac{\partial\left(r f_\phi\right)}{\partial r}\right]\mathbf{e}_\theta$$

$$+ \frac{1}{r}\left[\frac{\partial\left(r f_\theta\right)}{\partial r} - \frac{\partial f_r}{\partial\theta}\right]\mathbf{e}_\phi.$$

B.3   Random Variables and Random Fields

Definition B.3.1 A probability space is a mathematical model of an experiment characterised by a triple $(\Theta, \mathcal{F}, P)$, where Θ is the space of possible outcomes, or the sample space, with each outcome being the result of a single experimental test, $\mathcal{F}$ is a collection of events, with each event being a subset of Θ, and $P : \mathcal{F} \to [0, 1]$ is a function that assigns probabilities to individual events, such that events that almost certainly will not happen have probability 0, and events that will happen almost surely have probability 1.

Definition B.3.2 Given a probability space $(\Theta, \mathcal{F}, P)$, a random variable is a real-valued function, $X : \Theta \to \mathbb{R}$, mapping a sample space Θ into the real line $\mathbb{R}$, such that $(X \leq x) = \{\theta \in \Theta : X(\theta \leq x)\} \in \mathcal{F}$, for every real number $x \in \mathbb{R}$. For any fixed $\theta \in \Theta$, the deterministic value $X(\theta) \in \mathbb{R}$ is called a realisation, or a sample, of the random variable X. The corresponding (cumulative) distribution function is $F_X : \mathbb{R} \to [0, 1]$, defined by $F_X(x) = P(X \leq x)$.

- The distribution function F_X of the random variable X has the following properties:

  (a) $\lim_{x \to \infty} F_X(x) = 1$, $\lim_{x \to -\infty} F_X(x) = 0$.
  (b) If $y < x$, then $F_X(y) \leq F_X(x)$.
  (c) $\lim_{0 < \varepsilon \to 0} F_X(x + \varepsilon) = F_X(x)$.

- From the above properties, it follows that:

  (a) $P(X > x) = 1 - F_X(x)$.
  (b) $P(y < X \leq x) = F_X(x) - F_X y x)$.
  (c) $P(X = x) = F_X(x) - \lim_{y \to x, y \leq x} F_X(y)$.

- If X is a random variable and $g : \mathbb{R} \to \mathbb{R}$ is a given function, then $Y = g(X)$ is also a random variable.

Definition B.3.3 The random variable X is called continuous if there exists a non-negative function $f_X : \mathbb{R} \to [0, \infty)$, such that the corresponding cumulative distribution function takes the form

$$F_X(x) = P(X \leq x) = \int_{-\infty}^{x} f_X(u)du, \qquad x \in \mathbb{R}.$$

The function f_X is called the probability density function (pdf) of X.

- A probability density function has the following properties:

  (a) $P(y \leq X \leq x) = \int_{y}^{x} f_X(u)du$.
  (b) $P(-\infty < X < \infty) = \int_{-\infty}^{\infty} f_X(u)du = 1$.

Definition B.3.4 The random variable X is called discrete if the range of its possible values is a countable set $\{x_1, x_2, \ldots\} \subset \mathbb{R}$. Its distribution function takes the form

$$F_X(x) = \sum_{i\,:\,x_i \leq x} P(X = x_i).$$

Definition B.3.5 A random vector of length n is a vector $\mathbf{X} = (X_1, X_2, \ldots, X_n)$, where $X_i : \Theta \to \mathbb{R}$ is a random variable, for all $i = 1, 2, \ldots, n$. The corresponding joint distribution function is a function $F_{\mathbf{X}} : \mathbb{R}^n \to [0, 1]$, given by $F_{\mathbf{X}}(x_1, x_2, \ldots, x_n) = P(X_1 \leq x_1, X_2 \leq x_2, \ldots, X_n \leq x_n)$. Denoting $\mathbf{x} = (x_1, x_2, \ldots, x_n)$, the joint probability function can be written in the form $F_{\mathbf{X}}(\mathbf{x}) = P(\mathbf{X} \leq \mathbf{x})$.

- The joint distribution function $F_{\mathbf{X}}$ of the random vector $\mathbf{X}$ is characterised by the following properties:

   (a) $\lim_{x_1,x_2,\ldots,x_n \to \infty} F_{\mathbf{X}}(x_1, x_2, \ldots, x_n) = 1$, $\lim_{x_1,x_2,\ldots,x_n \to -\infty} F_{\mathbf{X}}(x_1, x_2, \ldots, x_n) = 0$.
   (b) if $(y_1, y_2, \ldots, y_n) < (x_1, x_2, \ldots, x_n)$, then $F_{\mathbf{X}}(y_1, y_2, \ldots, y_n) \leq F_{\mathbf{X}}(x_1, x_2, \ldots, x_n)$.
   (c) $\lim_{0 < \varepsilon_1, \varepsilon_2, \ldots, \varepsilon_n \to 0} F_{\mathbf{X}}(x_1 + \varepsilon_1, x_2 + \varepsilon_2, \ldots, x_n + \varepsilon_n) = F_{\mathbf{X}}(x_1, x_2, \ldots, x_n)$.

Definition B.3.6 The random variables $X_1, X_2, \ldots, X_n$ are called jointly continuous if there exists a joint non-negative function $f_{\mathbf{X}}(\mathbf{x})$, with $\mathbf{X} = (X_1, X_2, \ldots, X_n)$ and $\mathbf{x} = (x_1, x_2, \ldots, x_n)$, such that the corresponding distribution function takes the form

$$F_{\mathbf{X}}(\mathbf{x}) = \int_{-\infty}^{x_n} \cdots \int_{-\infty}^{x_2} \int_{-\infty}^{x_1} f_{\mathbf{X}}(u_1, u_1, \ldots, u_n)\,du_1 du_2 \ldots du_n, \qquad \mathbf{x} \in \mathbb{R}^n.$$

The function $f_{\mathbf{X}}$ is called the joint probability density function of $X_1, X_2, \ldots, X_n$.

Definition B.3.7 The random variables $X_1, X_2, \ldots, X_n$ are called jointly discrete if the range of its possible values of the random vector $\mathbf{X} = (X_1, X_2, \ldots, X_n)$ is a countable subset of $\mathbb{R}^n$. Its joint distribution function takes the form

$$F_{\mathbf{X}}(\mathbf{x}) = \sum_{i\,:\,\mathbf{x}^{(i)} \leq \mathbf{x}} P\left(\mathbf{X} = \mathbf{x}^{(i)}\right).$$

Definition B.3.8 The random variables $X_1, X_2, \ldots, X_n$ are independent if the events $(X_i = x_i)$ are independent for all $i = 1, 2, \ldots, n$, i.e.,

$$P(X_1 = x_1, X_2 = x_2, \ldots, X_n = x_n) = P(X_1 = x_1)P(X_2 = x_2)\ldots P(X_n = x_n),$$

for all sets $\{x_1, x_2, \ldots, x_n\} \in \mathbb{R}$ and all their finite subsets.

- If $X_1, X_2, \ldots, X_n$ are independent random variables and $g : \mathbb{R} \to \mathbb{R}$ is a given function, then $Y_1 = g(X_1)$, $Y_2 = g(X_2), \ldots, Y_n = g(X_n)$ are also independent random variables.

Definition B.3.9 The mean value, or expected value, or mathematical expectation of a continuous random variable X with probability function f_X is

$$E[X] = \int_{-\infty}^{\infty} x f_X(x) dx$$

provided that

$$\int_{-\infty}^{\infty} |x| f_X(x) dx < \infty.$$

- Equivalently, the expectation of a continuous random variable X is equal to

$$E[X] = \int_{\Theta} X dP,$$

where:

Definition B.3.10 The moment generating function of a random variable X is the function $M_X : \mathbb{R} \to [0, \infty)$, given by $M_X(t) = E[e^{tX}]$. This function may be approximated by the Taylor's expansion

$$M_X(t) = \sum_{k=0}^{\infty} E[X^k] \frac{t^k}{t!},$$

where $E[X^k]$, $k > 1$, is called the kth moment. X is a q-order random variable if $E[X^q] < \infty$.

Definition B.3.11 For a continuous random variable X, the kth central moment is $(X - E[X])^k$, $k > 1$, and its expected value is given by

$$E\left[(X - E[X])^k\right] = \int_{-\infty}^{\infty} (x - E[X])^k f_X(x) dx,$$

if

$$\int_{-\infty}^{\infty} \left|(x - E[X])^k\right| f_X(x) dx < \infty.$$

Definition B.3.12 For a random variable X, the second central moment is the variance, defined by

$$\text{Var}[X] = E\left[(X - E[X])^2\right] = E\left[X^2\right] - E[X]^2 = \sum_{i \geq 1}(x_i - E[X])^2 \, P(X_i = x_i).$$

The square root of the variance, $\sqrt{\text{Var}[X]}$, is called the standard deviation of X.

Definition B.3.13 The mean value, or expected value, or mathematical expectation of a discrete random variable X, with possible values in a countable set $\{x_1, x_2, \ldots\} \in \mathbb{R}$, is

$$E[X] = \sum_{i \geq 1} x_i P(X_i = x_i)$$

provided that

$$\sum_{i \geq 1} |x_i| P(X_i = x_i) < \infty.$$

Definition B.3.14 In general, for a discrete random variable X, the expected value of the kth central moment $(X - E[X])^k$, $k > 1$, is

$$E\left[(X - E[X])^k\right] = \sum_{i \geq 1}(x_i - E[X])^k \, P(X_i = x_i),$$

if

$$\sum_{i \geq 1}\left|(x_i - E[X])^k\right| P(X_i = x_i) < \infty.$$

- If X is a random variable and c is a deterministic constant, then:

  (a) $E[c] = c$, $\text{Var}[c] = 0$.
  (b) $E[cX] = cE[X]$, $\text{Var}[cX] = c^2\text{Var}[X]$.

- If X_1 and X_2 are random variables, and c_1 and c_2 are deterministic constants, then $E[c_1 X_1 + c_2 X_2] = c_1 E[X_1] + c_2 E[X_2]$.

Definition B.3.15 The covariance of two random variables X_1 and X_2 is defined by

$$\text{Cov}[X_1, X_2] = E\left[(X_1 - E[X_1])(X_2 - E[X_2])\right] = E[X_1 X_2] - E[X_1]E[X_2].$$

- The variance of the linear combination $c_1 X_1 \pm c_2 X_2$, where c_1 and c_2 are deterministic constants, is

$$\text{Var}\,[c_1 X_1 \pm c_2 X_2] = c_1^2 \text{Var}\,[X_1] + c_2^2 \text{Var}\,[X_2] \pm 2c_1 c_2 \text{Cov}[X_1, X_2].$$

- If the variables X_1 and X_2 are independent, then the variance of their product is

$$\text{Var}\,[X_1 X_2] = E[X_1]^2 \text{Var}\,[X_2] + E[X_2]^2 \text{Var}\,[X_1] + \text{Var}\,[X_1]\,\text{Var}\,[X_2].$$

- If $X_1 = X_2 = X$, then

$$\text{Cov}[X_1, X_2] = \text{Var}[X].$$

Definition B.3.16 A random field $U(\mathbf{x})$ is an uncountable family of random variables depending on a deterministic variable $\mathbf{x} = (x_1, \ldots, x_d)^T \in \Omega \subset \mathbb{R}^d$. When $d = 1$ and the deterministic variable is time, $x = t$, the random field is a random function of time, $U(t)$, known as a stochastic process.

Definition B.3.17 The mean function of a random field $U(\mathbf{x})$ is a real-valued function depending on a deterministic variable $\mathbf{x} \in \Omega \subset \mathbb{R}^d$, $E[U(\mathbf{x})] : \Omega \to \mathbb{R}$, where $E[\cdot]$ denotes the mathematical expectation (mean value).

B.4 Normal Distribution as Limiting Case of the Gamma Distribution

Theorem B.4.1 *The limiting distribution of the Gamma distribution with shape and scale parameters ρ_1 and ρ_2, respectively, such that $\rho_1 \to \infty$, is the normal (Gaussian) distribution with mean value $\rho_1 \rho_2$ and standard deviation $\sqrt{\rho_1}\rho_2$.*

Proof If X is a random variable following a Gamma probability distribution with shape parameter $\rho_1 > 0$ and scale parameter $\rho_2 > 0$, then its mean value and standard deviation, respectively, are equal to

$$\underline{X} = \rho_1 \rho_2, \qquad \|X\| = \sqrt{\rho_1}\rho_2.$$

Its probability density function takes the form (see Appendix B.6)

$$g_X(x; \rho_1, \rho_2) = \frac{x^{\rho_1 - 1} e^{-\frac{x}{\rho_2}}}{\rho_2^{\rho_1} \Gamma(\rho_1)}, \qquad x \geq 0, \tag{B.1}$$

where $\Gamma : \mathbb{R}_+^* \to \mathbb{R}$ is the complete Gamma function

$$\Gamma(z) = \int_0^\infty t^{z-1} e^{-t} dt. \tag{B.2}$$

The corresponding cumulative distribution function is equal to

$$G_X(x) = \int_0^x g_X(u; \rho_1, \rho_2) du. \tag{B.3}$$

The moment generating function of X is defined by

$$M_X(t) = E\left[e^{tX}\right] = (1 - \rho_2 t)^{-\rho_1}, \qquad t < \frac{1}{\rho_2}.$$

Subtracting the mean value $\underline{X}$ from X and dividing the result by the standard deviation $\|X\|$ give the following one-to-one transformation:

$$Y = \frac{X - \underline{X}}{\|X\|},$$

or equivalently,

$$Y = \frac{X}{\rho_2 \sqrt{\rho_1}} - \sqrt{\rho_1}.$$

The moment generating function of Y is then

$$M_Y(t) = E\left[e^{tY}\right] = e^{-t\sqrt{\rho_1}} E\left[e^{t\frac{X}{\rho_2\sqrt{\rho_1}}}\right] = e^{-t\sqrt{\rho_1}}\left(1 - \frac{t}{\sqrt{\rho_1}}\right)^{-\rho_1}, \qquad t < \sqrt{\rho_1}.$$

Thus, the limiting moment generating function of Y when $\rho_1 \to \infty$ takes the form

$$\lim_{\rho_1 \to \infty} M_Y(t) = \lim_{\rho_1 \to \infty} e^{-t\sqrt{\rho_1}}\left(1 - \frac{t}{\sqrt{\rho_1}}\right)^{-\rho_1}, \qquad -\infty < t < \infty.$$

The above limit can be calculated as follows:

$$\lim_{\rho_1 \to \infty} e^{-t\sqrt{\rho_1}}\left(1 - \frac{t}{\sqrt{\rho_1}}\right)^{-\rho_1} = \lim_{y=1/\sqrt{\rho_1} \to 0_+} e^{-(t/y)}\left(1 - ty\right)^{-1/y^2}$$

$$= e^{\lim_{y \to 0_+} \frac{-ty - \ln(1-ty)}{y^2}},$$

where, applying L'Hôspital's rule [321],

$$\lim_{y\to 0_+}\frac{-ty-\ln(1-ty)}{y^2}=\lim_{y\to 0}\frac{-t+t/(1-ty)}{2y}=\lim_{y\to 0_+}\frac{t^2}{2(1-ty)}=\frac{t^2}{2}.$$

Therefore,

$$\lim_{\rho_1\to\infty}M_Y(t)=e^{\frac{t^2}{2}},\qquad -\infty<t<\infty,$$

which is the moment generating function of a normally distributed random variable.

For a normally distributed random variable $Y \in (-\infty,\infty)$ with mean value $\underline{Y}=0$ and standard deviation $\|Y\|=1$, the probability density function takes the form

$$f_Y(y;0,1)=\frac{1}{\sqrt{2\pi}}e^{-\frac{y^2}{2}},\qquad -\infty<y<\infty. \tag{B.4}$$

The associated normal cumulative distribution function is

$$F_Y(y)=\int_{-\infty}^y f_Y(u;0,1)du$$
$$=\frac{1}{2}+\frac{1}{2}\mathrm{erf}\left(\frac{y}{\sqrt{2}}\right), \tag{B.5}$$

where

$$\mathrm{erf}(y)=2F_Y(y\sqrt{2})-1$$
$$=\frac{2}{\sqrt{\pi}}\int_0^y e^{-t^2}dt \tag{B.6}$$

is the *error function*, giving the probability that a random variable with normal distribution of mean 0 and variance 1/2 falls in the range $[-y,y]$ [219, pp. 114–115]. The error function has the following properties:

$$\mathrm{erf}(0)=0,\qquad \mathrm{erf}(\infty)=1,\qquad \mathrm{erf}(-\infty)=-1,\qquad \mathrm{erf}(-y)=-\mathrm{erf}(y). \tag{B.7}$$

For $-1<z<1$, there exists a unique $\mathrm{erf}^{-1}(z)$ satisfying $\mathrm{erf}\left(\mathrm{erf}^{-1}(z)\right)=z$. Thus, the limiting distribution of a Gamma distribution with shape parameter $\rho_1\to\infty$ is the normal distribution.   □

B.5 Linear Combination of Independent Gamma Distributions

The following theorem is applicable also to linear combinations of independent Gamma-distributed random variables by rescaling (see Theorem 1 of [387] for a proof).

Theorem B.5.1 *If* $\{R_1, R_2\}$ *are mutually independent Gamma-distributed random variables, with the corresponding shape and scale hyperparameters,* $\rho_1^{(i)}$ *and* $\rho_2^{(i)}$, $i = 1, 2$, *respectively, such that* $\rho_2^{(1)} \leq \rho_2^{(2)}$, *then the density of* $R = R_1 + R_2$ *can be expressed as follows:*

$$g(R) = C \sum_{k=0}^{\infty} \frac{\delta_k R^{\rho+k-1} e^{-R/\beta_1}}{\beta_1^{\rho+k} \Gamma(\rho+k)}, \qquad R > 0, \tag{B.8}$$

where

$$\rho = \rho_1^{(1)} + \rho_1^{(2)}, \qquad C = \left(\frac{\rho_2^{(1)}}{\rho_2^{(2)}} \right)^{\rho_1^{(2)}}, \tag{B.9}$$

and δ_k *satisfies*

$$e^{\sum_{k=1}^{\infty} \gamma_k \left(1 - \rho_1^{(1)} t \right)^{-k}} = \sum_{k=0}^{\infty} \delta_k \left(1 - \rho_1^{(1)} t \right)^{-k}, \tag{B.10}$$

with

$$\gamma_k = \frac{1}{k} \rho_1^{(2)} \left(1 - \frac{\rho_2^{(1)}}{\rho_2^{(2)}} \right)^k, \qquad k = 1, 2, \ldots. \tag{B.11}$$

B.6 Maximum Entropy Probability

The measure of entropy (uncertainty)

$$H(R_1, \ldots, R_n) = -\sum_{p=1}^{n} R_p \log R_p \geq 0 \tag{B.12}$$

of a discrete probability distribution $(R_1, \ldots, R_n)$ was first defined by Shannon [492] in the context of information theory.

The *principle of maximum entropy* was introduced by E.T. Jaynes (1957) [280, 281] and consists of explicitly constructing the probability distribution that maximises Shannon's entropy (B.12) under the following $m+1$ equality constraints:

$$\sum_{p=1}^{n} R_p = 1, \qquad \sum_{p=1}^{n} R_p f_{pq} = f_q, \qquad q = 1, \ldots, m, \qquad (B.13)$$

where f_q and f_{pq}, $p = 1, \ldots, n$, $q = 1, \ldots, m$, are given, usually by data [506]. Solving this constrained optimisation problem is equivalent to finding the maximum of the Lagrangian function

$$\mathcal{L}(R_1, \ldots, R_n) = H(R_1, \ldots, R_n) - \Lambda_0 \left(\sum_{p=1}^{n} R_p - 1 \right)$$

$$- \sum_{q=1}^{m} \Lambda_q \left(\sum_{p=1}^{n} R_p f_{pq} - f_q \right), \qquad (B.14)$$

where $\{\Lambda_q\}_{q=0,1,\ldots,m}$ are the Lagrange multipliers associated with the constraints (B.13), respectively. The general form solution of this problem is [280]

$$R_p = e^{-\Lambda_0 - \sum_{q=1}^{m} \Lambda_q f_{pq}}, \qquad p = 1, \ldots, n, \qquad (B.15)$$

and provides the most objective (unbiased) probability model under the given information. By substituting the explicit form (B.15) in (B.13), we can write

$$\Lambda_0 = \ln Z, \qquad f_q = -\frac{\partial}{\partial \Lambda_q} (\ln Z), \qquad q = 1, \ldots, m, \qquad (B.16)$$

where

$$Z(\Lambda_1, \ldots, \Lambda_m) = \sum_{p=1}^{n} e^{-\sum_{q=1}^{m} \Lambda_q f_{pq}}. \qquad (B.17)$$

Thus the entropy of the discrete distribution (B.15) is

$$H_{\max} = \Lambda_0 + \sum_{q=1}^{m} \Lambda_q f_q. \qquad (B.18)$$

Bibliography

1. Abramowitz M, Stegun IA. 1964. Handbook of Mathematical Functions with Formulas, Graphs, and Mathematical Tables, National Bureau of Standards, Applied Mathematics Series, vol. 55, U.S. Government Printing Office, Washington, D.C.
2. Adkins JE, Rivlin RS. 1952. Large elastic deformations of isotropic materials. IX. The deformation of thin shells, Philosophical Transactions of the Royal Society of London A 244, 505–531.
3. Agostiniani V, Dal Maso G, DeSimone A. 2015. Attainment results for nematic elastomers, Proceedings of the Royal Society of Edinburgh A 145, 669–701 (https://doi.org/10.1017/S0308210515000128).
4. Agostiniani V, DeSimone A. 2012. Ogden-type energies for nematic elastomers, International Journal of Non-Linear Mechanics 47(2), 402–412 (https://doi.org/10.1016/j.ijnonlinmec.2011.10.001).
5. Agrawal A, Yun TH, Pesek SL, Chapman WG, Verduzco R. 2014. Shape-responsive liquid crystal elastomer bilayers, Soft Matter 10(9), 1411–1415.
6. Aguiar AR, Lopes da Rocha G. 2018. On the number of invariants in the strain energy density of an anisotropic nonlinear elastic material with two material symmetry directions, Journal of Elasticity 131, 125–132.
7. Aguiar AR, Lopes da Rocha G. 2018. Erratum to: On the number of invariants in the strain energy density of an anisotropic nonlinear elastic material with two material symmetry directions, Journal of Elasticity 131, 133–136.
8. Ahamed T, Peattie RA, Dorfmann L, Cherry-Kemmerling EM. 2018. Pulsatile flow measurements and wall stress distribution in a patient specific abdominal aortic aneurysm phantom, Zeitschrift für Angewandte Mathematik und Mechanik (ZAMM) 98, 2258–2274 (https://doi.org/10.1002/zamm.201700281).
9. Ahn S-k, Ware TH, Lee KM, Tondiglia VP, White TJ. 2016. Photoinduced topographical feature development in blueprinted azobenzene-functionalized liquid crystalline elastomers, Advanced Functional Materials 26(32), 5819–5826 (https://doi.org/10.1002/adfm.201601090).
10. Akyüz U, Ertepinar A. 1998. Stability and asymmetric vibrations of pressurized compressible hyperelastic cylindrical shells, International Journal of Non-Linear Mechanics 34, 391–404.
11. Alijani F, Amabili M. 2014. Non-linear vibrations of shells: A literature review from 2003 to 2013, International Journal of Non-Linear Mechanics 58, 233–257.
12. Almansi E. 1911. Sulle deformazioni finite dei solidi elastici isotropi, I, Rendiconti Lincei-Matematica E Applicazioni (5A) 20, 705–714.

© Springer Nature Switzerland AG 2022
L. A. Mihai, *Stochastic Elasticity*, Interdisciplinary Applied Mathematics 55,
https://doi.org/10.1007/978-3-031-06692-4

13. Amabili M. 2008. Nonlinear Vibrations and Stability of Shells and Plates, Cambridge University Press, Cambridge.

14. Amabili M, Païdoussis MP. 2003. Review of studies on geometrically nonlinear vibrations and dynamics of circular cylindrical shells and panels, with and without fluid-structure interaction, Applied Mechanics Reviews 56, 349–381.

15. Ambulo CP, Tasmin S, Wang S, Abdelrahman MK, Zimmern PE, Ware TH. 2020. Processing advances in liquid crystal elastomers provide a path to biomedical applications, Journal of Applied Physics 128, 140901 (https://doi.org/10.1063/5.0021143).

16. Anand L, Govindjee S. 2020. Continuum Mechanics of Solids, Oxford University Press, Oxford, UK.

17. Anderson DR, Carlson DE, Fried E. 1999. A continuum-mechanical theory for nematic elastomers, Journal of Elasticity 56, 33–58 (https://doi.org/10.1023/A:1007647913363).

18. Anssari-Benam A, Bucchi A. 2021. A generalised neo-Hookean strain energy function for application to the finite deformation of elastomers, International Journal of Non-Linear Mechanics 128, 103626 (https://doi.org/10.1016/j.ijnonlinmec.2020.103626).

19. Anssari-Benam A, Bucchi A, Saccomandi G. 2021. Modelling the inflation and elastic instabilities of rubber-like spherical and cylindrical shells using a new generalised Neo-Hookean strain energy function, Journal of Elasticity (https://doi.org/10.1007/s10659-021-09823-x).

20. Anssari-Benam A, Horgan CO. 2022. Extension and torsion of rubber-like hollow and solid circular cylinders for incompressible isotropic hyperelastic materials with limiting chain extensibility, European Journal of Mechanics A Solids 92, 104443 (https://doi.org/10.1016/j.euromechsol.2021.104443).

21. Anssari-Benam A, Horgan CO. 2021. On modelling simple shear for isotropic incompressible rubber-like materials (https://doi.org/10.1007/s10659-021-09869-x).

22. Antman SS. 1973. Nonuniqueness of equilibrium states for bars in tension, Journal of Mathematical Analysis and Applications 44(2), 333–349 (https://doi.org/10.1016/0022-247X(73)90063-2).

23. Antman SS. 2005. Nonlinear Problems of Elasticity, 2nd ed, Springer, New York.

24. Antman SS, Carbone ER. 1977. Shear and necking instabilities in nonlinear elasticity, Journal of Elasticity 7(2), 125–151 (https://doi.org/10.1007/BF00041087).

25. Antman SS, Negro n-Marrero PV. 1987. The remarkable nature of radially symmetric equilibrium states of aleotropic nonlinearly elastic bodies, Journal of Elasticity 18, 131–164.

26. Aranda-Iglesias D, Vadillo G, Rodríguez-Martínez JA. 2015. Constitutive sensitivity of the oscillatory behaviour of hyperelastic cylindrical shells, Journal of Sound and Vibration 358, 199–216.

27. Aranda-Iglesias D, Ram'on-Lozano C, Rodríguez-Martínez JA. 2017. Nonlinear resonances of an idealized saccular aneurysm, International Journal of Engineering Science 121, 154–166.

28. Aranda-Iglesias D, Rodríguez-Martínez JA, Rubin MB. 2018. Nonlinear axisymmetric vibrations of a hyperelastic orthotropic cylinders, International Journal of Non-Linear Mechanics 99, 131–143.

29. Araújo FS, Nunes LCS. 2020. Experimental study of the Poynting effect in a soft unidirectional fiber-reinforced material under simple shear, Soft Matter (https://doi.org/10.1039/d0sm00745e).

30. Arruda EM, Boyce MC. 1993. A three-dimensional constitutive model for the large stretch behavior of rubber elastic materials, Journal of Mechanics and Physics of Solids 41, 389–412 (https://doi.org/10.1016/0022-5096(93)90013-6).

31. Atkin RJ, Fox N. 2005. An Introduction to the Theory of Elasticity, Dover Publications, New York.

32. Audoly B, Hutchinson JW. 2016. Analysis of necking based on a one-dimensional model, Journal of the Mechanics and Physics of Solids 97, 68–91 (https://doi.org/10.1016/j.jmps. 2015.12.018).

33. Babuška I, Nobile F, Tempone R. 2007. Reliability of computational science, Numerical Methods for Partial Differential Equations 23, 753–784.

34. Babuška I, Nobile F, Tempone R. 2008. A systematic approach to model validation based on Bayesian updates and prediction related rejection criteria, Computer Methods in Applied Mechanics and Engineering 197, 2517–2539.

35. Baker M, Ericksen JL. 1954. Inequalities restricting the form of stress-deformation relations for isotropic elastic solids and Reiner-Rivlin fluids, Journal of the Washington Academy of Sciences 44(2), 33–35 (www.jstor.org/stable/24533303).

36. Bai R, Bhattacharya K. 2020. Photomechanical coupling in photoactive nematic elastomers, Journal of the Mechanics and Physics of Solids 144, 104115 (https://doi.org/10.1016/j.jmps. 2020.104115).

37. Balakrishnan R, Shahinpoor M. 1978. Finite amplitude oscillations of a hyperelastic spherical cavity, International Journal of Non-Linear Mechanics 13, 171–176.

38. Balbi V, Trotta A, Destrade M, Annaidh AN. 2019. Poynting effect of brain matter in torsion, Soft Matter 15(25), 5147–5153 (https://doi.org/10.1039/c9sm00131j).

39. Ball JM. 1982. Discontinuous equilibrium solutions and cavitation in nonlinear elasticity, Philosophical Transactions of the Royal Society A 306(1496), 557–611 (https://doi.org/10. 1098/rsta.1982.0095).

40. Ball JM, James RD. 1987. Fine phase mixtures as minimizers of energy, Archive for Rational Mechanics and Analysis 100, 13–52.

41. Ball JM, James RD. 1992. Proposed experimental tests of a theory of fine microstructure, and the two-well problem, Philosophical Transactions of the Royal Society of London A 338, 389–450.

42. Ball JM, James RD. 2002. The scientific life and influence of Clifford Ambrose Truesdell III, Archive for Rational Mechanics and Analysis 161, 1–26.

43. Ball JM, Schaeffer DG. 1983. Bifurcation and stability of homogeneous equilibrium configurations of an elastic body under dead-load tractions, Mathematical Proceedings of the Cambridge Philosophical Society 94, 315–339.

44. Barnes M, Verduzco R. 2019. Direct shape programming of liquid crystal elastomers, Soft Matter 15(5), 870–879 (https://doi.org/10.1039/c8sm02174k).

45. Barney CW, Dougan CE, McLeod KR, Kazemi-Moridani A, Zheng Y, Ye Z, Tiwari S, Sacligil I, Riggleman RA, Cai S, Lee JH, Peyton SR, Tew GN, Crosby AJ. 2020. Cavitation in soft matter, Proceedings of the National Academy of Sciences of the United States of America (PNAS) 117(17), 9157–9165 (https://doi.org/10.1073/pnas.1920168117).

46. Batra RC. 1976. Deformation produced by a simple tensile load in an isotropic elastic body, Journal of Elasticity 6, 109–111.

47. Bayes T. 1763. An essay toward solving a problem in the doctrine of chances, Philosophical Transactions of the Royal Society 53, 370–418.

48. Beatty MF. 1967. A theory of elastic stability for incompressible hyperelastic bodies, International Journal of Solids and Structures 3, 23–37.

49. Beatty MF, Stalnaker DO. 1986. The Poisson function of finite elasticity, Journal of Applied Mathematics 53, 807–813.

50. Beatty MF. 1987. A class of universal relations in isotropic elasticity theory, Journal of Elasticity 17, 113–121.

51. Beatty MF. 2007. On the radial oscillations of incompressible, isotropic, elastic and limited elastic thick-walled tubes, International Journal of Non-Linear Mechanics 42, 283–297.

52. Beatty MF. 2011. Small amplitude radial oscillations of an incompressible, isotropic elastic spherical shell, Mathematics and Mechanics of Solids 16, 492–512.

53. Beck JV, Arnold KJ.1977. Parameter Estimation in Engineering and Science, John Wiley & Sons, New York.

54. Becker GW, Kruger O. 1972. On the nonlinear biaxial stress-strain behavior of rubberlike polymers, In: Kausch HH, Hessell JA, Jaffee RI (eds.), Deformation and fracture of high polymers, Plenum Press, New York, 115–130.

55. Belytschko T, Liu W, Moran B. 2000. Nonlinear Finite Elements for Continua and Structures, Wiley, New York.

56. Berger JO, Jefferys WH. 1992. The application of robust Bayesian analysis to hypothesis testing and Occam's razor, Journal of the Italian Statistical Society 1, 17–32.

57. Bertinetti L, Fischer FD, Fratzl P. 2013. Physicochemical basis for water-actuated movement and stress generation in nonliving plant tissues, Physical Review Letters 111, 238001.

58. Biggins JS, Warner M, Bhattacharya K. 2009. Supersoft elasticity in polydomain nematic elastomers, Physical Review Letters 103, 037802 (https://doi.org/10.1103/PhysRevLett.103. 037802).

59. Biggins JS, Warner M, Bhattacharya K. 2012. Elasticity of polydomain liquid crystal elastomers, Journal of the Mechanics and Physics of Solids 60, 573–590 (https://doi.org/10. 1016/j.jmps.2012.01.008).

60. Bigoni D. 2012. Nonlinear Solid Mechanics. Bifurcation Theory and Material Instability, Cambridge University Press, Cambridge, UK.

61. Biot MA. 1965. Mechanics of Incremental Deformation, John Wiley& Sons, Inc.. New York/London/Sydney.

62. Biscari P, Omati C. 2010. Stability of generalized Knowles solids, IMA Journal of Applied Mathematics 75, 479–491.

63. Bladon P, Terentjev EM, Warner M. 1994. Deformation-induced orientational transitions in liquid crystal elastomers, Journal de Physique II 4, 75–91 (https://doi.org/10.1051/jp2: 1994100).

64. Block HD. 1962. Introduction to Tensor Analysis, Charles E. Merrill Books, Inc., Columbus, Ohio.

65. Bogomolny A. 2020. Cut the Knot: Probability Riddles, Wolfram Media, Champaign, IL.

66. Bonet J, Wood RD. 1997. Nonlinear Continuum Mechanics for Finite Element Analysis, Cambridge University Press, Cambridge, UK.

67. Boothby JM, Kim H, Ware TH. 2017. Shape changes in chemoresponsive liquid crystal elastomers, Sensors and Actuators B-chemical 240, 511518 (https://doi.org/10.1016/j.snb. 2016.09.004).

68. Bouasse H, Carriére Z. 1903. Sur les courbes de traction du caoutchouc vulcanisé, Annales de la Faculté des Sciences de Toulouse: Mathématiques, Serie 2, 5 (3), 257–283.

69. Boyce MC, Arruda EM. 2000. Constitutive models of rubber elasticity: A review, Rubber Chemistry and Technology 73, 504–523.

70. Breslavsky I, Amabili M. 2018. Nonlinear vibrations of a circular cylindrical shell with multiple internal resonances under multi-harmonic excitation, Nonlinear Dynamics 93, 53–62.

71. Breslavsky I, Amabili M, Legrand M. 2016. Static and dynamic behaviors of circular cylindrical shells made of hyperelastic arterial materials, Journal of Applied Mechanics, American Society of Mechanical Engineers 83, 051002.

72. Brewick PT, Teferra K. 2018. Uncertainty quantification for constitutive model calibration of brain tissue, Journal of the Mechanical Behavior of Biomedical Materials 85, 237–255.

73. Brömmel F, Zou P, Finkelmann H, Hoffmann A. 2013. Influence of the mesogenic shape on the molecular dynamics and phase-biaxiality of liquid crystal main-chain polymer, Soft Matter 9(5), 1674–1677.

74. Bucchi A, Hearn EH. 2013. Predictions of aneurysm formation in distensible tubes: Part A - Theoretical background to alternative approaches, International Journal of Mechanical Sciences 71, 1–20.

75. Bucchi A, Hearn EH. 2013. Predictions of aneurysm formation in distensible tubes: Part B - Application and comparison of alternative approaches, International Journal of Mechanical Sciences 70, 155–170.

76. Buchheim A. 1884. Mathematical Notes, Messenger of Mathematics 13, 65–66.

77. Budday S, Sommer G, Birkl C, Langkammer C, Haybäck J, Kohnert J, Bauer M, Paulsen F, Steinmann P, Kuhl E, Holzapfel GA. 2017. Mechanical characterization of human brain tissue, Acta Biomaterialia, 48, 319–340 (https://doi.org/10.1016/j.actbio.2016.10.036).

78. Bui-Thanh T. 2021. The optimality of Bayes' theorem, SIAM News 54(6), 1–2.

79. Busse WF. 1939. Physics of rubber as related to the automobile, Journal of Applied Physics 9(7), 438–451 (https://doi.org/10.1063/1.1710439).

80. Bustamante R, Rajagopal KR. 2021. A new type of constitutive equation for nonlinear elastic bodies. Fitting with experimental data for rubber-like materials, Proceedings of the Royal Society A 477, 20210330 (https://doi.org/10.1098/rspa.2021.0330).

81. Buze M, Woolley TE, Mihai LA. 2021. A stochastic framework for atomistic fracture, SIAM Journal on Applied Mathematics 82(2), 526–548 (https://doi.org/10.1137/21M1416436)

82. Calderer C. 1983. The dynamical behaviour of nonlinear elastic spherical shells, Journal of Elasticity 13, 17–47.

83. Camacho-Lopez M, Finkelmann H, Palffy-Muhoray P, Shelley M. 2004. Fast liquid-crystal elastomer swims into the dark, Nature Materials 3, 307–310 (https://doi.org/10.1038/nmat1118).

84. Campbell L, Garnett W. 1882. The Life of James Clerk Maxwell: With Selections from His Correspondence and Occasional Writings, Macmillan and Co., London (https://archive.org/details/lifeofjamesclerk00camprich).

85. Carroll MM. 1987. Pressure maximum behavior in inflation of incompressible elastic hollow spheres and cylinders. Quarterly of Applied Mathematics 45, 141–154.

86. Carroll MM. 2009. Must elastic materials be hyperelastic?, Mathematics and Mechanics of Solids 14, 369–376 (https://doi.org/10.1177/1081286508099385).

87. Carroll MM. 2011. A strain energy function for vulcanized rubber, Journal of Elasticity 103, 173–187.

88. Carroll MM, Horgan CO. 1990. Finite strain solutions for a compressible elastic solid, Quarterly of Applied Mathematics 48, 767–780.

89. Caylak I, Penner E, Dridger A, Mahnken R. 2018. Stochastic hyperelastic modeling considering dependency of material parameters, Computational Mechanics 62, 1273–1285 (https://doi.org/10.1007/s00466-018-1563-z).

90. Cayley A. 1858. A memoir on the theory of matrices, Philosophical Transactions of the Royal Society of London 148, 17–37 (https://doi.org/10.1098/rstl.1858.0002).

91. Chadwick P. 1999. Continuum Mechanics: Concise Theory and Problems, 2nd ed, Dover, New York.

92. Chagnon G, Rebouah M, Favier D. 2014. Hyperelastic energy densities for soft biological tissues: a review, Journal of Elasticity 120, 129–160 (https://doi.org/10.1007/s10659-014-9508-z).

93. Chatelin S, Constantinesco A, Willinger R. 2010. Fifty years of brain tissue mechanical testing: from in vitro to in vivo investigations, Biorheology 47, 255–276.

94. Chen YC, Fried E. 2006. Uniaxial nematic elastomers: constitutive framework and a simple application, Proceedings of the Royal Society A 462, 1295–1314 (https://doi.org/10.1098/rspa.2005.1585).

95. Chen P, Guilleminot J. 2022. Spatially dependent material uncertainties in anisotropic nonlinear elasticity: stochastic modeling, identification, and propagation, Computer Methods in Applied Mechanics and Engineering.

96. Choi J, Chung H, Yun J-H, Cho M. 2014. Photo-isomerization effect of the azobenzene chain on the opto-mechanical behavior of nematic polymer: A molecular dynamics study, Applied Physics Letters 105(22), 221906 (https://doi.org/10.1063/1.4903247).

97. Chou-Wang M-S, Horgan CO. 1989. Void nucleation and growth for a class of incompressible nonlinearly elastic materials, International Journal of Solids and Structures 25, 1239–1254.

98. Chou-Wang M-S, Horgan CO. 1989. Cavitation in nonlinear elastodynamics for neo-Hookean materials, International Journal of Engineering Science 27, 967–973.

99. Chu S, Guilleminot J, Kelly C, Abar B, Gall K. 2021. Stochastic modeling and identification of material parameters on structures produced by additive manufacturing, Computer Methods in Applied Mechanics and Engineering 387, 114166 (https://doi.org/10.1016/j.cma.2021.114166).

100. Chui C, Kobayashi E, Chen X, Hisada T, Sakuma I. 2004. Combined compression and elongation experiments and non-linear modelling of liver tissue for surgical simulation, Medical & Biological Engineering & Computing 42, 787–798.

101. Ciarlet PG. 1996. Mathematical Elasticity: Three-Dimensional Elasticity, vol. 1, North Holland, Amsterdam.

102. Ciarlet PG. 1997. Mathematical Elasticity: Theory of Plates, vol. 2, North-Holland, Amsterdam.

103. Ciarlet PG. 2000. Mathematical Elasticity: Theory of Shells, vol. 3, North-Holland, Amsterdam.

104. Cirak F, Long Q, Bhattacharya K, Warner M. 2014. Computational analysis of liquid crystalline elastomer membranes: Changing Gaussian curvature without stretch energy, International Journal of Solids and Structures 51(1), 144–153 (https://doi.org/10.1016/j.ijsolstr.2013.09.019).

105. Clarke SM, Hotta A, Tajbakhsh AR, Terentjev EM. 2001. Effect of crosslinker geometry on equilibrium thermal and mechanical properties of nematic elastomers, Physical Reviews E 64, 061702 (https://doi.org/10.1103/PhysRevE.64.061702).

106. Clarke SM, Terentjev EM. 1998. Slow stress relaxation in randomly disordered nematic elastomers and gels, Physical Review Letters 81(20), 4436–4439 (https://doi.org/10.1103/PhysRevLett.81.4436).

107. Clarke SM, Terentjev EM, Kundler I, Finkelmann H. 1998. Texture evolution during the polydomain-monodomain transition in nematic elastomers, Macromolecules 31(15), 4862–4872 (https://doi.org/10.1021/ma980195j).

108. Coleman BD, Newman DC. 1988. On the rheology of cold drawing. I. Elastic materials, Journal of Polymer Science: Part B: Polymer Physics 26, 1801–1822 (https://doi.org/10.1002/polb.1988.090260901).

109. Comley KSC, Fleck NA. 2012. The compressive response of porcine adipose tissue from low to high strain rate, International Journal of Impact Engineering 46, 1–10.

110. Connolly F, Polygerinos P, Walsh CJ, Bertoldi K. 2015. Mechanical programming of soft actuators by varying fiber angle, Soft Robotics 2, 26–32.

111. Connolly F, Walsh CJ, Bertoldi K. 2017. Automatic design of fiber-reinforced soft actuators for trajectory matching, Proceedings of the National Academy of Sciences of the United States of America (PNAS) 14, 51–56.

112. Considére A. 1885. Mémoire sur l'emploi du fer et de l'acier dans les constructions, Annales des Ponts et Chaussées 6(9), 574–775.

113. Conti S, DeSimone A, Dolzmann G. 2002. Soft elastic response of stretched sheets of nematic elastomers: a numerical study, Journal of the Mechanics and Physics of Solids 50, 1431–1451 (https://doi.org/10.1016/S0022-5096(01)00120-X).

114. Conti S, DeSimone A, Dolzmann G. 2002. Semi-soft elasticity and director reorientation in stretched sheets of nematic elastomers, Physical Review E 60, 61710 (https://doi.org/10.1103/PhysRevE.66.061710).

115. Corbett D, Warner M. 2008. Bleaching and stimulated recovery of dyes and of photocantilevers, Physical Review E 77, 051710.

116. Crespo J, Latorre M, Montans FJ. 2017. WYPIWYG hyperelasticity for isotropic, compressible materials, Computational Mechanics 59, 73–92.

117. Criscione JC, Humphrey JD, Douglas AS, Hunter WC. 2000. An invariant basis for natural strain which yields orthogonal stress response terms in isotropic hyperelasticity, Journal of the Mechanics and Physics of Solids 48, 2445–2465.

118. Croci M, Vinje V, Rognes ME. 2019. Uncertainty quantification of parenchymal tracer distribution using random diffusion and convective velocity fields, Fluids and Barriers of the CNS 16, 32 (https://doi.org/0.1186/s12987-019-0152-7).

119. Cviklinski J, Tajbakhsh AR, Terentjev EM. 2002. UV isomerisation in nematic elastomers as a route to photo-mechanical transducer, The European Physical Journal E 9, 427–434 (https://doi.org/10.1140/epje/i2002-10095-y).

120. Dal H, Açikgöz K, Badienia Y. 2021 On the performance of isotropic hyperelastic constitutive models for rubber-like materials: a state of the art review, Applied Mechanics Review 73, 020802 (https://doi.org/10.1115/1.4050978).

121. Davidson EC, Kotikian A, Li S, Aizenberg J, Lewis JA. 2019. 3D printable and reconfigurable liquid crystal elastomers with light-induced shape memory via dynamic bond exchange, Advanced Materials, 1905682 (https://doi.org/10.1002/adma.201905682).

122. De Bellis I, Martella D, Parmeggiani C, Keller P, Wiersma DS, Li MH, Nocentini S. 2020. Color modulation in morpho butterfly wings using liquid crystalline elastomers, Advanced Intelligent Systems, 2000035 (https://doi.org/10.1002/aisy.202000035).

123. de Gennes PG. 1979. The Physics of Liquid Crystals, Clarendon Press, Oxford.

124. de Gennes PG. 1975. Physique moléculaire - réflexions sur un type de polymères némauques, Comptes rendus de l'Académie des Sciences B 281, 101–103.

125. de Gennes PG, Prost J. 1993. The Physics of Liquid Crystals, 2nd ed, Clarendon Press, Oxford.

126. de Jeu WH (ed). 2012. Liquid Crystal Elastomers: Materials and Applications, Springer, New York.

127. De Pascalis R, Parnell WJ, Abrahams ID, Shearer T, Daly DM, Grundy D. 2018. The inflation of viscoelastic balloons and hollow viscera, Proceedings of the Royal Society A 474, 20180102.

128. DeSimone A. 1999. Energetics of fine domain structures, Ferroelectrics 222(1), 275–284 (https://doi.org/10.1080/00150199908014827).

129. DeSimone A, Dolzmann G. 2000. Material instabilities in nematic elastomers, Physica D 136(1–2), 175–191 (https://doi.org/10.1016/S0167-2789(99)00153-0).

130. DeSimone A, Dolzmann G. 2002. Macroscopic response of nematic elastomers via relaxation of a class of SO(3)-invariant energies, Archive of Rational Mechanics and Analysis 161, 181–204 (https://doi.org/10.1007/s002050100174).

131. DeSimone A, Gidoni P, Noselli G. 2015. Liquid crystal elastomer strips as soft crawlers, Journal of the Mechanics and Physics of Solids 84, 254–272.

132. DeSimone A, Teresi L. 2009. Elastic energies for nematic elastomers, The European Physical Journal E 29, 191–204 (https://doi.org/10.1140/epje/i2009-10467-9).

133. Destrade M, Gilchrist MD, Motherway J, Murphy JG. 2012. Slight compressibility and sensitivity to changes in Poisson's ratio, International Journal for Numerical Methods in Engineering 2012, 403–411.

134. Destrade M, Gilchrist MD, Murphy JG, Rashid B, Saccomandi G. 2015. Extreme softness of brain matter in simple shear, International Journal of Non-Linear Mechanics 75, 54–58.

135. Destrade M, Goriely A, Saccomandi G. 2011. Scalar evolution equations for shear waves in incompressible solids: a simple derivation of the Z, ZK, KZK and KP equations, Proceedings of the Royal Society A 467, 1823–1834.

136. Destrade M, Murphy JG, Saccomandi G. 2011. Simple shear is not so simple, International Journal of Non-Linear Mechanics 47, 210–214.

137. Destrade M, Saccomandi G. 2010. On the rectilinear shear of compressible and incompressible elastic slabs, International Journal of Engineering Science 48, 1202–1211.

138. Destrade M, Saccomandi G, Sgura I. 2017. Methodical fitting for mathematical models of rubber-like materials, Proceedings of the Royal Society A 473, 20160811 (https://doi.org/10.1098/rspa.2016.0811).

139. de St Venant AJB. 1844. Sur les pressions qui se développent à l'intérieur des corps solides lorsque les déplacements de leurs points, sans alterer l'élasticité, ne peuvent cependant pas être considérés comme très-petits, Bulletin de la Société Philomathique 5, 26–28.

140. Diaconis P, Skyrms B. 2018. Ten Great Ideas about Chance, Princeton University Press, New Jersey.

141. Diani J, Fayolle B, Gilormini P. 2009. A review on the Mullins effect, European Polymer Journal 45, 601–612 (https://doi.org/10.1016/j.eurpolymj.2008.11.017).

142. Dinwoodie JM. 1981. Timber, its Nature and behavior, Van Nostrand Reinhold, New York.

143. Dobrynin AV, Carrillo J-MY. 2011. Universality in nonlinear elasticity of biological and polymeric networks and gels, Macromolecules 44, 140–146.

144. Dong YH, Zhu B, Wang Y, Li YH, Yang J. 2018. Nonlinear free vibration of graded graphene reinforced cylindrical shells: Effects of spinning motion and axial load, Journal of Sound and Vibration 437, 79–96.

145. Dorfmann A, Ogden RW. A constitutive model for the Mullins effect with permanent set in particle-reinforced rubber, International Journal of Solids and Structures 41(7), 1855–1878 (https://doi.org/10.1016/j.ijsolstr.2003.11.014).

146. Dorfmann L, Ogden RW. 2014. Nonlinear Theory of Electroelastic and Magnetoelastic Interactions, Springer, New York.

147. Dorfmann L, Ogden RW. 2020. Waves and vibrations in a finitely deformed electroelastic circular cylindrical tube, Proceedings of the Royal Society A 476, 20190701 (https://doi.org/10.1098/rspa.2019.0701).

148. Drass M, Bartels N, Schneider J, Klein D. 2019. Pseudo-elastic cavitation model: part II - extension to cyclic behavior of transparent silicone adhesives, Glass Structures & Engineering 5, 67–82 (https://doi.org/10.1007/s40940-019-00103-8).

149. Drass M, Du Bois PA, Schneider J, Killing S. 2020. Pseudo-elastic cavitation model: part I - finite element analyses on thin silicone adhesives in façades, Glass Structures & Engineering 5, 41–46 (https://doi.org/10.1007/s40940-019-00115-4).

150. Elishakoff I. 2017. Probabilistic Methods in the Theory of Structures: Strength of Materials, Random Vibrations, and Random Buckling, 3rd ed, World Scientific Publishing Co., Singapore.

151. Elishakoff I. 2017. Problems Book for Probabilistic Methods for the Theory of Structures with Complete Worked Through Solutions, World Scientific Publishing Co., Singapore.

152. Elishakoff I, Soize C (eds.). 2012. Nondeterministic Mechanics, Springer, New York.

153. Emery D, Fu Y. 2021. Elasto-capillary circumferential buckling of soft tubes under axial loading: existence and competition with localised beading and periodic axial modes, Mechanics of Soft Materials 3, 3 (https://doi.org/10.1007/s42558-021-00034-x).

154. Ericksen JL. 1953. On the propagation of waves in isotropic incompressible perfectly elastic materials, Journal of Rational Mechanics and Analysis 2, 329–337.

155. Ericksen JL. 1954. Deformations possible in every isotropic, incompressible, perfectly elastic body, Zeitschrift für angewandte Mathematik und Physik (ZAMP) 5, 466–489.

156. Ericksen JL. 1955. Deformation possible in every compressible isotropic perfectly elastic materials, Journal of Mathematics and Physics 34, 126–128.

157. Ericksen JL. 1975. Equilibrium of bars, Journal of Elasticity 5(3–4), 191–201 (https://doi.org/10.1007/BF00126984).

158. Ertepinar A, Akay HU. 1976. Radial oscillations of nonhomogeneous, thick-walled cylindrical and spherical shells subjected to finite deformations, International Journal of Solids and Structures 12, 517–524.

159. Esmailzadeh E, Younesian D, Askari H. 2018. Analytical Methods in Nonlinear Oscillations: Approaches and Applications, Springer, Dordrecht, Netherlands.

160. Evans SL. 2017. How can we measure the mechanical properties of soft tissues?, In: Avril S, Evans SL (eds.), Material Parameter Identification and Inverse Problems in Soft Tissue, Biomechanics, Springer, London, 67–83.

161. Farmer CL. 2017. Uncertainty quantification and optimal decisions, Proceedings of the Royal Society A 473, 20170115.

162. Ferrers NM (ed). 2014. Mathematical Papers of the Late George Green, Cambridge University Press, Cambridge, UK.

163. Finkelmann H, Greve A, Warner M. 2001. The elastic anisotropy of nematic elastomers, The European Physical Journal E 5, 281–293 (https://doi.org/10.1007/s101890170060).

164. Finkelmann H, Kock HJ, Rehage G. 1981. Investigations on liquid crystalline polysiloxanes 3, Liquid crystalline elastomers - a new type of liquid crystalline material, Die Makromolekulare Chemie, Rapid Communications 2, 317–322 (https://doi.org/10.1002/marc.1981.030020413).

165. Finkelmann H, Kundler I, Terentjev EM, Warner M. 1997. Critical stripe-domain instability of nematic elastomers, Journal de Physique II 7, 1059–1069 (https://doi.org/10.1051/jp2:1997171).

166. Finkelmann H, Nishikawa E, Pereita GG, Warner M. 2001. A new opto-mechanical effect in solids, Physical Review Letters 87(1), 015501 (https://doi.org/10.1103/PhysRevLett.87.015501).

167. Fitt D, Wyatt H, Woolley TE, Mihai LA. 2019. Uncertainty quantification of elastic material responses: testing, stochastic calibration and Bayesian model selection, Mechanics of Soft Materials 1, 13 (https://doi.org/10.1007/s42558-019-0013-1).

168. Fleischmann E-K, Liang H-L, Kapernaum N, Giesselmann F, Lagerwall JPF, Zentel R. 2012. One-piece micropumps from liquid crystalline core-shell particles, Nature Communications 3, 1178 (https://doi.org/10.1038/ncomms2193).

169. Flory PJ. 1961. Thermodynamic relations for high elastic materials, Transactions of the Faraday Society 57, 829–838 (https://doi.org/10.1039/TF9615700829).

170. Fond, C. 2001. Cavitation criterion for rubber materials: a review of void-growth models, Journal of Polymer Science: Part B 39, 2081–2096.

171. Ford MJ, Ambulo CP, Kent TA, Markvicka EJ, Pan C, Malen J, Ware TH, Majidi C. 2019. A multifunctional shape-morphing elastomer with liquid metal inclusions, Proceedings of the National Academy of Sciences 116(43), 21438–21444 (https://doi.org/10.1073/pnas.1911021116).

172. Fornasini P. 2008. The Uncertainty in Physical Measurements: An Introduction to Data Analysis in the Physics Laboratory, Springer, New York.

173. Fortes MA, Nogueira MT. 1989. The Poisson effect in cork, Materials Science and Engineering A 122.

174. Frank FC. 1958. I. Liquid crystals. On the theory of liquid crystals, Discussions of the Faraday Society 25, 19–28.

175. Freedman D, Pusani R, Perves R. Statistics, 4th ed, W.W. Norton & Company, New York.

176. Fried I, Johnson AR. 1988. A note on elastic energy density functions for largely deformed compressible rubber solids, Computer Methods in Applied Mechanics and Engineering 69, 53–64.

177. Fried E. 2002. An elementary molecular-statistical basis for the Mooney and Rivlin-Saunders theories of rubber elasticity, Journal of the Mechanics and Physics of Solids 50, 571–582 (https://doi.org/10.1016/S0022-5096(01)00086-2).

178. Fried E, Sellers S. 2004. Free-energy density functions for nematic elastomers, Journal of the Mechanics and Physics of Solids 52(7), 1671–1689 (https://doi.org/10.1016/j.jmps.2003.12.005).

179. Fried E, Sellers S. 2005. Orientational order and finite strain in nematic elastomers, The Journal of Chemical Physics 123(4), 043521 (https://doi.org/10.1063/1.1979479).

180. Fried E, Sellers S. 2006. Soft elasticity is not necessary for striping in nematic elastomers, Journal of Applied Physics 100, 043521 (https://doi.org/10.1063/1.2234824).

181. Friedel G. 1922. Les états mésomorphes de la matière, Annales de Physique 9(18), 272–474.

182. Frolich LM, LaBarbera M, Stevens WP. 1994. Poisson's ratio of a crossed fibre sheath: the skin of aquatic salamanders, Journal of Zoology 232(2), 231–252 (https://doi.org/10.1111/j.1469-7998.1994.tb01571.x).

183. Fu Y, Chui CK, Teo CL. 2013. Liver tissue characterization from uniaxial stress-strain data using probabilistic and inverse finite element methods, Journal of the Mechanical Behavior of Biomedical Materials 20, 105–112.

184. Fu Y, Liu JL, Francisco GS. 2016. Localized bulging in an inflated cylindrical tube of arbitrary thickness - the effect of bending stiffness, Journal of the Mechanics and Physics of Solids 90, 45–60.

185. Fu Y, Jin L, Goriely A. 2021. Necking, beading, and bulging in soft elastic cylinders, Journal of the Mechanics and Physics of Solids 147, 104250 (https://doi.org/10.1016/j.jmps.2020.104250).

186. Fung YC. 1993. Biomechanics: Mechanical Properties of Living Tissues, 2nd ed, Springer, New York.

187. Gao Z, Lister K, Desai J. 2010. Constitutive modeling of liver tissue: experiment and theory, Annals of Biomedical Engineering 38, 505–516.

188. Geethama VG, Thomas S. 1997. Why does a rubber ball bounce? The molecular origins of rubber elasticity, Resonance 2(4), 48–54.

189. Gelebart AH, Mulder DJ, Varga M, Konya A, Vantomme G, Meijer EW, Selinger RLB, Broer DJ. 2017. Making waves in a photoactive polymer film, Nature 546, 632.

190. Gent AN. 1991. Cavitation in rubber: a cautionary tale, Rubber Chemistry and Technology 63, G49-G53.

191. Gent AN. 1996. A new constitutive relation for rubber, Rubber Chemistry and Technology 69, 59–61 (https://doi.org/10.5254/1.3538357).

192. Gent AN, Hua KC. 2004. Torsional instability of stretched rubber cylinders, International Journal of Non-Linear Mechanics 39, 483–489 (https://doi.org/10.1016/S0020-7462(02)00217-2).

193. Gent AN, Lindley PB. 1959. Internal rupture of bonded rubber cylinders in tension, Proceedings of the Royal Society of London A 249, 195–205.

194. Gent AN, Thomas AG. 1958. Forms for the stored (strain) energy function for vulcanized rubber, Journal of Polymer Science 28, 625–628.

195. Ghanem R, Higdon D, Owhadi H (eds.). 2017. Handbook of Uncertainty Quantification, Springer, New-York, 2017.

196. Gibson LJ, Ashby MF, Harley BA. 2010. Cellular Materials in Nature and Medicine, Cambridge University Press, Cambridge, UK.

197. Gigerenzer G, Swijtink Z, Porter T, Daston L, Beatty J, Krüger L. 1989. The Empire of Chance: How Probability Changed Science and Everyday Life, Cambridge University Press, New York.

198. Giudici A, Biggins JS. 2021. Large deformation analysis of spontaneous twist and contraction in nematic elastomer fibres with helical director, Journal of Applied Physics 129(15), 154701 (https://doi.org/10.1063/5.0040721).

199. Giudici A, Biggins JS. 2020. Giant deformations and soft-inflation in LCE balloons, Europhysics Letters 132(3), 36001 (https://doi.org/10.1209/0295-5075/132/36001).

200. Gilchrist MD, Murphy JG, Pierrat B, Saccomandi G. 2017. Slight asymmetry in the winding angles of reinforcing collagen can cause large shear stresses in arteries and even induce buckling, Meccanica 52, 3417–3429.

201. Godoy LA. 2006. Historical sense in the historians of the theory of elasticity, Meccanica 41, 529–538 (https://doi.org/10.1007/s11012-006-9001-2).

202. Golubitsky M, Schaeffer D. 1979. A theory for imperfect bifurcation via singularity theory, Communications on Pure and Applied Mathematics 32, 21–98.

203. Goncalves PB, Pamplona D, Lopes SRX. 2008. Finite deformations of an initially stressed cylindrical shell under internal pressure, International Journal of Mechanical Sciences 50, 92–103.

204. Goodbrake C, Goriely A, Yavari A. 2021. The mathematical foundations of anelasticity: existence of smooth global intermediate configurations, Proceedings of the Royal Society A 477, 20200462 (https://doi.org/10.1098/rspa.2020.0462).

205. Goriely A. 2017. The Mathematics and Mechanics of Biological Growth, Springer-Verlag, New York.

206. Goriely A, Destrade M, Ben Amar M. 2006. Instabilities in elastomers and in soft tissues, The Quarterly Journal of Mechanics and Applied Mathematics 59, 615–630.

207. Goriely A, Mihai LA. 2021. Liquid crystal elastomers wrinkling, Nonlinearity 34(8), 5599–5629 (https://doi.org/10.1088/1361-6544/ac09c1).

208. Goriely A, Moulton DE, Mihai LA. 2022. A rod theory for liquid crystalline elastomers, Journal of Elasticity, (https://doi.org/10.1007/s10659-021-09875-z).

209. Goriely A, Moulton DE, Vandiver R. 2010. Elastic cavitation, tube hollowing, and differential growth in plants and biological tissues, Europhysics Letters 91, 18001.

210. Goriely A, Tabor M. 1998. Spontaneous helix-hand reversal and tendril perversion in climbing plants, Physical Review Letters 80, 1564–1567.

211. Goriely A, Tabor M. 2013. Rotation, inversion and perversion in anisotropic elastic cylindrical tubes and membranes, Proceedings of the Royal Society A 469, 2013001.

212. Gough J. 1805. Memoirs of the Literary and Philosophical Society of Manchester, Second Series, Volume I, 288–295.

213. Graban K, Schweickert E, Martin RJ, Neff P. 2019. A commented translation of Hans Richter's early work "The isotropic law of elasticity", Mathematics and Mechanics of Solids 24(8), 2649–2660.

214. Green G. 1828. An Essay on the Application of Mathematical Analysis to the Theories of Electricity and Magnetism, Printed for the author by T. Wheelhouse, Nottingham.

215. Green G. 1839. On the propagation of light in crystallized media, Transactions of the Cambridge Philosophical Society 7, 121–140.

216. Green AE, Adkins JE. 1970. Large Elastic Deformations (and Non-linear Continuum Mechanics), 2nd ed, Oxford University Press, Oxford.

217. Green AE, Shield RT. 1950. Finite elastic deformations in incompressible isotropic bodies, Proceeding of the Royal Society of London A 202, 407–419.

218. Green AE, Zerna W. 2012. Theoretical Elasticity, 2nd rev. ed, Dover, New York.

219. Gregory P. 2005. Bayesian Logical Data Analysis for the Physical Sciences, Cambridge University Press, New York.

220. Grigoriu M. 2000. Stochastic mechanics, International Journal of Solids and Structures 37(1–2), 197–214 (https://doi.org/10.1016/S0020-7683(99)00088-8).

221. Grimmett GR, Stirzaker DR. 2001. Probability and Random Processes, 3rd ed, Oxford University Press, Oxford.

222. Griniasty I, Mostajeran C, Cohen I. 2021. Multivalued inverse design: Multiple surface geometries from one flat sheet, Physical Review Letters 127, 128001 (https://doi.org/10.1103/PhysRevLett.127.128001).

223. Gross B, Harris J, Riehl E. 2019. Fat Chance! Probability from 0 to 1, Cambridge University Press, New York.

224. Guilleminot J. 2020. Modelling non-Gaussian random fields of material properties in multi-scale mechanics of materials, In: Y. Wang Y, McDowell FL (eds.), Uncertainty Quantification in Multiscale Materials Modeling, Elsevier Series in Mechanics of Advanced Materials, Woodhead Publishing.

225. Guilleminot J, Soize C. 2012. Generalized stochastic approach for constitutive equation in linear elasticity: a random matrix model, International Journal for Numerical Methods in Engineering 90, 613–635.

226. Guilleminot J, Soize C. 2013. On the statistical dependence for the components of random elasticity tensors exhibiting material symmetry properties, Journal of Elasticity 11, 109–130.

227. Guilleminot J, Soize C. 2017. Non-Gaussian random fields in multiscale mechanics of heterogeneous materials, in Encyclopedia of Continuum Mechanics, Springer, Berlin Heidelberg.

228. Gurtin ME. 1981. An Introduction to Continuum Mechanics, Academic Press, New York.

229. Haas PA, Goldstein RE. 2019. Nonlinear and nonlocal elasticity in coarse-grained differential-tension models of epithelia, Physical Review E 99, 022411 (https://doi.org/10.1103/PhysRevE.99.022411).

230. Haas PA, Goldstein RE. 2015. Elasticity and glocality: Initiation of embryonic inversion in Volvox, Journal of the Royal Society Interface 12, 20150671 (https://doi.org/10.1098/rsif.2015.0671).

231. Haghiashtiani G, Habtour E, Park SH, Gardea F, McAlpine MC. 2018. 3D printed electrically-driven soft actuators, Extreme Mechanics Letters 21, 1–8 (https://doi.org/10.1016/j.eml.2018.02.002).

232. Halsey LG. 2019. The reign of the p-value is over: what alternative analyses could we employ to fill the power vacuum?, Biology Letters 15(5), 20190174 (https://doi.org/10.1098/rsbl.2019.0174).
233. Hamel G. 1912. Elementare Mechanik, Leipzig & Berlin.
234. Hand DJ. 2020. Dark Data, Princeton University Press, New Jersey.
235. Hartmann S. 2001. Parameter identification with a direct search method using finite elements, In: Besdo D, Schuster RH, Ihlemann J (eds.), Constitutive Models for Rubber II, Balkerna Publ. Lisse, 249–256.
236. Hartmann S. 2001. Numerical studies on the identification of the material parameters of Rivlin's hyperelasticity using tension-torsion tests, Acta Mechanica 148, 129–155.
237. Hartmann S. 2001. Parameter estimation of hyperelasticity relations of generalized polynomial-type with constraint conditions, International Journal of Solids and Structures 38, 7999–8018.
238. Hartmann S, Gilbert RR. 2017. Identifiability of material parameters in solid mechanics, Archive of Applied Mechanics, https://doi.org/10.1007/s00419-017-1259-4.
239. Haslach H, J. Humphrey J. 2004. Dynamics of biological soft tissue and rubber: internally pressurized spherical membranes surrounded by a fluid, International Journal of Non-Linear Mechanics 39, 399–420.
240. Hay GE. 1942. The finite displacement of thin rods, Transactions of the American Mathematical Society 51, 65–102.
241. Hayes MA, Knops RJ. 1966. On universal relations in elasticity theory, Zeitschrift für Angewandte Mathematik und Physik (ZAMP) 17, 636–639.
242. He, L, Lou J, Du J. 2018. Voltage-driven torsion of electroactive thick tubes reinforced with helical fibers, Acta Mechanica 229, 2117–2131.
243. He Q, Wang Z, Wang Y, Song Z, Cai S. 2020. Recyclable and self-repairable fluid-driven liquid crystal elastomer actuator, Applied Materials & Interfaces 12, 35454–35474 (https://doi.org/10.1021/acsami.0c10021).
244. He X, Zheng Y, He Q, Cai S. 2020. Uniaxial tension of a nematic elastomer with inclined mesogens, Extreme Mechanics Letters 40, 100936 (https://doi.org/10.1016/j.eml.2020.100936).
245. He Q, Zheng Y, Wang Z, He X, Cai S. 2020. Anomalous inflation of a nematic balloon, Journal of the Mechanics and Physics of Solids 142, 104013 (https://doi.org/10.1016/j.jmps.2020.104013).
246. Hebner TS, Bowman CN, White TJ. 2021. Influence of orientational genesis on the actuation of monodomain liquid crystalline elastomers, Macromolecules 54, 4023–4029 (https://doi.org/10.1021/acs.macromol.1c00437).
247. Hejazi M, Hsiang Y, Srikantha Phani A. 2021. Fate of a bulge in an inflated hyperelastic tube: theory and experiment, Proceedings of the Royal Society A 477, 20200837 (https://doi.org/10.1098/rspa.2020.0837).
248. Hencky H. 1928. Über die Form des Elastizitätsgesetzes bei ideal elastischen Stoffen, Zeitschrift für technische Physik 9, 215–220.
249. Heng GZ, Solecki R. 1963. Free and forced finite amplitude oscillations of an elastic thick-walled hollow sphere made of incompressible material, Archiwum Mechaniki Stosowanej 3, 427–433.
250. Higaki H, Takigawa T, Urayama K. 2013. Nonuniform and uniform deformations of stretched nematic elastomers, Macromolecules 46, 5223–5231 (https://doi.org/10.1021/ma400771z).
251. Higham NJ, Noferini V. 2016. An algorithm to compute the polar decomposition of a 3×3 matrix, Numerical Algorithms 73, 349–369 (https://doi.org/10.1007/s11075-016-0098-7).
252. Hill R. 1957. On uniqueness and stability in the theory of finite elastic strain, Journal of Mechanics and Physics of Solids 5, 229–241.
253. Hill R. 1970. Constitutive inequalities for isotropic elastic solids under finite strain, Proceedings of the Royal Society of London A 314, 457–472.

254. Hill R. Hutchinson JW. 1975. Bifurcation phenomena in the plane tension test, Journal of the Mechanics and Physics of Solids 23(4–5), 239–264 (https://doi.org/10.1016/0022-5096(75)90027-7).

255. Hogan PM, Tajbakhsh AR, Terentjev EM. 2002. UV manipulation of order and macroscopic shape in nematic elastomers, Physical. Review E 65, 041720 (https://doi.org/10.1103/PhysRevE.65.041720).

256. Holzapfel GA. 2000. Nonlinear Solid Mechanics: A Continuum Approach for Engineering, John Wiley & Sons, New York.

257. Horgan CO, Murphy JG. 2017. Fiber orientation effects in simple shearing of fibrous soft tissues, Journal of Biomechanics 64, 131–135.

258. Horgan CO, Murphy JG. 2010. Simple shearing of incompressible and slightly compressible isotropic nonlinearly elastic materials, Journal of Elasticity 98, 205–221 (https://doi.org/10.1007/s10659-009-9225-1).

259. Horgan CO, Murphy JG. 2017. Poynting and reverse Poynting effects in soft materials, Journal of Soft Matter 13, 4916 (https://doi.org/10.1039/c7sm00992e).

260. Horgan CO, Murphy JG. 2019. Magic angles in the mechanics of fibrous soft materials, Mechanics of Soft Materials 1, 2 (https://doi.org/10.1007/s42558-018-0001-x).

261. Horgan CO, Murphy JG. 2021. Incompressible transversely isotropic hyperelastic materials and their linearized counterparts, Journal of Elasticity 143, 187–194 (https://doi.org/10.1007/s10659-020-09803-7).

262. Horgan HO, Murphy JG. 2022. On an angle with magical properties, Notices of the American Mathematical Society 69(1), 22–25 (https://doi.org/10.1090/noti2398).

263. Horgan CO, Ogden RW, Saccomandi G. 2004. A theory of stress softening of elastomers based on finite chain extensibility, Proceedings of the Royal Society A 460(2046), 1737–1754 (https://doi.org/10.1098/rspa.2003.1248).

264. Horgan CO, Pence TJ. 1989. Cavity formation at the center of a composite incompressible nonlinearly elastic sphere, Journal of Applied Mechanics 56, 302–308.

265. Horgan CO, Polignone DA. 1995. Cavitation in nonlinearly elastic solids: a review, Applied Mechanics Reviews 48, 471–485.

266. Horgan CO, Saccomandi G. 2002. A molecular-statistical basis for the Gent constitutive model of rubber elasticity, Journal of Elasticity 68, 167–176 (https://doi.org/10.1023/A:1026029111723).

267. Horgan CO, Saccomandi G. 2004. Constitutive models for compressible nonlinearly elastic materials with limiting chain extensibility, Journal of Elasticity 77, 123–138 (https://doi.org/10.1007/s10659-005-4408-x).

268. Hrapko M, van Dommelen JAW, Peters GWM, Wismans JSHM. 2008. Characterisation of the mechanical behavior of brain tissue in compression and shear, Biorheology 45, 663–676.

269. Huet C. 1990. Application of variational concepts to size effects in elastic heterogeneous bodies, Journal of the Mechanics and Physics of Solids 38, 813–841.

270. Hughes I, Hase TPA. 2010. Measurements and Their Uncertainties: A Practical Guide to Modern Error Analysis, Oxford University Press, Oxford.

271. Huilgol RR. 1967. Finite amplitude oscillations in curvilinearly aeolotropic elastic cylinder, Quarterly of Applied Mathematics 25, 293–298.

272. Hussain M, Jull EIL, Mandle RJ, Raistrick T, Hine PJ, Gleeson HF. Liquid crystal elastomers for biological applications, Nanomaterials 11, 813 (https://doi.org/10.3390/nano11030813).

273. Hutchens SB, Fakhouri S, Crosby AJ. 2016. Elastic cavitation and fracture via injection, Soft Matter 12, 2557.

274. Il'ichev AT, Fu Y. 2014. Stability of an inflated hyperelastic membrane tube with localized wall thinning, International Journal of Engineering Science 80, 53–61.

275. James RD. 1985. Displacive phase transformations in solids, Journal of Mechanics and Physics of Solids 34, 359–394.

276. James RD, Spector SJ. 1991. The formation of filamentary voids in solids, Journal of Mechanics and Physics of Solids 39, 783–813.

277. Jampani VSR, Mulder DJ, De Sousa KR, Gélébart AH, Lagerwall JPF, Schenning APHJ. 2018. Micrometer-scale porous buckling shell actuators based on liquid crystal networks, Advanced Functional Materials 28(31), 1801209 (https://doi.org/10.1002/adfm.201801209).

278. Jampani VSR, Volpe RH, De Sousa KR, Machado JF, Yakacki CM, Lagerwall JPF, Schenning APHJ. 2019. Liquid crystal elastomer shell actuators with negative order parameter, Science Advances 5(4), eaaw2476 (https://doi.org/10.1126/sciadv.aaw2476).

279. Janmey PA, McCormick ME, Rammensee S, Leight JL, Georges PC, MacKintosh FC. 2006. Negative normal stress in semiflexible biopolymer gels, Nature Materials 6, 48–51.

280. Jaynes ET. 1957. Information theory and statistical mechanics i, Physical Review 108, 171–190.

281. Jaynes ET. 1957. Information theory and statistical mechanics ii, Physical Review 106, 620–630.

282. Jaynes ET. 2003. Probability Theory: The Logic of Science, Cambridge University Press, Cambridge, New York.

283. Jefferys WH, Berger JO. 1992. Ockham's razor and Bayesian analysis, American Scientist 80, 64–72.

284. Jeffreys H. 1935. Some tests of significance, treated by the theory of probability, Mathematical Proceedings of the Cambridge Philosophical Society 31, 203–222.

285. Jeffreys H. 1961. Theory of Probability, 3rd ed, Oxford University Press, Oxford, UK.

286. Jiang ZC, Xiao YY, Zhao Y. 2019. Shining light on liquid crystal polymer networks: preparing, reconfiguring, and driving soft actuators, Advanced Optical Materials 7, 1900262 (https://doi.org/10.1002/adom.201900262).

287. Jin X, Zhu F, Mao H, Shen M, Yang KH. 2013. A comprehensive experimental study on material properties of human brain tissue, Journal of Biomechanics 46, 2795–2801.

288. Johnson NL, Kotz S, Balakrishnan N. 1994. Continuous Univariate Distributions, Vol 1, 2nd ed, John Wiley & Sons, New York.

289. Jones DF, Treloar LRG. 1975. The properties of rubber in pure homogeneous strain, Journal of Physics D: Applied Physics 8(11), 1285–1304 (https://doi.org/10.1088/0022-3727/8/11/007).

290. Junker P, Nagel J. 2019. Modeling of visco-elastic structures with random material properties using time-separated stochastic mechanics, International Journal for Numerical Methods in Engineering, 1–26 (https://doi.org/10.1002/nme.6210).

291. Kaiser A, Winkler M, Krause S, Finkelmann H, Schmidt AM. 2009. Magnetoactive liquid crystal elastomer nanocomposites, Journal of Materials Chemistry 19(4), 538–543.

292. Kaminski M, Lauke B. 2018. Probabilistic and stochastic aspects of rubber hyperelasticity, Meccanica 53, 2363–2378.

293. Kang J, Tang Y. 2021. Dynamic cavitation in soft solids under monotonically increasing pressure, International Journal of Mechanical Sciences 209, 106730 (https://doi.org/10.1016/j.ijmecsci.2021.106730).

294. Kang J, Wang C, Tan H. 2018. Cavitation in inhomogeneous soft solids, Soft Matter 14,7979–7986.

295. Kang J, Zhang Z, Wang C. 2020. Analytical study of cavitation in elastic solids with affinely varied mechanical property, Soft Materials 18(1), 38–45 (https://doi.org/110.1080/1539445X.2019.1655052).

296. Kanner LM, Horgan CO. 2008. On extension and torsion of strain-stiffening rubber-like elastic circular cylinders 93, 39–61 (https://doi.org/10.1007/s10659-008-9164-2).

297. Katz VJ. 1979. The history of Stokes' theorem, Mathematics Magazine 52(3), 146–156.

298. Kearsley EA, Zapas LJ. 1980. Some methods of measurement of an elastic strain-energy function of the Valanis-Landel type, Journal of Rheology 24, 483–500.

299. Kemper A, Santago A, Stitzel J, Sparks J, Duma S. 2012. Biomechanical response of human spleen in tensile loading, Journal of Biomechanics 45, 348–355.

300. Kloczkowski A. 2002. Application of statistical mechanics to the analysis of various physical properties of elastomeric networks - a review. Polymer 43, 1503–1525.

301. Knowles JK. 1960. Large amplitude oscillations of a tube of incompressible elastic material, Quarterly of Applied Mathematics 18, 71–77.
302. Knowles JK. 1962. On a class of oscillations in the finite-deformation theory of elasticity, Journal of Applied Mechanics 29, 283–286.
303. Knowles JK, Jakub MT. 1965. Finite dynamic deformations of an incompressible elastic medium containing a spherical cavity, Archive of Rational Mechanics and Analysis 18, 376–387.
304. Kowalski BA, Mostajeran C, Godman NP, Warner M, White TJ. 2017. Curvature by design and on demand in liquid crystal elastomers, Physical Review E 97, 012504.
305. Kotz S, Balakrishnan N, Johnson NL. 2000. Continuous Multivariate Distributions, Vol 1: Models and Applications, 2nd ed, Wiley, New York.
306. Krauss H. 1967. Thin Elastic Shells, John Wiley & Sons, New York - London - Sydney.
307. Kroeger M, Karl H, Simmler B, Singer P. 2018. Viability Test Device for anisakid nematodes, Heliyon 4, e00552.
308. Krüger L, Daston L, Heidelberger M. 1987. The Probabilistic Revolution (2 vols), MIT Press, Cambridge, USA.
309. Kuenstler AS, Hayward RC. 2019. Light-induced shape morphing of thin films, Current Opinion in Colloid & Interface Science 40, 70–86 (https://doi.org/10.1016/j.cocis.2019.01.009).
310. Kumar N, DasGupta A. 2013. On the contact problem of an inflated spherical hyperelastic membrane, International Journal of Non-Linear Mechanics 57, 130–139.
311. Kundler I, Finkelmann H. 1995. Strain-induced director reorientation in nematic liquid single crystal elastomers, Macromolecular Rapid Communications 16, 679–686 (https://doi.org/10.1002/marc.1995.030160908).
312. Kundler I, Finkelmann H. 1998. Director reorientation via stripe-domains in nematic elastomers: influence of cross-link density, anisotropy of the network and smectic clusters, Macromolecular Chemistry and Physics 199, 677–686.
313. Küpfer J, Finkelmann H. 1991. Nematic liquid single crystal elastomers, Die Makromolekulare Chemie, Rapid Communications 12, 717–726 (https://doi.org/10.1002/marc.1991.030121211).
314. Küpfer J, Finkelmann H. 1994. Liquid crystal elastomers: Influence of the orientational distribution of the crosslinks on the phase behaviour and reorientation processes, Macromolecular Chemistry and Physics 195, 1353–1367.
315. Landau LD, Lifshitz EM. 1986. Theory of Elasticity, Volume 7 of Course of Theoretical Physics, 3rd revised ed, Pergamon Press, New York.
316. Lazopoulos KA, Ogden RW. 1998. Nonlinear elasticity theory with discontinuous internal variables, Mathematics and Mechanics of Solids 3, 29–51 (https://doi.org/10.1177/108128659800300103).
317. Leblanc JL. 2010. Filled Polymers: Science and Industrial Applications, CRC Press, Taylor & Francis Group, Boca Raton, USA.
318. Lee V, Bhattacharya K. 2021. Actuation of cylindrical nematic elastomer balloons, Journal of Applied Physics 129, 114701 (https://doi.org/10.1063/5.0041288).
319. Lee T, Bilionis I, Tepole AB. 2020. Propagation of uncertainty in the mechanical and biological response of growing tissues using multi-fidelity Gaussian process regression, Computer Methods in Applied Mechanics and Engineering 359, 112724 (https://doi.org/10.1016/j.cma.2019.112724).
320. Le Tallec P. 1994. Numerical methods for three-dimensional elasticity, In: Ciarlet PG, Lions JL (eds.), Handbook of Numerical Analysis, v. III, North-Holland, 465–624.
321. L'Hôspital GFA. 1696. Analyse des infiniment petits, pour l'intelligence des lignes courbes, A Paris, de l'Imprimerie Royale, 145–146.
322. Li S, Bai H, Liu Z, Zhan X, Huang C, Wiesner LW, Silberstein M, Shepherd RF. 2021. Digital light processing of liquid crystal elastomers for self-sensing artificial muscles, Science Advances 23(7), 30, eabg3677 (https://doi.org/10.1126/sciadv.abg3677).

323. Li Z, Xu H, Xia X, Song Y, Zheng Q. 2019. Energy dissipation accompanying Mullins effect of nitrile butadiene rubber/carbon black nanocomposites, Polymer 171, 106–114 (https://doi.org/10.1016/j.polymer.2019.03.043).

324. Limbert G, Masen MA, Pond D, Graham HK, Sherratt MJ, Jobanputra R, McBride A. 2019. Biotribology of the ageing skin - Why we should care, Biotribology 17, 75–90.

325. Limpert E, Stahel WA, ABBT M. 2001. Log-normal distributions across the sciences: keys and clues, BioScience 51(5), 341–352.

326. Lopez-Pamies O. 2009. Onset of cavitation in compressible, isotropic, hyperelastic solids, Journal of Elasticity 94, 115–145.

327. Lopez-Pamies O. 2010. A new I_1-based hyperelastic model for rubber elastic materials. Comptes Rendus Mécanique 338, 3–11 (https://doi.org/10.1016/j.crme.2009.12.007).

328. Lopez-Pamies O, Idiart MI, Nakamura T. 2011. Cavitation in elastomeric solids: I - A defect-growth theory, Journal of the Mechanics and Physics of Solids 59, 1464–1487.

329. Love AEH. 1888. On the small free vibrations and deformations of thin elastic shells, Philosophical Transactions of the Royal Society A 179, 491–546.

330. Love AEH. 1944. A Treatise on the Mathematical Theory of Elasticity, 4th ed, Dover Publications, New York.

331. Lubarda VA. 2004. Constitutive theories based on the multiplicative decomposition of deformation gradient: thermoelasticity, elastoplasticity and biomechanics, Applied Mechanics Reviews 57(2), 95–108 (https://doi.org/10.1115/1.1591000).

332. Ly HB, Desceliers C, Le LM, Le TT, Pham BT, Nguyen-Ngoc L, Doan VT, Le M. 2019. Quantification of uncertainties on the critical buckling load of columns under axial compression with uncertain random materials, Materials 12, 1828 (https://doi.org/10.3390/ma12111828).

333. Machado G, Chagnon G, Favier D. 2012. Induced anisotropy by the Mullins effect in filled silicone rubber, Mechanics of Materials 50, 70–80 (https://doi.org/10.1016/j.mechmat.2012.03.006).

334. Madireddy S, Sista B, Vemaganti K. 2015. A Bayesian approach to selecting hyperelastic constitutive models of soft tissue, Computer Methods in Applied Mechanics and Engineering 291, 102–122.

335. Madireddy S, Sista B, Vemaganti K. 2016. Bayesian calibration of hyperelastic constitutive models of soft tissue, Journal of the Mechanical Behavior of Biomedical Materials 59, 108–127.

336. Mahimwalla Z, Yager KG, Mamiya J-i, Shishido A, Priimagi A, Barrett CJ. 2012. Azobenzene photomechanics: prospects and potential applications, Polymer Bulletin 69, 967–1006 (https://doi.org/10.1007/s00289-012-0792-0).

337. Mallock A. Note on the instability of India-rubber tubes and balloons when distended by fluid pressure, Proceedings of the Royal Society of London 49, 458 (https://doi.org/10.1098/rspl.1890.0116).

338. Malyarenko A, Ostoja-Sarzewski M. 2019. Tensor-Valued Random Fields for Continuum Physics, 1st ed, Cambridge University Press, New York.

339. Malyarenko A, Ostoja-Sarzewski M, Amiri-Hezaveh A. 2020. Random Fields of Piezoelectricity and Piezomagnetism. Correlation Structures, Springer, New York.

340. Mangan R, Destrade M. 2015. Gent models for the inflation of spherical balloons, International Journal of Non-Linear Mechanics 68, 52–58.

341. Mangan R, Destrade M, Saccomandi G. 2016. Strain energy function for isotropic non-linear elastic incompressible solids with linear finite strain response in shear and torsion, Extreme Mechanics Letters 9, 204–206.

342. Marckmann G, Verron E. 2006. Comparison of hyperelastic models for rubber-like materials, Rubber Chemistry and Technology 79, 835–858 (https://doi.org/10.5254/1.3547969).

343. Marsden JE, Hughes TJR. 1983. Mathematical Foundations of Elasticity, Dover Publications, New York.

344. Martínez-Frutos J, Ortigosa R, Pedregal P, Periago F. 2020. Robust optimal control of stochastic hyperelastic materials 88, 888–904 (https://doi.org/10.13140/RG.2.2.18544.61449).

345. Marzano M. 1983. An interpretation of Baker-Ericksen inequalities in uniaxial deformation and stress, Meccanica 18, 233–235 (https://doi.org/10.1007/BF02128248).
346. Mathai AM. 1982. Storage capacity of a dam with Gamma type inputs, Annals of the Institute of Statistical Mathematics 34, 591–597.
347. McCoy JJ. 1973. A statistical theory for predicting response of materials that possess a disordered structure, Technical report ARPA 2181, AMCMS Code 5911.21.66022, Army Materials and Mechanics Research Center, Watertown, Massachusetts.
348. McCracken JM, Donovan BR, White TJ. 2020. Materials as machines, Advanced Materials 32, 1906564 (https://doi.org/10.1002/adma.201906564).
349. McGrayne SB. 2012. The Theory That Would Not Die: How Bayes' Rule Cracked the Enigma Code, Hunted Down Russian Submarines, an Emerged Triumphant from Two Centuries of Controversy, Paperback ed., Yale University Press, New Haven.
350. McMahon J, Goriely A. 2010. Spontaneous cavitation in growing elastic membranes, Mathematics and Mechanics of Solids 15, 57–77.
351. McMillen T, Goriely A, 2002. Tendril perversion in intrinsically curved rods, Journal of Nonlinear Science 12, 241–281.
352. McLeish T. 2020. Soft Matter: A Very Short Introduction, Oxford University Press, Oxford.
353. Merodio J, Saccomandi G. 2006. Remarks on cavity formation in fiber-reinforced incompressible non-linearly elastic solids, European Journal of Mechanics A/Solids 25, 778–792.
354. Mihai LA, Alamoudi M. 2021. Likely oscillatory motions of stochastic hyperelastic spherical shells and tubes, International Journal of Non-Linear Mechanics130, 103671 (https://doi.org/10.1016/j.ijnonlinmec.2021.103671).
355. Mihai LA, Budday S, Holzapfel GA, Kuhl E, Goriely A. 2017. A family of hyperelastic models for human brain tissue, Journal of Mechanics and Physics of Solids 106, 60–79.
356. Mihai LA, Chin L, Janmey PA, Goriely A. 2015. A comparison of hyperelastic constitutive models applicable to brain and fat tissues, Journal of the Royal Society Interface 12, 20150486.
357. Mihai LA, Fitt D, Woolley TE, Goriely A. 2019. Likely equilibria of stochastic hyperelastic spherical shells and tubes, Mathematics and Mechanics of Solids 24(7), 2066–2082 (https://doi.org/10.1177/1081286518811881).
358. Mihai LA, Fitt D, Woolley TE, Goriely A. 2019. Likely oscillatory motions of stochastic hyperelastic solids, Transactions of Mathematics and Its Applications 3, 1–42 (https://doi.org/10.1093/imatrm/tnz003).
359. Mihai LA, Fitt D, Woolley TE, Goriely A. 2019. Likely cavitation in stochastic elasticity, Journal of Elasticity 137(1), 27–42 (https://doi.org/10.1007/s10659-018-9706-1).
360. Mihai LA, Goriely A. 2011. Positive or negative Poynting effect? The role of adscititious inequalities in hyperelastic materials, Proceedings of the Royal Society A 467, 3633–3646.
361. Mihai LA, Goriely A. 2013. Numerical simulation of shear and the Poynting effects by the finite element method: An application of the generalised empirical inequalities in non-linear elasticity, International Journal of Non-Linear Mechanics 49, 1–14.
362. Mihai LA, Goriely A. 2017. How to characterize a nonlinear elastic material? A review on nonlinear constitutive parameters in isotropic finite elasticity, Proceedings of the Royal Society A 473, 20170607 (https://doi.org/10.1098/rspa.2017.0607).
363. Mihai LA, Goriely A. 2020. Likely striping in stochastic nematic elastomers, Mathematics and Mechanics of Solids 25(10), 1851–1872 (https://doi.org/0.1177/1081286520914958).
364. Mihai LA, Goriely A. 2020. A plate theory for nematic liquid crystalline solids, Journal of the Mechanics and Physics of Solids 144, 104101 (https://doi.org/10.1016/j.jmps.2020.104101).
365. Mihai LA, Goriely A. 2020. A pseudo-anelastic model for stress softening in liquid crystal elastomers, Proceedings of the Royal Society A, 20200558 (https://doi.org/10.1098/rspa.2020.0558).
366. Mihai LA, Goriely A. 2021. Instabilities in liquid crystal elastomers, Material Research Society (MRS) Bulletin 46 (https://doi.org/10.1557/s43577-021-00115-2).

367. Mihai LA, Mistry D, Raistrick T, Gleeson HF, Goriely A. 2022. A mathematical model for the auxetic response of liquid crystal elastomers, Philosophical Transaction of the Royal Society A, in print.

368. Mihai LA, Neff P. 2017. Hyperelastic bodies under homogeneous Cauchy stress induced by non-homogeneous finite deformations, International Journal of Non-Linear Mechanics 89, 93–100.

369. Mihai LA, Neff P. 2017. Hyperelastic bodies under homogeneous Cauchy stress induced by three-dimensional non-homogeneous deformations, Mathematics and Mechanics of Solids (https://doi.org/10.1177/1081286516682556).

370. Mihai LA, Wang H, Guilleminot J, Goriely A. 2021. Nematic liquid crystalline elastomers are aeolotropic materials, Proceedings of the Royal Society A 477, 20210259 (https://doi.org/10.1098/rspa.2021.0259).

371. Mihai LA, Woolley TE, Goriely A. 2018. Stochastic isotropic hyperelastic materials: constitutive calibration and model selection, Proceedings of the Royal Society A 474, 20170858 (https://doi.org/10.1098/rspa.2017.0858).

372. Mihai LA, Woolley TE, Goriely A. 2019. Likely equilibria of the stochastic Rivlin cube, Philosophical Transactions of the Royal Society A 377, 20180068 (https://doi.org/10.1098/rsta.2018.0068).

373. Mihai LA, Woolley TE, Goriely A. 2019. Likely chirality of stochastic anisotropic hyperelastic tubes, International Journal of Non-Linear Mechanics 114, 9–20 (https://doi.org/10.1016/j.ijnonlinmec.2019.04.004).

374. Mihai LA, Woolley TE, Goriely A. 2020. Likely cavitation and radial motion of stochastic elastic spheres, Nonlinearity 33(5), 1987–2034 (https://doi.org/10.1088/1361-6544/ab7104).

375. Miller GA. 1956. The magical number seven, plus or minus two: Some limits on our capacity for processing information, Psychological Review 63(2): 81–97 (https://doi.org/10.1037/h0043158).

376. Misra S, Ramesh KT, Okamura AM. 2010. Modelling of non-linear elastic tissues for surgical simulation, Computer Methods in Biomechanics and Biomedical Engineering 13, 811–818.

377. Mistry D, Connell SD, Mickthwaite SL, Morgan PB, Clamp JH, Gleeson HF. 2018. Coincident molecular auxeticity and negative order parameter in a liquid crystal elastomer, Nature Communications 9, 5095 (https://doi.org/0.1038/s41467-018-07587-y).

378. Mistry D, Gleeson HF. 2019. Mechanical deformations of a liquid crystal elastomer at director angles between 0° and 90°: Deducing an empirical model encompassing anisotropic nonlinearity, Journal of Polymer Science 57, 1367–1377 (https://doi.org/0.1002/polb.24879).

379. Mistry D, Nikkhou M, Raistrick T, Hussain M, Jull EIL, Baker DL, Gleeson HF. 2020. Isotropic liquid crystal elastomers as exceptional photoelastic strain sensors, Macromolecules 53, 3709–3718 (https://doi.org/10.1021/acs.macromol.9b02456).

380. Mistry D, Traugutt NA, Yu K, Yakacki CM. 2021. Processing and reprocessing liquid crystal elastomer actuators, Journal of Applied Physics 129, 130901 (https://doi.org/10.1063/5.0044533).

381. Modes CD, Bhattacharya K, Warner M. 2010. Disclination-mediated thermo-optical response in nematic glass sheets, Physical Review E 81, 060701(R) (https://doi.org/10.1103/PhysRevE.81.060701).

382. Modes CD, Bhattacharya K, Warner M. 2011. Gaussian curvature from flat elastica sheets, Proceedings of the Royal Society A 467, 1121–1140 (https://doi.org/10.1098/rspa.2010.0352).

383. Modes CD, Warner M. 2011. Blueprinting nematic glass: Systematically constructing and combining active points of curvature for emergent morphology, Physical Review E 84, 021711 (https://doi.org/10.1103/PhysRevE.84.021711).

384. Moon H, Truesdell C. 1974. Interpretation of adscititious inequalities through the effects pure shear stress produces upon an isotropic elastic solid, Archive for Rational Mechanics and Analysis 55, 1–17.

385. Mooney M. 1940. A theory of large elastic deformation, Journal of Applied Physics 11, 582–592 (https://doi.org/10.1063/1.1712836).

386. Mori T, Cukelj R, Prévôt ME, Ustunel S, Story A, Gao Y, Diabre K, McDonough JA, Johnson Freeman E, Hegmann E, Clements RJ. 2020. 3D porous liquid crystal elastomer foams supporting long-term neuronal cultures, Macromolecular Rapid Communications, 1900585 (https://doi.org/10.1002/marc.201900585).

387. Moschopoulos PG. 1985. The distribution of the sum of independent Gamma random variables, Annals of the Institute of Statistical Mathematics 37(3), 541–544.

388. Mostajeran C. 2015. Curvature generation in nematic surfaces, Physical Review E 91, 062405 (https://doi.org/10.1103/PhysRevE.91.062405).

389. Müller I, Struchtrup H. 2002. Inflation of rubber balloon, Mathematics and Mechanics of Solids 7, 569–577.

390. Mullins L. 1948. Effect of stretching on the properties of rubber, Rubber Chemistry and Technology 21(2), 281–300 (https://doi.org/10.5254/1.3546914).

391. Mullins L. 1949. Permanent set in vulcanized rubber, Rubber Chemistry and Technology 22(4), 1036–1044 (https://doi.org/10.5254/1.3543010).

392. Mullins L. 1969. Softening of rubber by deformation, Rubber Chemistry and Technology 42(1), 339–362 (https://doi.org/10.5254/1.3539210).

393. Mullins L, Tobin NR. 1957. Theoretical model for the elastic behaviour of filler-reinforced vulcanized rubbers, Rubber Chemistry and Technology 30(2), 555–571 (https://doi.org/10.5254/1.3542705).

394. Mullins L, Tobin NR. 1965. Stress softening in rubber vulcanizates. part I. Use of a strain amplification factor to describe the elastic behavior of filler-reinforced vulcanized rubber, Journal of Applied Polymer Science 9(9), 2993–3009 (https://doi.org/10.1002/app.1965.070090906).

395. Murphy JG. 2013. Transversely isotropic biological soft tissue must be modeled using both anisotropic invariants, European Journal of Mechanics A/Solids 42, 90–96 (https://doi.org/10.1016/j.euromechsol.2013.04.003).

396. Murphy JG. 2014. Evolution of anisotropy in soft tissue, Proceedings of the Royal Society A 470, 20130548 (https://doi.org/10.1098/rspa.2013.0548).

397. Nayfeh AH, Mook DT. 1995. Nonlinear Oscillations, Wiley-VCH, Weinheim, Germany.

398. Neff P, Mihai LA. 2016. Injectivity of the Cauchy-stress tensor along rank-one connected lines under strict rank-one convexity condition, Journal of Elasticity 127, 309–315.

399. Nelson P. 2021. Physical Models of Living Systems, 2nd ed, Chiliagon Science.

400. Neubert D, Saunders DW. 1958. Some observations of the permanent set of cross-linked natural rubber samples after heating in a state of pure shear, Rheologica Acta 1, 151–157 (https://doi.org/10.1007/BF01968858).

401. Ni B, Liu G, Zhang M, Keller P, Tatoulian M, Li MH. 2021. Large-size honeycomb-shaped and iris-like liquid crystal elastomer actuators, CCS Chemistry 3, 1081–1088 (https://doi.org/10.31635/ccschem.021.202100818).

402. Nörenberg N, Mahnken R. 2013. A stochastic model for parameter identification, Archive of Applied Mechanics 83, 367–378 (https://doi.org/10.1007/s00419-012-0684-7).

403. Nörenberg N, Mahnken R. 2015. Parameter identification for rubber materials with artificial spatially distributed data, Computational Mechanics 56, 353–370 (https://doi.org/10.1002/pamm.201410201).

404. Nowinski JL. 1966. On a dynamic problem in finite elastic shear, International Journal of Engineering Science 4, 501–510.

405. Nowinski JL, Schultz AR. 1964. Note on a class of finite longitudinal oscillations of thick-walled cylinders, Proceedings of the Indian National Congress of Theoretical and Applied Mechanics, 31–44.

406. Nunes ICS, Moreira DC. 2013. Simple shear under large deformation: experimental and theoretical analyses, European Journal of Mechanics A/Solids 42, 315–322 (https://doi.org/10.1016/j.euromechsol.2013.07.002).

407. Oden JT. 2006. Finite Elements of Nonlinear Continua, 2nd ed, Dover, New York.

408. Oden JT. 2018. Adaptive multiscale predictive modelling, Acta Numerica 27, 353–450.

409. Oden JT, Prudencio EE, Hawkins-Daarud A. 2013. Selection and assessment of phenomeno-logical models of tumor growth, Mathematical Models and Methods in Applied Sciences 23, 1309–1338.

410. Oden JT, Moser R, Ghattas O. 2010. Computer predictions with quantified uncertainty, part I, SIAM News 43(9), 1–3.

411. Oden JT, Moser R, Ghattas O. 2010. Computer predictions with quantified uncertainty, part II, SIAM News 43(10), 1–4.

412. Ogden RW. 1972. Large deformation isotropic elasticity - on the correlation of theory and experiment for incompressible rubberlike solids, Proceedings of the Royal Society of London A 326, 565–584.

413. Ogden RW. 1972. Large deformation isotropic elasticity - on the correlation of theory and experiment for compressible rubberlike solids, Proceedings of the Royal Society of London A 328, 567–583.

414. Ogden RW. 1974. On isotropic tensors and elastic moduli, Mathematical Proceed-ings of the Cambridge Philosophical Society 75, 427–436 (https://doi.org/10.1017/S0305004100048635).

415. Ogden RW. 1997. Non-Linear Elastic Deformations, 2nd ed, Dover, New York.

416. Ogden RW, Roxburgh DG. 1998. A pseudo-elastic model for the Mullins effect in filled rubber, Proceedings of the Royal Society A 455, 2861–2877 (https://doi.org/10.1098/rspa.1999.0431).

417. Ogden RW, Saccomandi G, Sgura I. 2004. Fitting hyperelastic models to experimental data, Computational Mechanics 34, 484–502 (https://doi.org/10.1007/s00466-004-0593-y).

418. Okamoto S, Sakurai S, Urayama K. 2021. Effect of stretching angle on the stress plateau behavior of main-chain liquid crystal elastomers, Soft Matter (https://doi.org/10.1039/d0sm02244f).

419. Olmsted P. 1994. Rotational invariance and Goldstone modes in nematic elastomers and gels, Journal de Physique II, EDP Sciences 4(12), 2215–2230 (https://doi.org/10.1051/jp2:1994257).

420. Ostoja-Starzewski M. 2007. Microstructural Randomness and Scaling in Mechanics of Materials, Taylor & Francis Group, Boca Raton, FL.

421. Owen N. 1987. Existence and stability of necking deformations for nonlinearly elastic rods, Archive for Rational Mechanics and Analysis 98, 357–383 (https://doi.org/10.1007/BF00276914).

422. Pang X, Lv J-a., Zhu C, Qin L, Yu Y. 2019. Photodeformable azobenzene-containing liquid crystal polymers and soft actuators, Advanced Materials, 1904224 (https://doi.org/10.1002/adma.201904224).

423. Payne LE. 1963. Review of: C. Truesdell & R. Toupin, Static grounds for inequalities in finite strain of elastic materials, Archive for Rational Mechanics and Analysis, 12, 1–33.

424. Pearce SP, Fu Y. 2010. Characterization and stability of localized bulging/necking in inflated membrane tubes, IMA Journal of Applied Mathematics 75, 581–602.

425. Pearson K. 1900. On the criterion that a given system of deviations from the probable in the case of a correlated system of variables is such that it can be reasonably supposed to have arisen from random sampling, Philosophical Magazine, Series 5, 50(302), 157–175 (https://doi.org/10.1080/14786440009463897).

426. Pearson ES, Wishart J (eds.). 1942. "Student's" Collected Papers, Issued by the Biometrika Office, University College London, London.

427. Pence TJ, Tsai SJ. 2007. Bulk cavitation and the possibility of localized deformation due to surface layer swelling, Journal of Elasticity 87, 161–185.

428. Penn W. 1970. Volume changes accompanying the extension of rubber, Transactions of the Society of Rheology 14, 509–517.

429. Penn RW, Kearsley EA. 1976. The scaling law for finite torsion of elastic cylinders, Transactions of the Society of Rheology 20, 227–238.

430. Pereira GG, Warner M. 2001. Mechanical and order rigidity of nematic elastomers, The European Physical Journal E 5, 295–307 (https://doi.org/10.1007/s101890170061).

431. Perepelyuk M, Chin LK, Cao X, van Oosten A, Shenoy VB, Janmey PA, Wells RG. 2016. Normal and fibrotic rat livers demonstrate shear strain softening and compression stiffening: a model for soft tissue mechanics, PLoS ONE 11, e0146588.

432. Petelin A, Čopič M. 2009. Observation of a soft mode of elastic instability in liquid crystal elastomers, Physical Review Letters 103, 077801 (https://doi.org/10.1103/PhysRevLett.103.077801).

433. Petelin A, Čopič M. 2010. Strain dependence of the nematic fluctuation relaxation in liquid-crystal elastomers, Physical Review E 82, 011703 (https://doi.org/10.1103/PhysRevE.82.011703).

434. Pilz da Cunha M, Peeketi AR, Ramgopal A, Annabattula RK, Schenning APHJ. 2020. Light-driven continual oscillatory rocking of a polymer film, Chemistry Open 9(11), 1149–1152 (https://doi.org/0.1002/open.202000237).

435. Plucinsky P, Bhattacharya K. 2017. Microstructure-enabled control of wrinkling in nematic elastomer sheets, Journal of the Mechanics and Physics of Solids 102, 125–150 (https://doi.org/10.1016/j.jmps.2017.02.009).

436. Pogoda K, Chin LK, Georges PC, Byfield FRG, Bucki R, Kim R, Weaver M, Wells RG, Marcinkiewicz C, Janmey PA. 2014. Compression stiffening of brain and its effect on mechanosensing by glioma cells, New Journal of Physics 16, 075002.

437. Poincaré H. 2001. The Value of Science: Essential Writings of Henri Poincaré, paperback ed, The Modern Library, New York.

438. Polignone DA, Horgan CO. 1993. Cavitation for incompressible anisotropic nonlinearly elastic spheres, Journal of Elasticity 33, 27–65.

439. Polignone DA, Horgan CO. 1993. Effects of material anisotropy and inhomogeneity on cavitation for composite incompressible anisotropic nonlinearly elastic spheres, International Journal of Solids and Structures 30, 3381–3416.

440. Pollack HN. 2005. Uncertain Science... Uncertain World, paperback ed, Cambridge University Press, New York.

441. Poulain X, Lefèvre V, Lopez-Pamies O, Ravi-Chandar K. 2017. Damage in elastomers: nucleation and growth of cavities, micro-cracks, and macro-cracks, International Journal of Fracture 205, 1–21.

442. Poulain X, Lopez-Pamies O, Ravi-Chandar K. 2018. Damage in elastomers: Healing of internally nucleated cavities and micro-cracks, Soft Matter 14, 4633–4640.

443. Poynting JH. 1909. On pressure perpendicular to the shear planes in finite pure shears, and on the lengthening of loaded wires when twisted, Proceedings of the Royal Society A 82(557), 546–559 (https://doi.org/10.1098/rspa.1909.0059).

444. Poynting JH. 1912. On the changes in the dimensions of a steel wire when twisted, and on the pressure or distortional waves in steel, Proceedings of the Royal Society of London A 86(590), 534–561 (https://doi.org/10.1098/rspa.1912.0045).

445. Prathumrat P, Sbarski I, Hajizadeh E, Nikzad M. 2021. A comparative study of force fields for predicting shape memory properties of liquid crystalline elastomers using molecular dynamic simulations, Journal of Applied Physics 129, 155101 (https://doi.org/10.1063/5.0044197).

446. Prasad D, Kannan K. 2020. An analysis driven construction of distortional-mode-dependent and Hill-Stable elastic potential with application to human brain tissue, Journal of the Mechanics and Physics of Solids 134, 103752 (https://doi.org/10.1016/j.jmps.2019.103752).

447. Prévôt ME, Andro H, Alexander SLM, Ustunel S, Zhu C, Nikolov Z, Rafferty ST, Brannum MT, Kinsel B, Korley LTJ, Freeman EJ, McDonough JA, Clements RJ, Hegmann E. 2018. Liquid crystal elastomer foams with elastic properties specifically engineered as biodegradable brain tissue scaffolds, Soft Matter 14, 354–360 (https://doi.org/10.1039/c7sm01949a).

448. Prigogine I. 1997. The End of Certainty: Time, Chaos, and the New Laws of Nature, The Free Press, New York.

449. Pucci E, Saccomandi G. 1997. On universal relations in continuum mechanics, Continuum Mechanics and Thermodynamics 9, 61–72.

450. Pucci E, Saccomandi G. 2002. A note on the Gent model for rubber-like materials, Rubber Chemistry and Technology 75, 839–852.

451. Quarteroni A, Lassila T, Rossi S, Ruiz-Baier R. 2017. Integrated heart - Coupling multiscale and multiphysics models for the simulation of the cardiac function, Computer Methods in Applied Mechanics and Engineering 314, 345–407.
452. Raayai-Ardakani S, Earla DR, Cohen T. 2019. The intimate relationship between cavitation and fracture, Soft Matter 15, 4999.
453. Raistrick T, Zhang Z, Mistry D, Mattsson J, Gleeson HF. 2021. Understanding the physics of the auxetic response in a liquid crystal elastomer, Physical Review Research 3, 023191 (https://doi.org/10.1103/PhysRevResearch.3.023191).
454. Rajagopal KR, Wineman AS. 1987. New universal relations for nonlinear isotropic elastic materials. Journal of Elasticity 17, 75–83.
455. Rashid B, Destrade M, Gilchrist MD. 2012. Mechanical characterization of brain tissue in compression at dynamic strain rates, Journal of the Mechanical Behavior of Biomedical Materials 10, 23–38.
456. Rashid B, Destrade M, Gilchrist MD. 2013. Mechanical characterization of brain tissue in simple shear at dynamic strain rates, Journal of the Mechanical Behavior of Biomedical Materials 28, 71–85.
457. Rashid B, Destrade M, Gilchrist MD. 2014. Mechanical characterization of brain tissue in tension at dynamic strain rates, Journal of the Mechanical Behavior of Biomedical Materials 33, 43–54.
458. Reissner E. 1941. A new derivation of the equations for the deformation of elastic shells, American Journal of Mathematics 63, 177–184.
459. Ren J.-s.. 2008. Dynamical response of hyper-elastic cylindrical shells under periodic load, Applied Mathematics and Mechanics 29, 1319–1327.
460. Ren J.-s.. 2009. Dynamics and destruction of internally pressurized incompressible hyper-elastic spherical shells, International Journal of Engineering Science 47, 745–753.
461. Richter H. 1948. Das isotrope Elastizitätsgesetz, Zeitschrift für Angewandte Mathematik und Mechanik 28, 205–209.
462. Rickaby SR, Scott NH. 2013. A cyclic stress softening model for the Mullins effect, International Journal of Solids and Structures 50, 111–120 (https://doi.org/10.1016/j.ijsolstr.2012.09.006).
463. Riggs JD, Lalonde TL. 2017. Handbook for Applied Modeling: Non-Gaussian and Correlated Data, Cambridge University Press, Cambridge, UK.
464. Rivlin RS. 1948. Large elastic deformations of isotropic materials. II. Some uniqueness theorems for pure, homogeneous deformation, Philosophical Transactions of the Royal Society A 240, 491–508.
465. Rivlin RS. 1948. Large elastic deformations of isotropic materials. IV. Further developments of the general theory, Philosophical Transactions of the Royal Society of London. Series A, Mathematical and Physical Sciences 241, 379–397.
466. Rivlin RS. 1949. Large elastic deformations of isotropic materials. VI. Further results in the theory of torsion, shear and flexure, Philosophical Transactions of the Royal Society of London A 242(845), 173–195.
467. Rivlin RS. 1953. The solution of problems in second order elasticity theory, Journal of Rational Mechanics and Analysis 2, 53–81.
468. Rivlin RS. 1974. Stability of pure homogeneous deformations of an elastic cube under dead loading, Quarterly of Applied Mathematics 32, 265–271.
469. Rivlin RS. 1997. Collected papers of R.S. Rivlin, vol I, II, Barenblatt GI, Joseph DD (eds.), Springer, New York.
470. Rivlin RS, Saunders DW. 1951. Large elastic deformations of isotropic materials. VII. Experiments on the deformation of rubber, Philosophical Transactions of the Royal Society of London A 243(865), 251–288.
471. Roan E, Vemaganti K. 2007. The nonlinear material properties of liver tissue determined from no-slip uniaxial compression experiments, Journal of Biomechanical Engineering 129, 450–456.

472. Robert CP. 2007. The Bayesian Choice: From Decision-Theoretic Foundations to Computational Implementation, 2nd ed, Springer, New York.

473. Rodríguez-Martínez JA, Fernández-Sáez J, Zaera R. 2015. The role of constitutive relation in the stability of hyper-elastic spherical membranes subjected to dynamic inflation, International Journal of Engineering Science 93, 31–45.

474. Röntgen WC. 1986. Ueber das Verhältnis der Quercontraction zur Längendilatation bei Kautschuk, Annalen der Physik 235(12), 601–616.

475. Rothemund P, Kim Y, Heisser RH, Zhao X, Shepherd RF, Keplinger C. 2021. Shaping the future of robotics through materials innovation, Nature Materials 20, 1582–1587 (https://doi.org/10.1038/s41563-021-01158-1).

476. Saccomandi G, Vergori L. 2021. Some remarks on the weakly nonlinear theory of isotropic elasticity, Journal of Elasticity, (https://doi.org/10.1007/s10659-021-09865-1).

477. Sadik S, Yavari A. 2017. On the origins of the idea of the multiplicative decomposition of the deformation gradient, Mathematics and Mechanics of Solids 22(4), 771–772 (https://doi.org/10.1177/1081286515612280).

478. Saed MO, Gablier A, Terentjev EM. 2021. Exchangeable liquid crystalline elastomers and their applications, Chemical Reviews (https://doi.org/10.1021/acs.chemrev.0c01057).

479. Saed MO, Torbati AH, Starr CA, Visvanathan R, Clark NA, Yakacki CM. 2017. Thiol-acrylate main-chain liquid-crystalline elastomers with tunable thermomechanical properties and actuation strain, Journal of Polymer Science 55(2), 157–168 (https://doi.org/0.1002/polb.24249).

480. Safar A, Mihai LA. 2018. The nonlinear elasticity of hyperelastic models for stretch-dominated cellular structures, International Journal of Non-Linear Mechanics 106, 144–154.

481. Sawyers KN. 1976. Stability of an elastic cube under dead loading: two equal forces, International Journal of Non-Linear Mechanics 11, 11–23.

482. Schmidt G, Tondl A. 1986. Non-Linear Vibrations, Cambridge University Press, Cambridge.

483. Schuhladen S, Preller F, Rix R, Petsch S, Zentel R, Zappe H. 2014. Iris-like tunable aperture employing liquid-crystal elastomers, Advanced Materials 26(42), 7247–7251 (https://doi.org/10.1002/adma.201402878).

484. Schweickert E, Mihai LA, Martin RJ, Neff P. 2020. A note on non-homogeneous deformations with homogeneous Cauchy stress for a strictly rank-one convex energy in isotropic hyperelasticity, International Journal of Non-Linear Mechanics 119, 103282 (https://doi.org/10.1016/j.ijnonlinmec.2019.103282).

485. Scott NH. 2007. The incremental bulk modulus, Young's modulus and Poisson's ratio in nonlinear isotropic elasticity: Physically reasonable response, Mathematics and Mechanics of Solids 12, 526–542 (https://doi.org/10.1177/1081286506064719).

486. Serak S, Tabiryan N, Vergara R, Whie TJ, Vaia RA, Bunning TJ. 2010. Liquid crystalline polymer cantilever oscillators fueled by light, Soft Matter 6, 779–783 (https://doi.org/10.1039/B916831A).

487. Shahinpoor M. 1974. Exact solution to finite amplitude oscillation of an anisotropic thin rubber tube, The Journal of the Acoustical Society of America 56, 477–480.

488. Shahinpoor M. 1973. Combined radial-axial large amplitude oscillations of hyperelastic cylindrical tubes, Journal of Mathematical and Physical Sciences 7, 111–128.

489. Shahinpoor M, Balakrishnan R. 1978. Large amplitude oscillations of thick hyperelastic cylindrical shells, International Journal of Non-Linear Mechanics 13, 295–301.

490. Shahinpoor M, Nowinski JL. 1971. Exact solution to the problem of forced large amplitude radial oscillations of a thin hyperelastic tube, International Journal of Non-Linear Mechanics 6, 193–207.

491. Shahsavan H, Aghakhani A, Zengb H, Guo Y, Davidson ZS, Priimagi A, Sitti M. 2020. Bioinspired underwater locomotion of light-driven liquid crystal gels, Proceedings of the National Academy of Sciences of the United States of America (PNAS) 117(10), 5125–5133 (https://doi.org/0.1073/pnas.1917952117).

492. Shannon CE. 1948. A mathematical theory of communication, Bell System Technical Journal 27, 379–423, 623–659.

493. Sharma A, Stoffel AM, Lagerwall JPF. 2021. Liquid crystal elastomer shells with topological defect-defined actuation: Complex shape morphing, opening/closing, and unidirectional rotation, Journal of Applied Physics 129, 174701 (https://doi.org/10.1063/5.0044920).

494. Shield RT. 1971. Deformations possible in every compressible, isotropic, perfectly elastic material, Journal of Elasticity 1, 91–92.

495. Shield RT. 1972. On the stability of finitely deformed elastic membranes. II: Stability of inflated cylindrical and spherical membranes, Zeitschrift für Angewandte Mathematik und Physik (ZAMP) 23, 16–34.

496. Šilhavý M. 2007. Ideally soft nematic elastomers, Networks Heterogeneous Media 2, 279–311.

497. Singh M, Pipkin AC. 1965. Note on Ericksen's problem, Zeitschrift für angewandte Mathematik und Physik (ZAMP) 16, 706–709.

498. Sivaloganathan, I. 1991. Cavitation, the incompressible limit, and material inhomogeneity, Quarterly of Applied Mathematics 49, 521–541.

499. Sivaloganathan J. 1999. On cavitation and degenerate cavitation under internal hydrostatic pressure, Proceedings of the Royal Society A 455, 3645–3664.

500. Sivaloganathan J, Spector SJ. 2011. On the stability of incompressible elastic cylinders in uniaxial extension, Journal of Elasticity 105(1–2), 313–330 (https://doi.org/10.1007/s10659-011-9330-9).

501. Skačej G, Zannoni C. 2012. Molecular simulations elucidate electric field actuation in swollen liquid crystal elastomers, Proceedings of the National Academy of Sciences of the United States of America (PNAS) 109(26), 10193–10198 (https://doi.org/10.1073/pnas.1121235109).

502. Skačej G, Zannoni C. 2014. Molecular simulations shed light on supersoft elasticity in polydomain liquid crystal elastomers, Macromolecules 47, 8824–8832 (https://doi.org/10.1021/ma501836j).

503. Smith PH. 2004. The Body of the Artisan: Art and Experience in the Scientific Revolution, The University of Chicago Press, Chicago, London.

504. Soares RM, Amaral PFT, Silva FMA, Gonçalves PB. 2019. Nonlinear breathing motions and instabilities of a pressure-loaded spherical hyperelastic membrane, Nonlinear Dynamics 99(1), 351–372 (https://doi.org/10.1007/s11071-019-04855-4).

505. Sobczyk K, Kirkner DJ. 2012. Stochastic Modeling of Microstructures, Springer, New York.

506. Soize C. 2000. A nonparametric model of random uncertainties for reduced matrix models in structural dynamics, Probabilistic Engineering Mechanics 15, 277–294 (https://doi.org/10.1016/S0266-8920(99)00028-4).

507. Soize C. 2001. Maximum entropy approach for modeling random uncertainties in transient elastodynamics, Journal of the Acoustical Society of America 109, 1979–1996 (https://doi.org/0.1121/1.1360716).

508. Soize C. 2006. Non-Gaussian positive-definite matrix-valued random fields for elliptic stochastic partial differential operators, Computer Methods in Applied Mechanics and Engineering 195, 26–64 (https://doi.org/10.1016/j.cma.2004.12.014).

509. Soize C. 2013. Stochastic modeling of uncertainties in computational structural dynamics - Recent theoretical advances, Journal of Sound and Vibration 332, 2379–2395 (https://doi.org/10.1016/j.jsv.2011.10.010).

510. Soize C. 2017. Uncertainty Quantification: An Accelerated Course with Advanced Applications in Computational Engineering, Interdisciplinary Applied Mathematics Book 47, Springer, New York.

511. Soldatos KP. 2006. On the stability of a compressible Rivlin's cube made of transversely isotropic material, IMA Journal of Applied Mathematics 71, 332–353 (https://doi.org/10.1093/imamat/hxh114).

512. Soltani M, Raahemifar K, Nokhosteen A, Kashkooli FM, Zoudani EL. 2021. Numerical methods in studies of liquid crystal elastomers, Polymers 13, 1650 (https://doi.org/10.3390/polym13101650).

513. Sommer G, Eder M, Kovacs L, Pathak H, Bonitz L, Mueller C, Regitnig P, Holzapfel GA. 2013. Multiaxial mechanical properties and constitutive modeling of human adipose tissue: a basis for preoperative simulations in plastic and reconstructive surgery, Acta Biomaterialia 9, 9036–9048.

514. Soni J, Goodman R. 2017. A Mind at Play: How Claude Shannon Invented the Information Age, Simon & Schuster, New York.

515. Soni H, Pelcovits RA, Powers TR. 2016. Wrinkling of a thin film on a nematic liquid-crystal elastomer, Physical Review E 94, 012701 (https://doi.org/10.1103/PhysRevE.94.012701).

516. Souri H, Banerjee H, Jusufi A, Radacsi N, Stokes AA, Park I, Sitti M, Amjadi M. 2020. Wearable and stretchable strain sensors: materials, sensing mechanisms, and applications, Advanced Intelligent Systems 2000039 (https://doi.org/10.1002/aisy.202000039).

517. Spencer AJM. 1971. Theory of invariants. In: Eringen, A.C. (ed.) Continuum Physics 1, 239–253, Academic Press, New York.

518. Spencer AJM. 2015. George Green and the foundations of the theory of elasticity, Journal of Engineering Mathematics 95, 5–6 (https://doi.org/10.1007/s10665-015-9791-0).

519. Staber B, Guilleminot J. 2015. Stochastic modeling of a class of stored energy functions for incompressible hyperelastic materials with uncertainties, Comptes Rendus Mécanique 343, 503–514 (https://doi.org/10.1016/j.crme.2015.07.008).

520. Staber B, Guilleminot J. 2016. Stochastic modeling of the Ogden class of stored energy functions for hyperelastic materials: the compressible case, Journal of Applied Mathematics and Mechanics/Zeitschrift für Angewandte Mathematik und Mechanik 97, 273–295 (https://doi.org/10.1002/zamm.201500255).

521. Staber B, Guilleminot J. 2017. Stochastic hyperelastic constitutive laws and identification procedure for soft biological tissues with intrinsic variability, Journal of the Mechanical Behavior of Biomedical Materials 65, 743–752 (https://doi.org/10.1016/j.jmbbm.2016.09.022).

522. Staber B, Guilleminot J. 2018. A random field model for anisotropic strain energy functions and its application for uncertainty quantification in vascular mechanics, Computer Methods in Applied Mechanics and Engineering 333, 94–113 (https://doi.org/10.1016/j.cma.2018.01.001).

523. Staber B, Guilleminot J, Soize C, Michopoulos J, Iliopoulos A. 2019. Stochastic modeling and identification of an hyperelastic constitutive model for laminated composites, Computer Methods in Applied Mechanics and Engineering 347, 425–444 (https://doi.org/10.1016/j.cma.2018.12.036).

524. Steigmann DJ. 1992. Cavitation in elastic membranes - an example, Journal of Elasticity 28, 277–287.

525. Steigmann DJ. 2017. Finite Elasticity Theory, Oxford University Press, Oxford, UK.

526. Steinmann P, Hossain M, Possart G. 2012. Hyperelastic models for rubber-like materials: consistent tangent operators and suitability for Treloar's data, Archive of Applied Mechanics 82, 1183–1217.

527. Stuji S, van Doorn JM, Kodger T, Sprakel J, Coulais C, Schall P. 2019. Stochastic buckling of self-assembled colloidal structures, Physical Review Research 1, 023033.

528. Sugerman GP, Kakaletsis S, Thakkar P, Chokshi A, Parekh SH, Rausch MK. 2020. A whole blood thrombus mimic: Constitutive behaviour under simple shear, Biorxiv (https://doi.org/10.1101/2020.07.19.210732).

529. Sullivan TJ. 2015. Introduction to Uncertainty Quantification, Springer-Verlag, New York.

530. Tabrizi M, Ware TH, Shankar MR. 2019. Voxelated molecular patterning in three-dimensional free forms, ACS Applied Materials & Interfaces 11, 28236–28245 (https://doi.org/10.1021/acsami.9b04480).

531. Tajbakhsh AR, Terentjev EM. 2001. Spontaneous thermal expansion of nematic elastomers, The European Physical Journal E 6, 181–188 (https://doi.org/10.1007/s101890170020).

532. Taleb NN. 2007. The Black Swan: The Impact of the Highly Improbable, Penguin Books, London, UK.

533. Talroze RV, Zubarev ER, Kuptsov SA, Merekalov AS, Yuranova TI, Plate NA, Finkelmann H. 1999. Liquid crystal acrylate-based networks: polymer backbone-LC order interaction, Reactive and Functional Polymers 41, 1–11 (https://doi.org/10.1016/S1381-5148(99)00032-2).

534. Tanner RI, Tanner E. 2003. Heinrich Hencky: A rheological pioneer, Rheologica Acta 42, 93–101 (https://doi.org/10.1007/s00397-002-0259-6).

535. Tarantino AM. 2008. Homogeneous equilibrium configurations of a hyperelastic compressible cube under equitriaxial dead-load tractions, Journal of Elasticity 92, 227–254.

536. Teferra K, Brewick PT. 2019. A Bayesian model calibration framework to evaluate brain tissue characterization experiments, Computer Methods in Applied Mechanics and Engineering 357, 112604.

537. Terentjev EM. 1999. Liquid-crystalline elastomers, Journal of Physics: Condensed Matter 11(24), R239-R257 (https://doi.org/10.1088/0953-8984/11/24/201).

538. Thiel C, Voss J, Martin RJ, Neff P. 2019. Shear, pure and simple, International Journal Non-Linear Mechanics 112, 57–72 (https://doi.org/10.1016/j.ijnonlinmec.2018.10.002).

539. Thorburn WM. 1918. The myth of Occam's razor, Mind 27(107), 345–353 (https://doi.org/10.1093/mind/XXVII.3.345).

540. Tian H, Wang Z, Chen Y, Shao J, Gao T, Cai S, 2018. Polydopamine-coated main-chain liquid crystal elastomer as optically driven artificial muscle, ACS Applied Materials & Interfaces 10(9), 8307–8316 (https://doi.org/10.1021/acsami.8b00639).

541. Timoshenko SP. 1983. History of Strength of Materials, Dover, New York.

542. Titterington DM. 1982. Irreverent Bayes, Journal of Applied Statistics 9 (1): 16–18 (https://doi.org/10.1080/02664768200000003).

543. Tobolsky AV. 1960. Properties and Structures of Polymers, John Wiley and Sons, New York, Chap. 5, 223–265.

544. Tobolsky AV, Prettyman IB, Dillon JH. 1944. Stress relaxation of natural and synthetic rubber stocks, Journal of Applied Physics 15, 380–395 (https://doi.org/10.1063/1.1707442).

545. Torras N, Zinoviev KE, Marshall JE, Terentjev EM, Esteve J. 2011. Bending kinetics of a photo-actuating nematic elastomer cantilever, Applied Physics Letters 99, 254102 (https://doi.org/10.1063/1.3670502).

546. Tottori S, Zhang L, Qiu FM, Krawczyk KK, Franco-Obregón A, Nelson BJ. 2012. Magnetic helical micromachines: fabrication, controlled swimming, and cargo transport, Advanced Materials 24(6), 811–816.

547. Traugutt NS, Mistry D, Luo C, Yu K, Ge Q, Yakacki CM. 2020. Liquid-crystal-elastomer-based dissipative structures by digital light processing 3D printing, Advanced Materials 2000797 (https://doi.org/10.1002/adma.202000797).

548. Traugutt NA, Volpe RH, Bollinger MS, Saed MO, Torbati AH, Yu K, Dadivanyan N, Yakacki CM. 2017. Liquid-crystal order during synthesis affects main-chain liquid-crystal elastomer behavior, Soft Matter 13, 7013 (https://doi.org/10.1039/c7sm01405h).

549. Treloar LRG. 1944. Stress-strain data for vulcanized rubber under various types of deformation, Transactions of the Faraday Society 40, 59–70 (https://doi.org/10.1039/TF9444000059).

550. Treloar LRG. 2005. The Physics of Rubber Elasticity, 3rd ed, Oxford University Press, Oxford, UK.

551. Treloar LRG, Hopkins HG, Rivlin RS, Ball JM. 1976. The mechanics of rubber elasticity [and discussions], Proceedings of the Royal Society of London A 351, 301–330 (https://doi.org/10.1098/rspa.1976.0144).

552. Truesdell CA. 1952. The mechanical foundations of elasticity and fluid dynamics, Journal of Rational Mechanics and Analysis 1, 125–171.

553. Truesdell CA. 1952. A programme of physical research in classical mechanics, Zeitschrift für Angewandte Mathematik und Physik (ZAMP) 3, 79–95.

554. Truesdell C. 1956. Das ungelöste Hauptproblem der endlichen Elastizitätstheorie, ZAMM, Journal of Applied Mathematics and Mechanics / Zeitschrift für Angewandte Mathematik und Mechanik 36, 97–103.

555. Truesdell C. 1962. Solutio generalis et accurata problematum quamplurimorum de motu corporum elasticorum incomprimibilium in deformationibus valde magnis, Archive of Rational Mechanics and Analysis 11, 106–113.

556. Truesdell C. 1966. The Elements of Continuum Mechanics, Springer, New York.

557. Truesdell C. 1968. Essays in the History of Mechanics, Springer, New York.

558. Truesdell C, Noll W. 2004. The Non-Linear Field Theories of Mechanics, 3rd ed, Springer-Verlag, New York.

559. Twizell EH, Ogden RW. 1983. Non-linear optimization of the material constants in Ogden's stress-deformation function for incompressible isotropic elastic materials, The ANZIAM Journal 24, 424–434.

560. Ube T, Ikeda T. 2014. Photomobile polymer materials with crosslinked liquid-crystalline structures: molecular design, fabrication, and functions, Angewandte Chemie International Edition 53(39), 10290–10299 (https://doi.org/10.1002/anie.201400513).

561. Ula SW, Traugutt NA, Volpe RH, Patel RP, Yu K, Yakacki CM. 2018. Liquid crystal elastomers: an introduction and review of emerging technologies, Liquid Crystals Review 6, 78–107 (https://doi.org/10.1080/21680396.2018.1530155).

562. Urayama K, Honda S, Takigawa T. 2005. Electrically driven deformations of nematic gels, Physical Review E 71, 051713.

563. Urayama K, Honda S, Takigawa T. 2006. Deformation coupled to director rotation in swollen nematic elastomers under electric fields, Macromolecules 39, 1943–1949.

564. Urayama K, Kohmon E, Kojima M, Takigawa T. 2009. Polydomain-monodomain transition of randomly disordered nematic elastomers with different cross-linking histories, Macromolecules 42, 4084–4089 (https://doi.org/10.1021/ma9004692).

565. van Oosten ASG, Chen X, Chin L, Cruz K, Patteson AE, Pogoda K, Vivek B. Shenoy VB, Janmey PA. 2019. Emergence of tissue-like mechanics from fibrous networks confined by close-packed cells, Nature 573, 96–101 (https://doi.org/10.1038/s41586-019-1516-5).

566. Valanis KC, Landel RF. 1967. The strain-energy function of a hyperelastic material in terms of the extension ratios, Journal of Applied Physics 38, 2997–3002 (https://doi.org/10.1063/1.1710039).

567. Vangerko H, Treloar LRG. 1978. The inflation and extension of rubber tube for biaxial strain studies, Journal of Physics D: Applied Physics 11, 1969–1978 (https://doi.org/10.1088/0022-3727/11/14/009).

568. van Oosten CL, Harris KD, Bastiaansen CWM, Broer DJ. 2007. Glassy photomechanical liquid-crystal network actuators for microscale devices, The European Physical Journal E 23(3), 329-33 (https://doi.org/10.1140/epje/i2007-10196-1).

569. Veronda DR, Westmann RA. 1970. Mechanical characterization of skin-finite deformations, Journal of Biomechanics 3(1), 111–124 (https://doi.org/10.1016/0021-9290(70)90055-2).

570. Verron E, Khayat RE, Derdouri A, Peseux B. 1999. Dynamic inflation of hyperelastic spherical membranes, Journal of Rheology 43, 1083–1097.

571. Verwey GC, Warner M. 1995. Soft rubber elasticity, Macromolecules 28, 4303–4306.

572. Verwey GC, Warner M, Terentjev EM. 1996. Elastic instability and stripe domains in liquid crystalline elastomers, Journal de Physique II France 6(9), 1273–1290 (https://doi.org/10.1051/jp2:1996130).

573. Verwey GC, Warner M. 1997. Compositional fluctuations and semisoftness in nematic elastomers, Macromolecules 30, 4189–4195.

574. Vogel S. 1998. Cat's Paws and Catapults, WW Norton and Company, New York, London.

575. Volokh K. 2019. Mechanics of Soft Materials, 2nd ed, Springer, New York.

576. von Mises R. 1982. Probability, Statistics and Truth, 2nd ed, Dover, New York (https://archive.org/details/in.ernet.dli.2015.189506/page/n1).

577. Wan G, Jin C, Trase I, Zhao S, Chen Z. 2018. Helical structures mimicking chiral seedpod opening and tendril coiling, Sensors 18(9), 2973 (https://doi.org/10.3390/s18092973).

578. Wang ASD. 1969. On free oscillations of elastic incompressible bodies in finite shear, International Journal of Engineering Science 7, 1199–1212.

579. Wang CC. 1965. On the radial oscillations of a spherical thin shell in the finite elasticity theory, Quarterly of Applied Mathematics 23, 270–274.

580. Wang CC, Ertepinar A. 1972. Stability and vibrations of elastic thick-walled cylindrical and spherical shells subjected to pressure, International Journal of Non-Linear Mechanics 7, 539–555.

581. Wang M, Fu Y. 2021. Necking of a hyperelastic solid cylinder under axial stretching: Evaluation of the infinite-length approximation, International Journal of Engineering Science 159, 103432 (https://doi.org/10.1016/j.ijengsci.2020.103432).

582. Wang Z, He Q, Wang Y, Cai S. 2019. Programmable actuation of liquid crystal elastomers via living exchange reaction, Soft Matter 15(13), 2811–2016.

583. Wang M, Hu XB, Zuo B, Huang S, Chen XM, Yang H. 2020. Liquid crystal elastomer actuator with serpentine locomotion, Chemical Communications 56, 7597 (https://doi.org/10.1039/d0cc02823a).

584. Wang SB, Guo GM, Zhou L, Li LA, Fu Y. 2019. An experimental study of localized bulging in inflated cylindrical tubes guided by newly emerged analytical results, Journal of the Mechanics and Physics of Solids 124, 536–554.

585. Wang C, Sim K, Chen J, Kim H, Rao Z, Li Y, Chen W, Song J, Verduzco R, Yu C. 2018. Soft ultrathin electronics innervated adaptive fully soft robots, Advanced Materials 30(13), 1706695 (https://doi.org/10.1002/adma.201706695).

586. Wang Z, Tian H, He Q, Cai S. 2017. Reprogrammable, reprocessable, and self-healable liquid crystal elastomer with exchangeable disulfide bonds, ACS Applied Materials & Interfaces 9(38), 33119–33128 (https://doi.org/10.1021/acsami.7b09246).

587. Wang Z, Wang Z, Zheng Y, He Q, Wang Y, Cai S. 2020. Three-dimensional printing of functionally graded liquid crystal elastomer, Science Advances 6(39), eabc0034 (https://doi.org/10.1126/sciadv.abc0034).

588. Ware TH, McConney ME, Wie JJ, Tondiglia VP, White TJ. 2015. Voxelated liquid crystal elastomers, Science 347, 982–984.

589. Warner M. 2020. Topographic mechanics and applications of liquid crystalline solids, Annual Review of Condensed Matter Physics 11, 125–145 (https://doi.org/10.1146/annurev-conmatphys-031119-050738).

590. Warner M, Bladon P, Terentjev E. 1994. "Soft elasticity" - deformation without resistance in liquid crystal elastomers, Journal de Physique II France 4, 93–102 (https://doi.org/10.1051/jp2:1994116).

591. Warner M, Gelling KP, Vilgis TA. 1988. Theory of nematic networks, The Journal of Chemical Physics 88, 4008–4013 (https://doi.org/10.1063/1.453852).

592. Warner M, Modes CD, Corbett D. 2010. Curvature in nematic elastica responding to light and heat, Proceedings of the Royal Society A 466, 2975–2989 (https://doi.org/10.1098/rspa.2010.0135).

593. Warner M, Terentjev EM. 1996. Nematic elastomers - a new state of matter?, Progress in Polymer Science 21, 853–891.

594. Warner M, Terentjev EM. 2007. Liquid Crystal Elastomers, paper back, Oxford University Press, Oxford, UK.

595. Warner M, Wang, XJ. 1991. Elasticity and phase behavior of nematic elastomers, Macro-molecules 24, 4932–4941 (https://doi.org/10.1021/ma00017a033).

596. Waters JT, Li S, Yao Y, Lerch MM, Aizenberg M, Aizenberg J, Balazs AC. 2020. Twist again: Dynamically and reversibly controllable chirality in liquid crystalline elastomer microposts, Science Advances 6(13), 5349 (https://doi.org/10.1126/sciadv.aay5349).

597. Weaire D, Fortes MA. 1994. Stress and strain in liquid and solid foams, Advances in Physics 43, 685–738.

598. Wei J, Yu Y. 2012. Photodeformable polymer gels and crosslinked liquid-crystalline poly-mers, Soft Matter 8(31), 8050–8059.

599. Weissenberg K. 1947. A continuum theory of rheological phenomena, Nature 159, 310–311.

600. Weissenberg K. 1950. Rheology of hydrocarbon gel, Proceedings of the Royal Society 200(1061), 183–188 (https://doi.org/10.1098/rspa.1950.0008).

601. Wen Z, Yang K, Raquez JM. 2020. A review on liquid crystal polymers in free-standing reversible shape memory materials, Molecules 25, 1241 (https://doi.org/10.3390/molecules25051241).

602. Wex C, Arndt S, Stoll A, Bruns C, Kupriyanova Y. 2015. Isotropic incompressible hyper-elastic models for modelling the mechanical behaviour of biological tissues: A review, Biomedizinische Technik/Biomedical Engineering 60, 577–592.

603. White TJ. 2018. Photomechanical effects in liquid crystalline polymer networks and elastomers, Journal of Polymer Science, Part B: Polymer Physics 56, 695–705 (https://doi.org/10.1002/polb.24576).

604. White TJ, Broer DJ. 2015. Programmable and adaptive mechanics with liquid crystal polymer networks and elastomers, Nature Materials 14, 1087–1098 (https://doi.org/10.1038/nmat4433).

605. Whitmer JK, Roberts TF, Shekhar R, Abbott NL, de Pablo JJ. 2013. Modeling the polydomain-monodomain transition of liquid crystal elastomers, Physical Review E 87, 020502 (https://doi.org/10.1103/PhysRevE.87.020502).

606. Wie JJ, Shankar MR, White TJ. 2016. Photomotility of polymers, Nature Communications 7, 13260.

607. Wiener JH. 2002. Statistical Mechanics of Elasticity, 2nd ed, Dover, New York.

608. Willcox KE, Ghattas O, Heimbach P. 2021. The imperative of physics-based modeling and inverse theory in computational science, Nature Computational Science 1, 166–168 (https://doi.org/10.1038/s43588-021-00040-z).

609. Winkler M, Kaiser A, Krause S, Finkelmann H, Schmidt AM. 2010. Liquid crystal elastomers with magnetic actuation, Macromolecular Symposia 291–292(1), 186–192.

610. Wriggers P. 2007. Mixed Finite-Element-Methods, Springer, New York.

611. Xia Y, Honglawan A, Yang S. 2019. Tailoring surface patterns to direct the assembly of liquid crystalline materials, Liquid Crystals Reviews 7(1), 30–59 (https://doi.org/10.1080/21680396.2019.1598295).

612. Yavari A. 2021. Universal deformations in inhomogeneous isotropic nonlinear elastic solids, Proceedings of the Royal Society A 477, 20210547 (https://doi.org/10.1098/rspa.2021.0547).

613. Ye Y, Liu Y, Fu Y. 2020. Weakly nonlinear analysis of localized bulging of an inflated hyperelastic tube of arbitrary wall thickness, Journal of the Mechanics and Physics of Solids 135, 103804 (https://doi.org/10.1016/j.jmps.2019.103804).

614. Ye S, Yin SF, Li B, Feng XQ. 2019. Torsion instability of anisotropic cylindrical tissues with growth, Acta Mechanica Solida Sinica 32(5), 621–632.

615. Yeoh OH. 1990. Characterization of elastic properties of carbon-black-filled rubber vulcan-izates, Rubber Chemistry & Technology 63, 792–805 (https://doi.org/10.5254/1.3538289).

616. Yeoh OH. 1993. Some forms of the strain energy function for rubber, Rubber Chemistry & Technology 66, 754–771.

617. Yerzley FL. 1939. Adhesion of neoprene to metal, Industrial & Engineering Chemistry 31(8), 950–956 (https://doi.org/10.1021/ie50356a007).

618. Yu Y, Nakano M, Ikeda T. 2003. Directed bending of a polymer film by light, Nature 425, 145–145 (https://doi.org/10.1038/425145a).

619. Yuan X, Zhang R, Zhang H. 2008. Controllability conditions of finite oscillations of hyperelastic cylindrical tubes composed of a class of Ogden material models, Computers, Materials and Continua 7, 155–166.

620. Zamani V, Pence TJ. 2017. Swelling, inflation, and a swelling-burst instability in hyperelastic spherical shells, International Journal of Solids and Structures 125, 134–149.

621. Zannoni C. 2000. Computer simulation and molecular design of model liquid crystals, In: Fabbrizzi L, Poggi A (eds.), Chemistry at the Beginning of the Third Millennium: Molecular Design, Supramolecules, Nanotechnology and Beyond, Springer: Berlin/Heidelberg, Germany, 329–342.

622. Zellner A. 1988. Optimal information processing and Bayes's theorem, The American Statistician 42(4), 278–280 (https://doi.org/10.1080/00031305.1988.10475585).

623. Zeng H, Wasylczyk P, Parmeggiani C, Martella D, Burresi M, Wiersma DS. 2015. Light-fueled microscopic walkers, Advanced Materials 27 (26), 3883–3887.

624. Zentel R. 1986. Shape variation of cross-linked liquid-crystalline polymers by electric fields, Liquid Crystals 1, 589–592.

625. Zhang Y, Xuan C, Jiang Y, Huo Y. 2019. Continuum mechanical of liquid crystal elastomers as dissipative ordered solids, Journal of the Mechanics and Physics of Solids 126, 285–303 (https://doi.org/10.1016/j.jmps.2019.02.018).

626. Zhang YS, Jiang SA, Lin JD, Lee CR. 2020. Bio-inspired design of active photo-mechano-chemically dual-responsive photonic film based on cholesteric liquid crystal elastomers, Journal of Material Chemistry C 8(16), 5517–5524 (https://doi.org/10.1039/C9TC05758G).

627. Zhao D, Liu Y. 2020. Light-induced spontaneous bending of a simply supported liquid crystal elastomer rectangular plate, Physical Review E 101, 042701.

628. Zhou L, Wang Y, Li K. 2022. Light-activated elongation/shortening and twisting of a nematic elastomer balloon, Polymers 14, 1249 (https://doi.org/10.3390/polym14061249).

629. Zhu F, Wu B, Destrade M, Chen W. 2020. Electrostatically tunable axisymmetric vibrations of soft electro-active tubes, Journal of Sound and Vibration 115467 (https://doi.org/10.1016/j.jsv.2020.115467).

630. Zubarev ER, Kuptsov SA, Yuranova TI, Talroze RV, Finkelmann H. 1999. Monodomain liquid crystalline networks: reorientation mechanism from uniform to stripe domains, Liquid Crystals 26, 1531–1540 (https://doi.org/10.1080/026782999203869).

631. Zurlo G, Blackwell J, Colgan N, Destrade M. 2020. The Poynting effect, American Journal of Physics 88, 1036 (https://doi.org/0.1119/10.0001997).

Index

A

Acceleration potential, 46, 115, 155
Adscititious, 16–18
Algorithm, 9
Almansi–Hamel strain tensor, 12
Analytic, 21
Anisotropic, 2, 4, 11, 20, 25, 28–30, 42–45, 49,
68, 72, 86, 98–110, 112
Anssari–Benam and Bucchi model, 36
Arruda–Boyce model, 35
Auxeticity, 186, 208–215

B

Baker–Ericksen (BE) inequalities, 17, 18, 23,
24, 33, 50, 67, 68, 76, 89–93, 157, 158,
192
Bayes' factor, 5, 62, 64, 65
Bayes theorem, 49, 60, 61
Beta distribution, 54, 55, 71, 83, 85, 94, 95,
202, 203, 206, 208
Biaxiality, 208–215
Bifurcation, 50, 78–84, 86, 89, 90, 92–95, 124,
127, 133, 134, 138, 139, 147, 150, 154,
195
Biot strain tensor, 12
Bulk modulus, 1, 21, 29–30, 220

C

Calibration, 4, 5, 49, 51–57, 59, 212
Carroll model, 34, 35
Cauchy–Green, 10, 11, 13, 16–18, 22, 24,
31–33, 38, 42, 43, 96, 99, 115, 156,
177, 196, 210, 213, 220

Cauchy's laws of motion, 45
Cauchy stress, 12–14, 17, 18, 21, 23–25, 29,
32, 33, 38, 41, 43, 44, 46, 47, 50, 57, 73,
97, 99, 116, 117, 124, 126, 137, 147,
156, 178, 179, 185, 191–193, 199, 205,
210, 214, 215, 220
Cavitation, 30, 67, 86–95, 112–154
Circulation preserving, 46, 112
Coaxial, 13, 16, 18
Compressible, 2, 11, 16, 21, 23, 32, 33, 47, 72,
112, 114
Configuration, 1, 7, 8, 11, 24, 29, 31, 38, 43,
46, 57, 68, 69, 72–74, 76, 79–81, 85,
87, 88, 95, 99, 112, 113, 115, 123,
126, 128, 132, 137, 147, 154, 157,
167, 177, 185–189, 194, 205, 209, 212,
220
Contact, 8
Continuum, vii–ix, 5, 7, 57, 99, 185, 186,
193–195, 217
Controllable, 19, 20
Coordinates, 7, 19, 24, 31, 38, 43, 69, 73, 74,
78, 87, 95, 96, 115, 154, 155, 177, 195,
205, 208, 228–231
Covariance, 53, 235
Critical value, 68, 86, 90, 94, 95, 103, 113,
122, 133, 134, 139, 147, 149, 163, 165,
166, 202–204
Cuboid, vii, 23, 31, 33, 42–45, 114,
177–181
Current configuration, 8, 24, 29, 31, 38, 43,
69, 73, 74, 87, 113, 115, 137, 154, 185,
209, 212
Cylinder, 19, 37–42, 86, 97, 98, 111

© Springer Nature Switzerland AG 2022
L. A. Mihai, *Stochastic Elasticity*, Interdisciplinary Applied Mathematics 55,
https://doi.org/10.1007/978-3-031-06692-4

D

Dead-load traction, 67, 72–75, 78, 79, 82, 85, 89, 92, 95, 112–114, 117–124, 128, 131–135, 141–150, 172, 173, 176
Deformation, 1, 7, 50, 68, 111, 183, 219
Deformed configuration, 8, 95, 112, 188, 189, 214
Deformed state, 8, 88, 112, 128, 154, 155, 167, 177
de Gennes, P.G., 183
Determinant, 8, 223–225
Deterministic, vii, viii, 3, 4, 49, 52, 57, 68, 70, 73, 80, 85, 87, 95, 99, 103–105, 109, 115, 122, 123, 127, 133, 140, 148, 149, 173, 201–208, 213, 215, 217, 219, 232, 235, 236
Director, 184–188, 191–196, 199, 200, 202–205, 207–209, 211, 212, 214, 220
Displacement, 1, 7, 9, 22, 57, 58, 68, 86, 177, 179–181, 220
Dobrynin–Carrillo model, 36

E

Eigenvalues, 10, 13, 14, 18, 31, 195, 210, 223–227
Eigenvectors, 13, 195, 225
Elastic, 1, 7, 49, 67, 111, 187
Empirical inequalities, 18, 24
Entropy, vii, 4, 51, 54, 239–240
Equilibrium, 46, 47, 72–74, 76, 77, 80, 85–87, 97, 116, 140, 156, 197–199, 201, 204, 229
Ericksen's theorem, 23
Eulerian, 8, 12, 45, 177
Expectation, ix, 54, 55, 94, 234, 236
Experimental data, viii, 5, 19, 49–51, 57, 58, 60, 64, 67, 81, 211, 212, 217
Extension, 2, 4, 17, 19, 23, 24, 40, 68–70, 72, 140, 173, 185, 187, 199, 211

F

Fibre, 42, 45, 68, 99–100, 108, 110
Finite elasticity, 1, 5, 7–47, 49, 50, 67, 188, 217
Finite element modelling, 49
Frame-indifference, 11, 188
Function, 2, 11, 50, 67, 117, 187
Fung model, 34, 36

G

Gamma distribution, 54, 55, 57, 59, 71, 79, 81, 83, 85, 93–95, 102–104, 107, 108, 115, 121–124, 126–128, 131, 133, 134, 137, 139–141, 144, 147–150, 154, 160, 161, 163–167, 171–174, 176, 180, 202–206, 208, 236, 238
Gaussian, 4, 59, 67, 113, 236
Gent–Gent model, 34, 64
Gent model, 34
Gent–Thomas model, 207
Gradient, 8, 9, 11, 12, 19, 21, 22, 24, 29, 31, 38, 43, 73–77, 87, 96, 115, 155, 177, 186, 187, 194, 196, 199, 205, 207, 209, 212, 220, 228, 230, 231
Green elastic, 11
Green, G., viii, 11, 20
Green–St Venant strain tensor, 12

H

Hadamard jump condition, 199
Hencky strain tensor, 12, 20, 25
Heterogeneous, 3, 4
Homogeneous, 1, 2, 5, 11, 17–20, 24, 29, 30, 40, 42, 46, 47, 50, 67–69, 72–79, 82, 85, 86, 112–139, 145, 153, 167, 171, 176, 199–202, 204
Hydrostatic pressure, 14–16, 29, 40, 75, 116
Hyperelastic, vii, 2, 4, 11, 17–21, 23, 25, 29, 30, 35, 36, 40, 42, 49–57, 67–69, 73–75, 78, 86, 87, 89, 99–102, 104, 108, 111–115, 140, 151, 173, 177, 184, 185, 187, 192–194, 199, 206–208, 222
Hyperparameters, 54, 57, 71, 79, 80, 84, 103, 140, 141, 148, 149, 174, 180, 202, 221, 239

I

Impulse traction, 113, 114, 124–127, 135–139, 151–154, 159–176
Incompressible, 2, 12–15, 18, 23, 25–27, 29, 32–35, 37, 40, 41, 46, 47, 67–69, 72, 74, 75, 86, 87, 111–116, 140, 151
Incremental, 20–23, 26–28
Infinitesimal, 28, 33, 34, 37, 39, 44, 45, 101, 157, 194, 195, 213
Inflation, 68, 95–112, 114, 140, 155, 166, 173, 185, 186, 204–206
Information theory, 4, 239
Inhomogeneous, 113–115, 140–154, 173–176, 199

Instability, vii, 5, 40, 67–111, 114, 134, 166,
    185, 195–207, 218
Invariants, 10, 11, 13–16, 32, 42, 52, 116, 141,
    156, 174, 178, 188, 189, 191, 192, 194,
    220, 223, 224, 227
Isotropic, 1, 2, 4, 11, 16–42, 50, 67, 68, 72,
    74, 75, 86, 99, 111–113, 115, 184–189,
    193, 199, 227

J

Jaynes, E.T., 4, 51, 240

L

Lagrange multiplier, 12, 33, 43, 74, 75, 99,
    116, 156, 178, 191, 193, 240
Lagrangian, 8, 12, 177, 240
Likelihood, 61, 64
Linear elastic, 2, 19–21, 26, 37, 40, 44, 45, 50,
    52, 56
Linear strain tensor, 12
Liquid crystal elastomer, viii, 5, 29, 183–215
Liquid crystals, 183, 184, 210
Load, viii, 23–25, 47, 49, 67–95, 113, 115,
    121, 123, 126, 132, 134, 137, 138, 143,
    145, 146, 152, 153, 185, 199, 200, 205,
    207, 208, 211, 218
Logarithmic strain tensor, 21
Lopez–Pamies model, 34, 35
Lower bound, 5, 64, 65, 130, 136, 160, 161,
    166, 171, 201, 203

M

Matrix, 9, 68, 76, 99, 101, 102, 188, 191,
    223–226, 228
Maximum entropy, vii, 4, 54, 239–240
Maxwell, J.C., 49
Mean value, 4, 50–57, 62–64, 68, 71, 72, 79,
    80, 93, 94, 103, 122–124, 126, 127,
    131, 133, 134, 137, 139–141, 145,
    147–151, 154, 161, 163, 164, 166,
    172–174, 176, 180, 189, 202–206, 208,
    214, 221, 234–238
Mechanical response, viii, 5, 186
Mesoscopic, 3
Model, 3, 11, 49, 67, 111, 184, 219
Modulus, 1, 7, 50, 67, 115, 191
Molecular dynamics simulation, 184, 186
Molecular frame, 211, 212
Monodomain, 184, 185, 187, 196, 200, 202,
    203

Mooney–Rivlin, 34, 35, 41, 63, 64, 69, 72,
    73, 77, 78, 81–86, 90, 112–115, 153,
    159–167, 185, 194, 204, 206
Multiplicative decomposition, 185, 187, 191,
    226

N

Nematic, 5, 183–188, 191–196, 199, 200,
    202–212, 220
Neo-Hookean (NH), 34, 35, 42, 69, 72, 78–80,
    85, 90, 112–115, 122–140, 147, 149,
    150, 153, 167–173, 178–181, 185, 194,
    206, 207
Nested square roots, 31
Nondeterministic, vii, viii, ix, 4, 217
Nonlinear, vii, viii, ix, 2, 4, 5, 7, 19–45, 50–54,
    57, 58, 61–63, 67, 72, 86, 111–114,
    156, 157, 174, 178, 185, 214, 217, 220
Normal distribution, 4, 59, 60, 141, 174,
    236–238
Normal stress effect, 40
Numerical procedure, 49

O

Objectivity, 11, 50, 188, 192
Odds, 61, 62, 64
Ogden, viii, 34, 35, 63, 64, 185
Order parameter, 186, 210–212
Oscillation, 111–114, 120, 122, 153, 178–181

P

Periodic orbits, 120
Piola–Kirchhoff stress, 14–16, 22, 26, 27, 32,
    33, 57, 61, 62, 69, 74, 88, 192, 193,
    207, 208, 210, 211, 220
Point mass, 121
Poisson function, 20, 25, 28–29, 36, 220
Poisson ratio, 28, 187
Polar coordinates, 7, 87, 95, 96, 115, 154,
    229–231
Polar decomposition, 9, 226, 227
Polydomain, 184, 185
Polymer, 1, 3, 185, 186, 194, 209
Posterior probability, 60
Potential, 11, 46, 115, 120, 155
Poynting effect, 40, 41, 45, 50, 186
Poynting modulus, 40–42, 220
Pressure-compression (PC) inequalities,
    16–18, 24, 28
Prigogine, I., 111–181

Principal stretches, 10, 11, 14, 17, 31, 35, 50, 69, 75, 76, 189, 191, 194, 195
Prior probability, 61, 62, 64, 67
Probability, vii, viii, ix, 3–5, 49, 51, 52, 59–62, 64, 67–71, 73, 77–80, 83–86, 93–95, 99, 102–104, 106–108, 111, 114, 120–122, 124, 127, 130, 134, 140, 149, 150, 154, 164, 172, 173, 200–204, 206, 208, 213, 217, 221, 232–234, 236, 238–240
p-value, 59

Q

Quasi-equilibrated motion, 5, 45–47, 111, 112, 115–153, 155, 167
Quasi-static, 5

R

Random variable, vii, 3, 4, 49, 50, 53–55, 67–69, 72, 73, 83, 85, 87, 94, 99, 102, 113, 115, 120, 128, 141, 148, 149, 167, 172, 200, 201, 205, 218, 219, 221, 232–236, 238, 239
Rank-one connected, 199
Reference configuration, 8, 11, 57, 72–74, 76, 79, 87, 88, 99, 113, 123, 128, 132, 145, 167, 185, 187, 194, 212, 220
Reference state, 8, 14, 72, 73, 76–78, 80, 81, 85, 88, 128, 155, 167, 177
Relative error, 5, 61, 62, 64
Rigid-body transformation, 11, 50, 188
Rivlin cube, 67, 72–78
Rivlin, R.S., 2, 3, 40, 51, 72
Röntgen, W.C., 28
Rotation, 9–11, 78, 81, 186–188, 196, 207, 224
Rubber, viii, 1–3, 29, 40, 57–65, 81, 86, 111, 183–186

S

Saunders, D.W., 2, 3, 51
Scalar, 11–16, 28, 188, 189, 191, 193, 211, 212, 219, 223, 224, 228–231
Shannon, C.E., 4, 239, 240
Shear modulus, 1, 2, 4, 19, 21, 30, 32–34, 37, 44, 45, 50–57, 59–64, 67, 69–71, 73, 79, 80, 91, 94, 103, 104, 107, 115, 120, 122–124, 127, 132, 134, 137, 140, 141, 144, 147–149, 151, 157, 173, 174, 176, 194, 195, 201–203, 205–208, 213, 220
Shell, 100, 111–115, 140, 173, 185, 186, 204–206

Skew-symmetric, 76, 192
Snap cavitation, 87, 94, 147, 148, 153
Sphere, vii, 50, 67, 86–95, 111–154, 205
Standard deviation, 51–57, 63, 64, 141, 214, 219, 221, 235–238
Stochastic, 3, 7, 49, 67, 113, 185, 220
Strain, 1, 7, 49, 67, 127, 185, 220
Strain-energy density, vii, 11, 74, 75, 99, 185, 186, 188, 193–195, 209, 220
Stress, 2, 12, 50, 69, 114, 184, 220
Stretch modulus, 2, 20, 23–28, 57, 61, 62, 213, 215, 220
Striping, 186, 195–204
Student's t-test, 59
Subcritical bifurcation, 50, 90, 91, 93, 94
Supercritical bifurcation, 50, 90, 92–94
Symmetric, 10–12, 14, 15, 18, 46, 62, 86–95, 111–115, 124, 188, 189, 191–193, 205, 210, 225–227

T

Tension, 5, 16, 17, 23–25, 28, 30, 51, 68, 69, 73, 88, 126, 130, 136, 185, 195, 207
Tensor, 8–28, 31–33, 38, 42–44, 46, 69, 73–77, 88, 96, 99, 100, 115, 155, 156, 177, 178, 185–188, 191–196, 199, 205, 207, 209–210, 213, 219–220, 223–229
Titterington, D.M., 62
Torque, 39–40, 68
Torsion modulus, 37–40, 220
Treloar, L.R.G., viii, 2, 3
Truesdell, C., viii, 7–47
Tube, vii, 68, 95–114, 153–176
Twist, 38, 39, 41, 42, 105, 106

U

Uncertainty, vii, viii, ix, 3, 4, 51, 77, 104, 217, 239
Uniaxial, 5, 17, 23–25, 28, 30, 49, 57, 59, 68, 69, 185–191, 195, 207, 208, 211–212
Universal relation, 2, 19, 21, 36–37, 40

V

Variable, vii, 3, 4, 8, 49, 50, 53–55, 67–69, 72, 73, 83, 85, 87, 89, 94, 98, 99, 102, 113, 115, 120, 128, 141, 149, 154, 167, 172, 174, 191, 193, 200, 201, 205, 218, 219, 221, 223, 227–229, 232–236, 238, 239
Variance, 4, 50–56, 68, 140, 141, 150, 151, 174, 181, 189, 235, 236, 238

Vector, 7, 8, 33, 54, 55, 186, 187, 194, 199,
    219, 223, 226, 228–231, 233
Velocity, 46, 122, 123, 149, 150, 164, 172, 173,
    176, 179, 230, 231
Volume, 2, 8, 16, 25, 29, 30, 46, 141, 186, 187,
    193, 209, 217, 229
von Mises, R., 51

W
Weissenberg effect, 40

Y
Yeoh model, 34, 35, 113
Young's modulus, 1, 19, 20, 25–27, 30, 37, 220

Printed by Printforce, the Netherlands